In relation to the origin and spread of grasses, domestication is a recent event confined to about the last ten thousand years and to relatively few grasses. Part I of *Grass evolution and domestication* considers, from an evolutionary viewpoint, grass taxonomy, the origin and diversification of C_4 photosynthesis, S–Z self-incompatibility and apomixis. It also includes a discussion of how the grass inflorescence and the spikelet could have originated. In Part II the origins of domestication are explored, both for cereals and for grasses which have latterly come to have either amenity or ecological significance. For the major cereals, domestication now involves not only classical plant breeding, but also the application of molecular techniques to obtain new varieties with desirable characteristics. The world's three most important cereals, wheat, maize and rice, are therefore presented as model systems in an attempt to explore the interaction of plant breeding, cytogenetics and molecular biology.

GRASS EVOLUTION
AND DOMESTICATION

GRASS EVOLUTION AND DOMESTICATION

Edited by

G. P. Chapman BSc, PhD, FLS

Wye College
University of London

CAMBRIDGE
UNIVERSITY PRESS

CAMBRIDGE UNIVERSITY PRESS
Cambridge, New York, Melbourne, Madrid, Cape Town, Singapore, São Paulo, Delhi

Cambridge University Press
The Edinburgh Building, Cambridge CB2 8RU, UK

Published in the United States of America by Cambridge University Press, New York

www.cambridge.org
Information on this title: www.cambridge.org/9780521107273

First published 1992
This digitally printed version 2009

A catalogue record for this publication is available from the British Library

Library of Congress Cataloguing in Publication data
Grass evolution and domestication/edited by G.P. Chapman.
   p.   cm.
Includes indexes.
ISBN 0–521–41654–X (hardback)
1. Grasses – Evolution.   2. Grain.   3. Plants, Cultivated – Origin.
I. Chapman, G. P. (Geoffrey Peter)
QK495.G74G734 1992
584'.9—dc20   92–2797 CIP

ISBN 978-0-521-41654-2 hardback
ISBN 978-0-521-10727-3 paperback

And the earth brought forth grass . . .

Genesis 1 v. 12

Man does not attempt to cause variability; though he unintentionally effects this by exposing organisms to new conditions of life, and by crossing breeds already formed. But variability being granted, he works wonders.

C. Darwin
Plants and animals under domestication, London, 1890

I am convinced that the 8 billion people projected to be living 40 to 50 years from now will continue to find most of their sustenance from the same plant species that supply most of our food needs now.

N. E. Borlaug
Science 1983

Contents

x   *Contents*

Colour plates are between pp. 158 and 159

Contributors

R. Appels
CSIRO
Division of Plant Industry
Institute of Plant Production & Processing
GPO Box 1600
Canberra
ACT 2601, Australia

G. P. Chapman
Division of Biochemistry & Biological Sciences
Wye College
(University of London)
Ashford
Kent TN25 5AH, UK

W. D. Clayton
The Herbarium
Royal Botanic Gardens
Kew
Richmond
Surrey TW9 3AB, UK

M. S. Davies
School of Pure and Applied Biology
University of Wales
PO Box 915
Cardiff, UK

J. M. J. de Wet
ICRISAT
Patancheru
Andhra Pradesh 502 324 India

J. R. Harlan
1016 North Hagan Street
New Orleans
Louisiana, LA 70119, USA

P. W. Hattersley
Australian Biological Resources Study
Australian National University
PO Box 636
Canberra
ACT 2601, Australia

D. L. Hayman
Department of Genetics
R. A. Fisher Laboratories
GPO Box 498, Adelaide
South Australia 5001, Australia

G. Hillman
Department of Archaeology
University College
Gower Street
London WC1E 6BT, UK

D. Hoisington
Dept. of Applied Molecular Genetics
CIMMYT, Lisboa 27 Colonia Juarez
Apdo. Postal 6-641
06600 Mexico, DF Mexico

G. Kochert
Department of Botany
College of Arts and Sciences
University of Georgia
Athens
Georgia 30602, USA

E. S. Lagudah
CSIRO
Division of Plant Industry
Institute of Plant Production & Processing
GPO Box 1600
Canberra
ACT 2601, Australia

S. A. Renvoize
The Herbarium
Royal Botanic Gardens
Kew
Richmond
Surrey TW9 3AB, UK

L. Watson
Australian Biological Resources Study
Australian National University
PO Box 636
Canberra
ACT 2601, Australia

Preface

Reproductive versatility in the grasses was published in 1990. Its emphasis was upon grasses as they now are but contemporary grasses have antecedents and will leave descendants. These are two aspects of the current volume, though we must add a third – the human involvement that seeks not only to understand but to direct grass evolution. Evolution is not merely 'reproductive versatility in action'. Mechanisms such as self-incompatibility and apomixis have themselves evolved and we make some effort here to grapple with the problem of how these might have come about.

Among the recent growth points of grass science has been the study of photosynthesis and it is thought-provoking that even so fundamental a process as this has further evolved among the Poaceae. Considerable emphasis therefore is given in this volume to photosynthesis. Behind its complex terminology are the issues of whether, in response to environmental change, grasses, at this level, having evolved might yet again do so and what implications there are for technology.

An assumption, widely current, is that evolution is well studied as a cytogenetic phenomenon. And so it is, but we have to recognize the patchiness of cytogenetic investigation across the Poaceae. Aside from a vast literature on chromosome counts, meiotic analyses have tended to be confined to relatively few genera and conspicuous among them are the north temperate grasses with larger and more conveniently examined chromosomes. There is a further consideration. The sophisticated cytogenetic studies in wheat, for example, continue to stimulate interest in prospective developments there as much as they discourage as premature any generalized view of grass cytogenetics. I have therefore preferred to concentrate attention on three 'model' systems: wheat, maize and rice. They are, I believe, in some ways complementary and represent this book's contribution to the cytogenetic debate.

Traditional plant breeding has a record of which its practitioners are justly proud. How then are the newer techniques of molecular biology to relate to it? Clearly, it is not a question of either/or since both offer insights into our crop plants. One of the aims of this book is to explore the emerging synthesis as it affects cereals.

Agrostologists are fortunate in the level of taxonomic interest that grasses have attracted. Work in Canberra and Kew particularly is not only highly accomplished, it is also different in approach and both in this book and its companion, I have sought to utilize each. I am grateful to each group for the summary of their systems. That for Canberra forms an appendix to the earlier volume and for Kew to the present one, so as to facilitate comparison and cross reference. Although chapter 2, for example, is appropriate to the present volume, Hattersley & Watson use the system of classification outlined in the Appendix to *Reproductive Versatility in the Grasses*. In preparing the organism index to the present volume, I have taken the opportunity to indicate there the genera and higher taxa individual to each system of classification bearing in mind that each is continually subject to revision.

Domestication in relation to the origin and spread of grasses is, of course, a 'recent' event confined both to about the last 10 000 years and to relatively few grasses. The choices made by Neolithic farmers have proved remarkably durable and we need to appreciate their originality sufficiently. Without their achievement, the settled societies in which art and science began to flourish could scarcely have been possible. Indeed, it is time, I believe, to recognize that to invent agriculture is to be part of that creative tradition rather than merely to induce an unwitting change that made it possible. Probably, agriculture arose under pressure of need and as our global requirement for food becomes more problematic, so agriculture has increasingly drawn in intellectual resources and it is realistic to see domestication, not as a once-for-all event, but as a process that continually refines and extends its control.

In the concluding chapter, there surfaces the element of social concern. It must because, in the face of a deteriorating situation, the problems addressed here do not merely indulge scientific curiosity. Those eight billion of which N. E. Borlaug writes – how are they to be fed and their environment sustained? Self-evidently, it will require an immense commitment of both pure and applied science that must not be blown off course by short-term political or economic considerations. At the centre of this concern are not merely our cereals but increasingly too, other grasses of more 'ecological' consequence.

As Watson remarked in the earlier volume, the grass family is a 'magnificent resource' though one we have far from sufficiently explored. This present volume extends the exploration and, having considered the matter, the reader can make his or her own judgement as to whether humankind inhabits the earth as host to, or guest of, the grass family.

G. P. Chapman
Wye College

Acknowledgements

My thanks are due to the various contributors, especially in view of their commitments elsewhere. I should like to record my appreciation of the way in which they sought to fulfil the concept this book embodies.

While the book was in preparation I had, through the generosity of the Royal Society and the Chinese Association for Science and Technology, the opportunity to travel widely in China. In Guanzhou (Canton) I saw something of Mrs G. Z. Zhang's work on bamboo hybridization and at the Nanjing Foresty University the bamboo collection of Professor C. S. Chou and his colleagues. I also visited Professor Ren Jizhou and his colleagues and saw the use of *Puccinellia* in combatting salinization in the Gobi desert. In Beijing I met Professor D. H. S. Chang whose work has concerned the remarkably cold-tolerant bamboos of Tibet.

Much nearer home, it is a pleasure to thank colleagues for support, including Professor D. A. Baker. My thanks are due to Dr W. E. Peat for his continuing steady enthusiasm for grasses. Mr Jeff Brooks has, where necessary, redrawn diagrams and maintained an inventive approach to photographing the grasses whose portraits embellish these pages.

I thank Mrs Susan Briant whose secretarial skills have contributed so much at all stages, Mrs Margaret Critchley and Mrs Terry Dinsdale, and the staff of Cambridge University Press who have, for this as for previous books, been helpful throughout.

Finally, I thank my wife, Sheila, for her support and encouragement.

G.P.C.

Part I

NATURAL DIVERSITY

1

Classification and evolution of the grasses

S. A. Renvoize and W. D. Clayton

Introduction

A classification is constructed by clustering taxa according to the characters they have in common, and expressing the result as a hierarchy of discrete groups. An associated activity is the search for diagnostic characters: those which are constant within a group but rare among its neighbours, and which therefore serve as good predictors of group membership.

The situation is vastly complicated by the occurrence of parallel evolution: the development of similar characters, such as awned lemmas, among plants which have little else in common. It is prevalent in grasses, and the resolution of the complications for taxonomy which are associated with it lies at the heart of a successful classification. The problem of distinguishing between sequential and parallel trends has been addressed in various ways, which fall into four broad categories.

(1) Seeking consensus among a large number of characters;

(2) Comparing the consensus derived from the different organs, such as spikelet structure and leaf-blade anatomy, on the assumption that they are unlikely to be subject to the same selection pressures and are therefore unlikely to evolve in a parallel manner;

(3) Searching for conservative characters: those which are either insensitive to transient selection pressure, or which have reached a local optimum from which they cannot progress because any change must initially be deleterious;

The earliest family name, Gramineae, is sanctioned by the International Code of Botanical Nomenclature even though it is not based upon a legitimate included genus; the later name, Poaceae, is an allowable alternative.

3

(4) Looking for clues about the way in which evolution works, and the direction in which it is moving, on the assumption that kinship can be presented as a rational sequence of transformations.

The application of these methods is often as much an art as a science, but considerable efforts have been made to analyse the processes involved and to represent them by algorithms which can be tested on a computer. These numerical techniques have done much to advance both theory and practice, but they still entail many subjective decisions and have difficulty with the large data sets required for an overall classification. However, the breath of critical reasoning which they have introduced is reflected in growing agreement about the core of our classification, and by a flurry of controversy over its many loose ends.

Inevitably the variability of the group increases as one moves up the hierarchy, and therefore the number of constant characters available for diagnosis drops until at the summit there are very few of them. The characters lost first are those most affected by parallel evolution. Thus, vegetative characters are extremely labile and seldom of much diagnostic value above species level. Spikelet characters are generally more stable, but even they become of limited significance above tribal level. The remaining ultraconservative characters are often cryptic expressions of internal physiology, which brings its own problems for they are expensive, in terms of time and skill, to observe and are often imperfectly recorded. Even they are not entirely free from the effects of parallel evolution, and so the search continues for the elusive touchstone that will infallibly map observable patterns of diversity and put them in congruence with the evolutionary tree from which they sprang.

Conservative characters

A practical taxonomy, intended as a ready means of communication, must recognize the limitations of our own thought processes, which do not store information as density functions of a multivariate continuum or as probabilistic measures of association, but which cling tenaciously to discrete groups and exclusive diagnoses. It has, therefore, to be constructed from characters which are conservative, exclusive and comprehensively recorded. At the summit of the classification, there are not many of these, and none of them is an infallible arbiter of group membership. The most important characters were discussed by Clayton & Renvoize (1986), and are briefly outlined below.

1. Photosynthetic pathways and their associated anatomical syndromes provide a framework for the classification, though they are not free from the complicating effects of parallel evolution, for the C_4 version seems to have evolved 11 times at least, on the evidence of present day members of the family.
2. Embryo strucure is a difficult character to use, but appears to be one of the most reliable diagnostic characters at subfamily level.
3. Fusoid cells are virtually confined to the Bambusoideae, though not possessed by all its members. They also appear anomalously in *Homolepis* (Paniceae).
4. Microhairs are generally slender. The swollen and absent variants are good predictors for the Eragrostideae and Pooideae, respectively.
5. Lodicules, whether thin or fleshy, are usefully diagnostic at subfamily level, though imperfectly recorded.
6. Large chromosomes with a basic number of $x=7$ characterize the Pooideae.
7. Woody culms characterize the Bambuseae.
8. One-flowered spikelets are an obvious, but unreliable, character. However, the variant with a male or barren floret below the bisexual floret is diagnostic for the Panicoideae.
9. Trimerous flowers have some diagnostic value, though they are mainly of interest for their phylogenetic implications.
10. Ligules are too erratic to be of much diagnostic significance, but are sometimes characteristic of a major group.

Direction of evolution

The taxonomic process is primarily phenetic – based on observable characters. There is much debate (Cronquist, 1987; Kellogg & Campbell, 1987) over the extent to which it should be influenced by phyletic theory, but the issue of phylogenetic polarity cannot be ignored. The conservative characters provide a few firm pointers when compared with their counterparts in other monocotyledonous families. Thus C_4 photosynthesis, fusoid cells, large chromosomes and woody culms are all rare among monocotyledons and can confidently be designated as derived; on rather flimsy evidence, it is conjectured that C_4 photosynthesis seldom reverts to C_3. Likewise trimerous floral symmetry is usual in monocotyledons and can be regarded as the ancestral state. The retention of trimerous flowers links the bambusoid grasses to the base of the phylogenetic tree, but their proximity to the origin is

6   S. A. Renvoize and W. D. Clayton

Fig. 1.1. First approximation to a classification of grasses, with superimposed grids for major differences in anatomy (double lines) and embryo structure (thin lines). (From Clayton & Renvoize, 1986.)

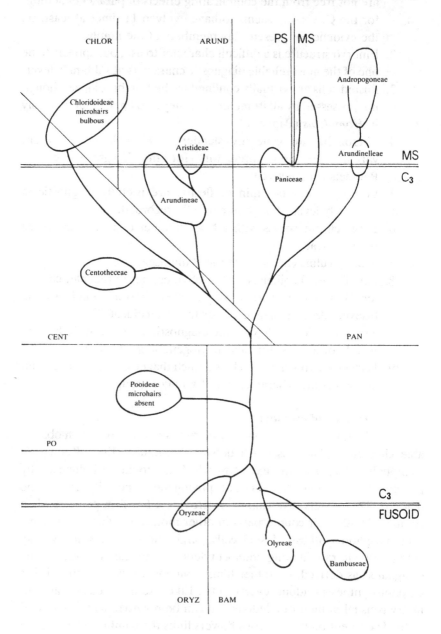

challenged by the concurrence of derived states. Confirmation of this link, and a similar distancing juxtaposition of derived states, can be found in the inflorescence.

Origin of the grass inflorescence

The putative origin of the grass inflorescence has been discussed previously (Clayton & Renvoize, 1986), but further study indicates that this interpretation erred in treating the spathate inflorescences of the Andropogoneae and Bambuseae as homologous. The spatheoles of the Andropogoneae terminate branches and subtend ebracteate racemes; they are clearly derived from simple racemes by multiplication of axillary branches and modification of their subtending leaves. By contrast, the inflorescence bracts of the Bambuseae are axillary to a continuing axis and subtend single spikelets. They are clearly not comparable to those of the Andropogoneae, so how did these structures arise?

There is no extant archetype of the grass spikelet, but it was suggested (Clayton, 1990) that an analogue could be found in the Marantaceae (Andersson, 1976). In *Thalia* the distal inflorescence branches become reduced to a distichous sequence of imbricate bracts and prophylls (the spikelet analogue); at the same time the proximal branches become modified to bracts (the panicle analogue) by the reduction of their subtending leaves. This structure is echoed in the bracteate quasi-panicle of many bamboos, and would seem to represent the primitive grass inflorescence. Subsequent transformations developed along two lines. The minor line, adopted by many of the giant bamboos, shortened the branches until the whole inflorescence became tightly compacted about the culm node. The major line lost first the prophylls and then the bracts subtending its branches to become the familiar panicle of many-flowered spikelets.

The situation is complicated by another minor line, exemplified by the bamboo *Phyllostachys*. Here the spikelets, subtended by glumaceous bracts, are borne on a short axis, the whole being enveloped in a spatheole and aggregated into fascicles. This transformation mimics the compound inflorescence of the Andropogoneae, wherein lay the seeds of error. In fact the *Phyllostachys* inflorescence appears to be a direct transformation from the primitive bracteate form; although sometimes displaying a reduced complement of prophylls and bracts, it does not progress to a true ebracteate raceme and lies outside the mainstream evolution of the grass inflorescence.

The prevalence of primitive inflorescence structures among tropical

8   *S. A. Renvoize and W. D. Clayton*

Fig. 1.2. Second approximation to a classification of grasses showing relationships between the tribes. (From Clayton & Renvoize, 1986.)

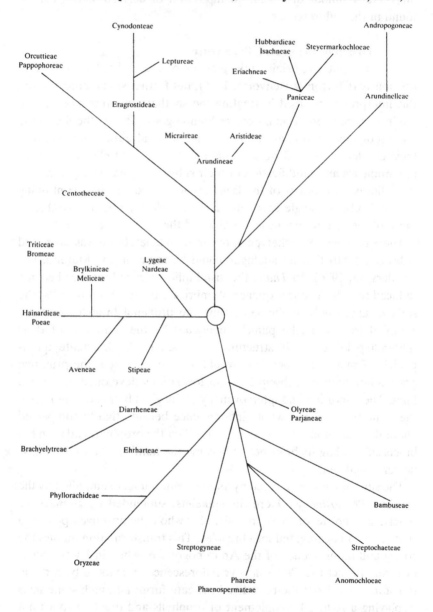

forest-dwelling grasses would seem to imply a forest origin for the family. However, this is belied by their universally anemophilous pollination, the very few recorded cases of entomophily being evidently secondary adaptations. Certainly their enormous success in open places, with a strongly seasonal climate that inhibits the development of a continuous tree cover, points to evolution driven by the selection pressures of seasonality and change, with their concomitant ecosystems involving fire, grazing and the activities of humans. The paradox can be resolved by postulating an origin in the tropical forest margin ecotone, where no commitment is made and forest or savanna are equal possibilities for evolution. In the end the mainstream of evolution led to the exploitation of the more stressful environment of the open country beyond the ecotone with an early offshoot resorting to the forest.

Photosynthetic alternatives

The earliest grasses almost certainly exhibited C_3 photosynthesis, in common with the majority of the monocotyledons. The C_3 pathway is perpetuated by the four most primitive groups, the bamboos, the arundinoids (with the exception of *Aristida*, *Stipagrostis* and *Centropodia*), the centothecoids and the stipoids, as well as by the more advanced pooid group. The subsequent evolution of the C_4 pathway increased the adaptive power of the family and enabled two advanced groups, the panicoids and chloridoids, to exploit more effectively the open habitats of the tropics and subtropics, ousting their forerunners, the uncompetitive arundinoids, and restricting them almost completely to the southern hemisphere. The C_3 pathway, disadvantaged in open tropical environments but competitive in the moist or shady environments of forest and swamp, enabled the bambusoids to survive unchallenged. It also persisted in some primitive panicoids and in the pooids when this group evolved from the early arundinoid line and expanded in the temperate zone. For a review of this subject, see Hattersley (chapter 2, this volume).

The evolutionary course of the family has been in three principal directions: (1) from an unspecialized ancestor of forest margins, a route to a forest environment was taken by the bamboos; (2) cool-temperate open steppe or meadows was favoured by the pooids; and (3) open savannas and semi-arid regions of the tropics and subtropics provided the ideal environments for the expansion of the chloridoids and panicoids.

The successful exploitation of open habitats provided the greatest

(a) Thalia geniculata

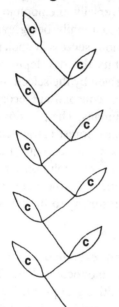

(b) Glaziophyton type

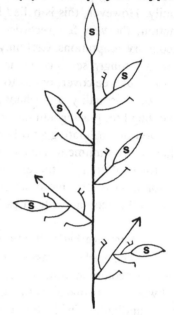

(c) Bambusa type

(d) Phyllostachys type

prophyll   bract

Fig. 1.3. (*a*) *Thalia geniculata* distal inflorescence branches in which each cyme (c) is reduced to a single flower enclosed by a bract and a prophyll. Sterilization of the 2 lowest cymes and their reduction to a single bract, plus the condensation of the branch segments, would result in a structure similar to the grass spikelet. (*b*) *Glaziophyton* type. This represents the simplest arrangement, with each spikelet (s) subtended by a bract and prophyll and borne in an open paniculate inflorescence. (*c*) *Bambusa* type, the result of further condensation of the branches into a compact cluster. (*d*) *Phyllostachys* type, an alternative arrangement combining condensation of the branches with the introduction of large intermittent bracts enclosing the spikelet clusters.

stimulus for the expansion of the family. This resulted largely from the quasi-symbiotic relationship with grazing animals which is associated with the refinement of the perennial habit, the strategically situated meristems at the base of the leaf blade and above each node and the absence of poisonous secondary compounds, commonly found in many other plant families.

A second influence encouraging the exploitation of open habitats, especially in the tropics, was the ability to withstand fire, an annual event in many environments with a pronounced dry season.

The forest grasses, subfamily Bambusoideae

The least amount of change from a monocotyledon ancestor is shown by the subfamily Bambusoideae, whose floral morphology includes one to six stamens, multinerved paleas, two to three lodicules and one to three stigmas. Their adaptation to forest habitats precluded them from the main evolutionary line of the whole family, and therefore they are perceived as a branch which diverged at an early stage from the ancestral grasses.

In contrast to their primitive floral structure, features such as woodiness, a strategy for competing in the forest environment, and a complex inflorescence indicate considerable evolutionary advancement.

Although the bambusoids are clearly undergoing adaptive radiation in groups such as the tribe Olyreae, they still include some taxa with very strange morphology which exist in isolation and provide little indication of their origins. Thus the subfamily appears as an assemblage of diverse forms which bear little resemblance to each other but which are united by cryptic characters of leaf-blade anatomy and embryo structure.

On the basis of their habit, three major tribes are identified: the woody Bambuseae, semiwoody Olyreae and herbaceous Oryzeae. Within these groups the various genera exhibit a mixture of primitive and advanced features.

BAMBUSEAE

The tribe Bambuseae (49 genera) is the pre-eminent group in the subfamily and comprises the woody-stemmed genera – the typical bamboos.

Bamboo taxonomy has been much neglected and the subdivision of the tribe remains controversial, though currently a revival of interest in the tribe is making some progress.

Undoubtedly the woody bamboos hold the key to primitive spikelet structures, though their woody culms, complex inflorescences and branch architecture and their possession of fusoid cells are all derived states which represent considerable advancements. The tribe can be seen as a line which has adapted to forest ecosystems and is competing with dicotyledonous trees. In this environment, selection pressure on the inflorescence has differed from that in the herbaceous grasses, leading to adaptations which have preserved many primitive features.

OLYREAE

The tropical tribe Olyreae (16 genera) represents a line of evolution which has adapted to the environment of the forest floor, developing lanceolate to ovate leaf blades which are supported by semiwoody culms and held horizontally and at right angles to each other in order to intercept what little light filters through the canopy. This group includes plants of relatively small stature, 20–30 cm high as well as plants with canes 200–300 cm high. Although pantropical in distribution, the Olyreae are most numerous and varied in tropical South America, where they appear to be undergoing adaptive radiation, as evidenced by the rather narrowly defined genera. The tropical forest floor is apparently an environment where there are selection pressures and opportunities to which, in South and Central America at least, the Olyreae are responding.

The origins of the Olyreae may lie at the beginning of the bambusoid evolutionary line, as evidenced by their semiwoody culms and primitive floral structure with three lodicules and three stigmas.

PARIANEAE AND PHAREAE

Allied to the Olyreae are two other tribes, the Parianeae (two genera) in which the male and female spikelets are separate, but on the same inflorescence, and the Phareae (three genera).

ORYZEAE, STREPTOGYNEAE AND PHYLLORACHIDEAE

The herbaceous habit, common to the rest of the family, is an anomaly in a subfamily which is principally woody and shrub-like. However, the tribe Oryzeae (12 genera) has adopted it in the course of its exploitation of aquatic habitats. Its membership of the subfamily is confirmed, however, by its leaf-blade anatomy, which is bambusoid.

Allied to the Oryzeae are two minor tribes, Streptogyneae (one genus) and Phyllorachideae (two genera).

EHRHARTEAE, DIARRHENEAE, BRACHYELYTREAE, PHAENOSPERMATEAE, STREPTOCHAETEAE AND ANOMOCHLOEAE

The remaining six tribes present some bizarre examples of grass morphology, especially in spikelet structure, and although qualifying for inclusion in the subfamily on the basis of embryo structure and leaf-blade anatomy, they occupy rather isolated positions, giving no hint of their evolutionary relationship with the three principal tribes described above or with each other.

The tribe Ehrharteae comprises a single genus *Ehrharta* (35 species) in which the lodicules are membranous and the embryo bambusoid. The spikelets, although three-flowered, are reduced to one fertile floret and two sterile florets, thus bearing a strong similarity to the situation in the Oryzeae. Unlike other minor bambusoid tribes, which comprise few species, the Ehrharteae has found a niche in the semi-arid, winter rainfall zone of the western Cape of South Africa, where there is a burst of adaptive radiation; other species, however, are found in more mesic environments.

The four tribes Diarrheneae, Brachyelytreae, Phaenospermateae and Streptochaeteae are monotypic and of uncertain affinities; their inclusion in the subfamily rests primarily on cryptic characters.

The final tribe, the monotypic Anomochloeae, is remarkable for its spathate inflorescence and spikelet, which are highly unorthodox and have no parallel among the bamboos, or the rest of the family for that matter.

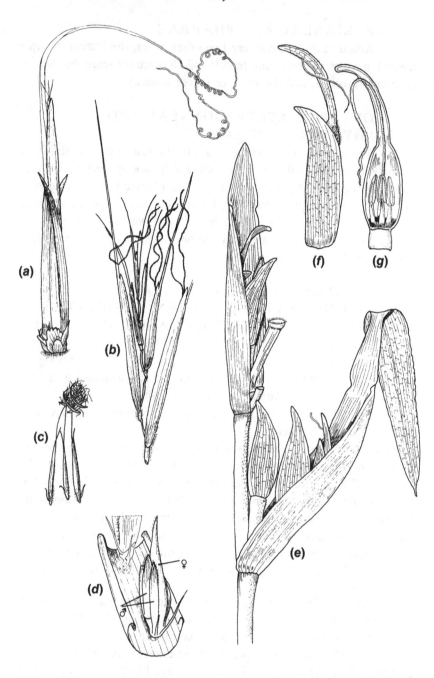

THE BAMBOO ACHIEVEMENT

The achievement of the bambusoids has been to adapt success-fully to the competitive tropical forest-shade environment. The variations in inflorescence structure and in habit which have evolved appear to be far greater than in any other grass group. However, many of these variations are represented only in the minor tribes, which often comprise solitary taxa, and the impression is that they are mostly relicts which survive in isolated ecological backwaters, where stable environments have exerted little pressure on them and thus provided no stimulus for further speciation.

The Bambusoideae has remained outside the mainstream of taxonomic research for almost a century and although the work of the late Thomas Soderstrom (1987; Soderstrom & Zuloaga, 1988) did much to promote an interest in the subfamily, the enigmatic complexity of their habit and floral biology remains imperfectly understood. The interpretation of their complex inflorescences is a confusing issue and the unconventionally long flowering cycles of many of the woody members continues to frustrate efforts to understand relationships between the taxa. However, the recent achievements of *in vitro* culture methods in inducing flowering in laboratory plants (Nadgauda, Parasharmi &

Fig. 1.4. (a) *Streptochaeta spicata* spikelet, showing long awn on the lemma, 3 very large lodicules and large palea split to the base with reflexed tips which serve as a hook mechanism; the basal glumes are very short. (b) *Streptogyna crinita* spikelet with the glumes removed to show the filamentous, retrorsely scabrid stigmas and the springy rachilla internode which remains on the floret and acts as a hook mechanism, shown in (c). In both examples (a) and (b), the long awns or stigmas become entangled on the inflorescence, serving to retain the florets after they have disarticulated and to dangle them in the manner of a fish-hook until caught in the hair or fur of a passing animal and consequently being swept off 'en masse' and dispersed. (d) *Phyllorachis sagittata*. Section of the foliaceous central axis bearing short, cuneate racemes of 1–4 unisexual spikelets. (e) *Anomochloa marantoidea*. Portion of the inflorescence showing the lowest and next to lowest partial inflorescences enclosed by large bracts, the lower bract bearing a reduced blade. (f) Spikelet in side view, showing the lemma enclosed by a bract; the palea and glumes are absent (note the exerted stigma). (g) Longitudinal section of the spikelet with the bract removed and the stamens at a pre-emergence stage.

Mascarenhas, 1990) may be the breakthrough which helps to overcome this obstacle.

Of all the grass groups, the bamboos are the least known and certainly the least understood in terms of their evolutionary progress. They still present the taxonomist with a challenge for the future, to understand the adaptive significance of their habit and of their inflorescence and spikelet forms.

The earliest grasses, the subfamily Arundinoideae

If the Bambusoideae represents an early offshoot into forest habitats, then the subfamily Arundinoideae, with the primitive C_3 photosynthetic pathway and unspecialized spikelets, appears to be the descendant of an ancestral line closest to the earliest true grasses. To a limited extent, almost every feature found elsewhere in the herbaceous members of the family is manifested in the Arundinoideae, which is further evidence of its ancestral position.

It is most likely that members of the Arundinoideae are the direct descendants of the grasses which first moved into the open 'savanna ecosystem' from the savanna/forest ecotone. The Arundinoideae represents a line of evolution which originated in the tropical zone but, from a position of relative predominance at an early stage in the evolution of the family, appears to have been subsequently ousted by the more efficient C_4 tribes that arose from it and which now dominate the tropics and subtropics. Those members of the subfamily which have survived have mostly retreated into southern hemisphere refugia in South Africa and Australia. In another direction, descendants of the early arundinoids spread into the temperate northern hemisphere, adapting to the rigours of a seasonal climate which, during part of the year was very cold. These species, which now constitute the subfamily Pooideae, retained the C_3 pathway, as the temperate environment offered no stimulus for the evolution of the C_4 pathway. The members of the Arundinoideae which occur in the northern hemisphere are limited to a few species, possibly again the survivors of a much larger group which has been decimated as a result of competition from its better adapted descendants.

The majority of the genera in this subfamily are grouped in the tribe Arundineae, to which are linked two satellite tribes, the Aristideae and Micraireae.

ARUNDINEAE

The genera in the Arundineae (41 genera) are weakly linked or form isolated groups, which is indicative of a long evolutionary history, the tribe having reached an advanced stage of decline with gaps in the continuity of genera resulting from extinctions. There are some exceptions however, for genera such as *Pentaschistis* (65 species) and *Rytidosperma* (90 species) have found a new lease of life in the southern hemisphere. There are also a few small genera which have been successful: *Phragmites* (three species) is a significant component in many temperate and tropical swamp ecotypes, and tropical *Gynerium* (one species) and warm–temperate *Arundo* (three species) likewise dominate, or are prominent members of, the communities in which they occur.

The Arundineae, in spite of the rather weak affinities among its genera or generic groups, is unified by the characteristic occurrence of the C_3 photosynthetic pathway and the possession of a distinctive embryo structure. It is difficult to categorize features of the tribe as either advanced or primitive, and therefore to infer which direction evolution has taken within the tribe but the following sequence has some credibility.

The most primitive species appear to be in a heterogeneous group (nine genera), unified by short glumes and multinerved lemmas with entire tips.

A possible advancement is towards long glumes and bilobed lemmas bearing a geniculate central awn. Sixteen closely related and narrowly defined genera, centred on *Chionochloa*, fit here, and mostly comprise few species, the exceptions being *Pentaschistis* and *Rytidosperma*. *Centropodia* belongs with this group and is unique in the tribe in having the C_4 photosynthetic pathway. Whether this is a relict or an advanced genus is debatable, but it nevertheless shows the potential within the tribe to evolve the C_4 pathway. To these may be added four genera with 1-flowered spikelets and a wrap-around lemma, two of which, *Diplopogon* and *Amphipogon*, are intermediate with the satellite tribe Aristideae.

A group of four genera, centred on *Crinipes*, have 3-nerved lemmas, but they appear to be unrelated to the tribe Eragrostideae, for which this character is diagnostic.

Cortaderia (which includes *Lamprothyrsus*) is an isolated genus which has adapted successfully to a lifestyle colonizing unstable slopes. It is unusual in being dioecious, but apart from this it is very closely allied to *Chionochloa* in the main group.

Six genera displaying a reed habit form an artificial assemblage; they comprise few species and, along with the group with short glumes and the group with 3-nerved lemmas, may represent the remnants of the earliest members of the tribe.

MICRAIREAE

The monotypic tribe Micraireae is almost unique within the family in the spiral phyllotaxy of its leaves, the only other occurrence of this arrangement being in *Arundoclaytonia* (Steyermarkochloeae, Panicoideae).

With its curious moss-like habit, *Micraira* (13 species), has found a niche on the tops of inselbergs in northern and north eastern Australia. This is a habitat in which soil is almost totally lacking and competition is limited to those plants which can adapt to alternating periods of drought and flood. The tribe occupies an isolated position, with no obvious relatives and, although it is included in the subfamily on the basis of its arundinoid leaf-blade anatomy, its position can only be explained as an advancement, possibly derived from the *Pentaschistis* line.

ARISTIDEAE

If the Arundineae represent the declining remnants of the earliest grasses, then the tribe Aristideae (three genera), which is probably derived from the Arundineae, has reversed the decline by successfully adapting in several ways to the demanding environments of arid and semi-arid regions.

The smallest and most primitive genus of this tribe, *Sartidia* (four species), is confined to South Africa and Madagascar and utilizes the C_3 pathway for photosynthesis. The more advanced *Stipagrostis* (40 species) is spread throughout the Old World and utilizes the C_4 pathway. *Aristida* (250 species) is pantropical and also utilizes the C_4 pathway. However, its anatomical structure comprises inner and outer bundle sheaths, both of which contain chloroplasts. In all other C_4 grasses with a double bundle sheath, only the outer is chlorenchymatous (but see Neurachninae, Paniceae). Hattersley & Browning (1981) and Hattersley (1988) suggested that the outer sheath serves as the site of a secondary Calvin cycle, fixing CO_2 'leaking' from the site of the primary Calvin cycle in the inner sheath. Such a refinement as this, resulting in increased conservation of CO_2 within the leaf blade, can only be viewed as an advancement. *Aristida* has been biochemically typed as NADP-malic enzyme (NADP-ME; Hattersley, 1988). Thus within a single tribe a transition has been achieved from C_3 to C_4 photosynthesis.

Although most of the genera in the subfamily appear to be rather unsuccessful, *Pentaschistis*, *Rytidosperma*, *Aristida* and *Stipagrostis* are apparently actively evolving. The combination of (1) morphological and physiological adaptation to arid environments, (2) specialized floret structure resulting in efficient dispersal of the seeds and (3) the C_4 pathway enabling them to exploit the tropical environment with maximum efficiency, has proved successful and has resulted in the remarkable expansion of the tribe, especially for *Aristida*.

The picture of the Arundinoideae is of a subfamily in which most of the genera are unspecialized in their spikelet structure and, to a large extent, in their physiology and photosynthesis, and in which there are isolated genera or groups of genera which are widely distributed both geographically and ecologically and yet are uncompetitive; all of which suggests evolutionary age. However, the presence of the C_4 pathway in three genera, *Aristida*, *Stipagrostis* and *Centropodia*, and specific trends in spikelet structure, indicate that the ancestral line which culminates in this subfamily as we know it is a potential source of the subfamilies Panicoideae, Chloridoideae and Pooideae which are so outstandingly successful in the present day.

The subfamily Centothecoideae

The subfamily Centothecoideae is an odd group which has adapted to forest shade in the tropics and in so doing it has adopted the obviously advantageous, broad leaf blades. The members of the subfamily thus bear a strong similarity to the Olyreae. This, however, is an example of parallel evolution as the leaf-blade anatomy is quite distinctive and bears no relation to that of the Bambusoideae. The cuneate lodicules suggest an alliance with the Arundinoideae, and *Megastachya*, which has the least modified spikelets, bears some similarity to primitive members of the Arundineae such as *Tribolium* and to primitive members of the Eragrostideae such as *Uniola*.

CENTOTHECEAE

The subfamily comprises a single tribe, the Centotheceae (10 genera), which is distributed in the Old and New World tropics and North America. Such a wide distribution suggests a lengthy evolutionary history, which is further supported by the very variable spikelets which have the effect of leaving the genera isolated morphologically. The subfamily does not have an obvious identity and possibly, like the Arundinoideae, it was larger in former times but now comprises a mosaic of evolutionary survivors and relict genera in widely dispersed niches.

The Centothecoideae appears to be an early C_3 offshoot from the arundinoid line that has remained in isolation and has no apparent descendants.

Expansion into the temperate zone – the subfamily Pooideae

The premier group of the cool temperate zone is the subfamily Pooideae, for which the main line of evolution has been divided into a physiological adaptation to the temperate environment and, to a lesser extent, into the annual habit typically associated with the winter rainfall régime of the mediterranean zone. The inflorescence and spikelets are unspecialized, which suggests that response to selection pressures have been related to internal physiology rather than to external morphology.

Modification of the endosperm storage material, liquid in some Aveneae or soft in some other tribes, and the almost universally large chromosomes are peculiar to the Pooideae.

In their adaptation to cold climates, the Pooideae are considered advanced. However, the primitive C_3 pathway has been retained consistently throughout the subfamily; the C_4 pathway, although considered an evolutionary advancement, is only advantageous in a tropical environment.

NARDEAE AND LYGEAE

Two monotypic tribes, Nardeae and Lygeae, are included in the subfamily, although they have no obvious affinities with the other members. A link to the Arundineae is suggested by the presence, in these tribes, of microhairs, which are otherwise absent from the subfamily, but their embryo and ligule are pooid. The two tribes appear to be intermediate and, although it seems most likely that they are primitive members of the Pooideae, their presence in the subfamily is somewhat arbitrary.

STIPEAE

The tribe Stipeae (nine genera) combines bambusoid features with pooid, which indicates a possible direct line from very primitive grasses and places them, with the Arundinoideae, as the direct descendants of those grasses which first moved into the savanna ecosystem from the forest margins. The lack of arundinoid features suggests that this tribe diverged before the arundinoids had evolved. Unlike the other primitive groups, Centotheceae, Nardeae, Lygeae and most Arundineae, the Stipeae have adapted successfully to the steppe environment of Eurasia and the Americas, where they now form a significant component.

The 3 lodicules and the rounded, enclosed, keel-less palea, are primitive features which indicate a possible link to the bamboo ancestral line. In contrast, the embryo is pooid, the specialized 1-flowered spikelets are far from primitive and the 1-celled pylidial hairs of the leaf blades are unique.

Although possibly pre-dating the Arundinoideae as an early divergent line, the Stipeae have subsequently undergone considerable evolution in isolation. In *Stipa* this has resulted not only in a large number of species but also in the development of numerous infrageneric groups.

The principal genus *Stipa* (300 species) is one of the largest in the family. It parallels *Aristida* in ecological preferences, habit and spikelet structure but any close phyletic relationship is ruled out by the C_3 pathway and lack of microhairs.

The centres of diversity in the subfamily lie with the tribes Poeae and Aveneae.

POEAE

The Poeae (47 genera) follows three main lines of evolution, centred on the genera *Festuca*, *Poa* and *Sesleria*. In many ways they parallel the Eragrostideae in spikelet structure, but have failed to develop the racemose inflorescence to the same extent. The tribe Triticeae, which is closely related to the Poeae, is characterized by a solitary, bilateral raceme, but this is simple compared to the inflorescence of eragrostoid genera such as *Dinebra* and *Acrachne*, which have multiple racemes. Throughout the tribe, in all the lines of evolution, there is a progression from perennial to annual habit. Closely related to the Poeae are the tribes Hainardieae, Meliceae and Brylkinieae.

HAINARDIEAE

Members of the Hainardieae (six genera) are almost all annual; one genus is perennial. They are adapted to saline soils and have a distinctive inflorescence in which the raceme is cylindrical and the

glumes collateral. Their annual habit and other features suggest a possible relationship with the group of annual species clustered around
Desmazeria in the Poeae.

MELICEAE, BRYLKINIEAE AND BROMEAE

Members of the Meliceae (eight genera), another tribe which
appears to be derived from the Poeae, have unique lodicules which are
connate, short, fleshy and truncate. Their chromosomes are small and
$n=9$ or 10, in contrast to the members of the Poeae, which have large
chromosomes and $n=7$. Allied to the Meliceae is the monotypic tribe
Brylkinieae, an oddment whose members possess the distinctive combination of large, free, hyaline lodicules and an umbonate ovary cap and
are therefore excluded from any other tribe.

Simple rounded starch grains are an unusual feature in grasses, occurring only in the tribes Bromeae and Triticeae which, although closely
related to the Poeae, are separated on this basis. Members of the
Bromeae (three genera, *Bromus* 150 species) bear a close similarity in
spikelet structure to some *Festuca* species and would appear to be an
offshoot from the ancestral members of this group in the Poeae.

TRITICEAE

The Triticeae (18 genera), like the Bromeae, represents a
separate line from the Poeae. This tribe, which is probably the most
advanced in the subfamily, presents some of the most vexatious problems of systematics in the whole family, as the genera have a propensity
to hybridize to a degree which is unparalleled by any other group. The
inflorescence is a solitary, bilateral raceme, which is an advancement
over the simple panicle which predominates in the Poeae.

Brachypodium, sometimes placed in a separate tribe, is included here
because it also has simple starch grains and intergrades morphologically
with *Elymus*, through the intermediate species *E. serpentini*.

AVENEAE

The other major group within the subfamily Pooideae is the
tribe Aveneae. For convenience it is divided into four subtribes, but
several lines of evolution and centres of diversity can be identified. The
long papery glumes and elaborate dorsal awn which characterize the
tribe, suggest an advancement over the Poeae where there is very little
elaboration of glumes and the awns are simple.

The most primitive group within the tribe is probably the subtribe

Duthieinae (four genera), characterized by three lodicules and three stigmas. Although the spikelets imitate the Arundineae in their bidentate, awned lemmas, their inclusion in that tribe is incompatible with their membranous ligules and thin lodicules, which are typical of the Pooideae.

The two main subtribes are Aveninae (23 genera) and Alopecurinae (27 genera). They follow several lines of evolution centred on *Helictotrichon*, *Trisetum* and *Deschampsia* in the Aveninae, and *Agrostis* and *Calamagrostis* in the Alopecurinae, the latter subtribe representing a progression from the *Deschampsia* group.

The subtribe Phalaridinae (three genera) appears to be an offshoot from the Aveninae, possibly linked through *Deschampsia*.

The subfamily Pooideae is more or less limited to the temperate zones, especially of the northern hemisphere, its occurrence in the tropics being limited to the tops of high mountains. The large number of genera in Eurasia implies an origin in the northern hemisphere, but alternatively this may be a reflection of the larger area available for exploitation and the consequent greater variety of possible habitats, the southern hemisphere temperate zone being virtually confined to New Zealand and southern South America. The tribes of the Pooideae appear to have advanced in parallel. They show little adaptive specialization, except in their tolerance of cold climates and the retention of the primitive C_3 photosynthetic pathway.

SUCCESS IN THE TROPICS – THE SUBFAMILIES CHLORIDOIDEAE AND PANICOIDEAE

The remaining members of the grass family, the subfamilies Chloridoideae and Panicoideae, have come to dominate the open environments of the tropical and subtropical zone. They are the descendants of the early arundinoid grasses which, through the evolution of the C_4 photosynthetic pathway, gained a competitive edge not only over their C_3 grass ancestors but over any C_3 plant which was likely to challenge their progress.

From a putative base in the arundinoid ancestral line, these grasses have diverged in the tropics along two principal routes. Through the chloridoids they have adapted to a wide range of ecotypes, especially stressful arid habitats, and through the panicoids they have adapted primarily to stable mesic situations and climax grasslands.

Chloridoideae

The trademark of the subfamily Chloridoideae is the unspecialized, usually many-flowered spikelets with (1-)3-nerved lemmas, distinctive leaf-blade anatomy and C_4 photosynthetic pathway. The uniformity which runs through the subfamily suggests strong evolutionary coherence of the genera and probably reflects the 'suitability' of the C_4 pathway to the environment to which the subfamily is adapted. This is in contrast to the other principal tropical subfamily the Panicoideae which is remarkably diverse in inflorescence type and photosynthetic pathway.

In this subfamily it appears that there has only been one route to the C_4 pathway, but subsequent evolution has resulted in two structural/ biochemical variants (phosphoenolpyruvate carboxykinase (PCK) and NAD-ME types). The adaptive significance of these variants is not entirely clear, although suggestions that they correspond to varying degrees of habitat aridity were discussed by Hattersley (1988, 1992).

The typical suites of anatomical/biochemical/ultrastructure features for PCK and NAD-ME types in chloridoids were outlined by Hattersley (1988) and additional data were given on variations found in *Triodia*, *Enneapogon*, *Triraphis* and *Eragrostis*.

The central tribes of the Chloridoideae are the Eragrostideae and Cynodonteae, to which are added three minor tribes, the Leptureae, Pappophoreae and Orcuttieae.

ERAGROSTIDEAE

The tribe Eragrostideae is divided into five subtribes on the basis of morphological characteristics, with each group often spanning a range of habitat types from grassland to semidesert, sand dunes and saline situations.

The principal subtribe is the Eleusininae (53 genera), in which there are several lines of evolution and centres of diversity based on *Triraphis*, *Tridens*, *Leptochloa*, *Eleusine* and *Eragrostis*. An evolutionary extension of the Eleusininae is the subtribe Sporobolinae (eight genera), in which the inflorescence is consistently a panicle and spikelets are reduced to a single floret; it includes the large and important genera *Sporobolus* and *Muhlenbergia*.

The Triodiinae (four genera), Uniolinae (four genera) and Monanthochloinae (seven genera) are subtribes which represent early offshoots in the evolution of the Eragrostideae. The Triodiinae and Uniolinae, in the morphological similarity of their spikelets with

Tribolium, *Spartochloa* and *Styppeiochloa*, provide evidence of a possible origin of the tribe from the subfamily Arundinoideae.

CYNODONTEAE

An evolutionary course different from that of the Eragrostideae has been taken by the tribe Cynodonteae, although its origins probably lie with the ancestral eragrostoid line. Both tribes are commonly associated with seral or disclimax communities but whereas the Eragrostideae show a tendency towards the weedy or the xerophytic lifestyle, the Cynodonteae are more likely to be components of closed communities.

The tribe is readily identified by the characteristic facies of its cuneate spikelets resulting from the reduction of the fertile florets to one per spikelet. This development has led to a new range of evolutionary trends which includes participation of the sterile florets in the construction of the disseminule, retreat of the floret within enclosing glumes, shedding of the spikelet as a whole, abbreviation of the raceme and conversion of the raceme to a disseminule. Panicles have been 'discarded' and racemose inflorescences have been adopted throughout the tribe.

Four subtribes are recognized. The Pommereullinae (three genera) is an early offshoot in which the spikelets contain two to four fertile florets, although the spikelet shape allies it to *Tetrapogon* in the Chloridinae rather than to any genus in the Eragrostidae.

Members of the Boutelouinae (16 genera) are, with one exception, all New World and are remarkable for their success in extending into temperate North America where they are an important component of the prairie vegetation. This subtribe appears to be an early offshoot from the main line, with many of the genera narrowly defined. The trend here is towards solitary or scattered, unilateral, deciduous racemes.

The Chloridinae (27 genera), forms the core of the tribe, with most of the genera centred on *Chloris*.

The Zoysiinae (13 genera), which has probably evolved from *Chloris*, is an advanced Old World group in which the highly modified spikelets are borne on deciduous reduced racemes. The genera occupy a variety of habitats from dry bushland to coastal sand dunes.

LEPTUREAE

The genus *Lepturus*, tribe Leptureae, is an oddment which has no strong link to any of the groups so far outlined. The one-flowered spikelets arranged in spikes suggest an affinity with the tribe Cynodonteae, but its nearest relatives are *Oropetium*, an anomalous member of the tribe Eragrostideae, with one-flowered spikelets, which appears to be a derivative from *Tripogon*, and *Lepturopetium*, which is tentatively assigned to the Cynodonteae. In the absence of any character which provides positive evidence of affinities to either the Eragrostideae or the Cynodonteae, *Lepturus* is kept in a separate tribe, albeit a closely related one.

PAPPOPHOREAE AND ORCUTTIEAE

The other two tribes included in the subfamily, the Pappophoreae (five genera) and Orcuttieae (three genera), comprise mostly New World species, identified by their many-nerved glumes and many-nerved or many-awned lemmas. Their inclusion is based on their chloridoid leaf-blade anatomy, but their affinities are obscure and they stand apart on a separate evolutionary line.

Subfamily Panicoideae

Whereas the Chloridoideae is a strongly coherent group, with only two biochemical variants of the C_4 pathway and limited diversity in their spikelet and inflorescence structure, the Panicoideae in contrast display a range of variation in both photosynthetic pathway and inflorescence structure which exceeds that of any other major group. The C_3 pathway and all three biochemical variants of the C_4 pathway are present, in addition to an inflorescence structure which ranges from solitary to paired spikelets arranged in panicles, racemes or spikes, with changes in sexuality of spikelets and modifications of the pedicels and rachis internodes.

The retention of the C_3 pathway has allowed some species successfully to exploit the mesic habitats of the tropics, especially forest margins and clearings, habitats in which the Chloridoideae are markedly absent. The greatest morphological photosynthetic diversity, however, is found amongst the C_4 species which have adapted to, and in most cases come to dominate, the climax grasslands of the tropics.

The subfamily Panicoideae has followed two principal lines of evolution: the first is centred on the tribe Paniceae and its associated minor tribes Isachneae, Hubbardieae, Eriachneae and Steyermarkochloeae; the second is centred on the tribe Andropogoneae and the associated tribe Arundinelleae.

The Arundinelleae and their descendants the Andropogoneae

represent an early divergence from the ancestral panicoid line and share with the other half of the subfamily the unique feature of the 2-flowered spikelets in which the florets are dimorphic. The Andropogoneae, however, have taken a very different line and developed a remarkable sequence of evolutionary advancements based upon the pairing of the spikelets, the selective sterilization of the florets and the modification of the internode and pedicel.

ISACHNEAE
The tribe Isachneae (five genera) has two-flowered spikelets in which both florets are fertile, a condition which probably prevailed in primitive panicoids. In this case, the tribe is most likely to be derived from *Panicum* sect. *Verruculosa*, with which it has many similarities, and to have reverted to bisexual spikelets in the course of its evolution.

Isachneae is maintained as a separate tribe on the strength of its florets and unusual leaf-blade anatomy, in which the chlorenchyma is a mixture of oblong radially arranged cells in the adaxial part of the leaf-blade and equidimensional, irregularly arranged cells in the abaxial part. The genera mostly favour marshy or aquatic habitats, although some species of *Isachne* have adapted to a forest shade environment, which is compatible with the retention of the C_3 photosynthetic pathway throughout the tribe. *Isachne* (100 species) is the central genus, the other four genera accounting for no more than 10 species.

HUBBARDIEAE
The monotypic tribe Hubbardieae, which is characterized by the absence of paleas, is apparently derived from the Isachneae, sharing similarities in spikelet structure, the disarticulation of the florets above the glumes and the C_3 photosynthetic pathway; it also favours damp habitats.

STEYERMARKOCHLOEAE
The tribe Steyermarkochloeae comprises two extraordinary genera which are found in isolated localities in the savannas and white-sand caatingas of the northern and south-central Amazon region. *Steyermarkochloa* has a rachilla extension, which is very rare in panicoids and led the original authors to consider it to be arundinoid, but features such as its spiciform panicle and two-flowered spikelets which fall entire are more in accord with the Paniceae. In fact this genus bears a strong resemblance in its spikelet and panicle structure to *Hymenachne*.

However, the segregated male and female spikelets are unusual in the subfamily Panicoideae, and the placing of the tribe is still controversial. *Arundoclaytonia*, although very different in habit, leaf and inflorescence morphology is included in this tribe on the basis of spikelet similarity. It is nevertheless a very weird member of the subfamily.

Both genera are of the C_3 photosynthetic type, on the evidence of leaf-blade anatomy (Davidse & Ellis, 1984, 1987), which suggests an early divergence from the ancestral panicoid line, as does their morphological and geographical isolation.

ERIACHNEAE

The tribe Eriachneae (two genera) has 2-flowered spikelets which disarticulate between the florets, indurated lemmas with inrolled margins and an embryo structure which allies it to Isachneae and Paniceae, but there is no direct link and the relationship is a very loose one. The bisexual florets and awned lemmas place it in isolation, as does the C_4 photosynthetic pathway which comprises an atypical suite of characters: NADP-ME biochemistry but double bundle sheaths and, in some species, centripetal chloroplasts (Prendergast & Hattersley, 1987).

PANICEAE

In the tribe Paniceae the 2-flowered spikelet is shed whole, the florets are dimorphic with the lemma and palea of the upper floret indurated and tightly enclosing the caryopsis, taking over the role of the glumes as protective bracts. Within this tribe there is the complete range of C_3 and C_4 photosynthetic pathways, which suggests that the tribe is in an early stage of evolution, responding to numerous different environmental pressures and not committed to any particular one, as seems to be the case with the Eragrostideae and Cynodonteae.

The most primitive group is the subtribe Neurachninae (three genera) of Australian semi-arid environments. The tough glumes, unspecialized upper lemma and 3 stigmas suggest an early divergence from the main Paniceae line. This subtribe makes the transition from the primitive C_3 photosynthetic pathway in the genus *Thyridachne* to the advanced C_4 pathway in the genera *Neurachne* and *Paraneurachne*. In this case, because all the species concerned are found in arid habitats, the C_3 pathway is most likely to be a relict feature from a C_3 ancestor, rather than a secondary adaptation; in tropical or subtropical groups, the C_3 pathway is usually associated with an aquatic or mesic shade environment.

Although the C_4 members of the subtribe are of the NADP-ME type biochemically, the anatomy is not completely in accord with the usual suite of characters associated with this pathway (Hattersley, 1988). In the genus *Neurachne*, *N. minor* is of an intermediate C_3–C_4 type, whereas *N. munroi* is of the C_4 NADP-ME type but has a double chlorenchymatous sheath similar to the arrangement recorded for *Aristida*. *Paraneurachne* is similarly of the C_4 NADP-ME type, with a double chlorenchymatous sheath. The atypical photosynthetic suites in this tribe are discussed at length in papers by Prendergast & Hattersley (1985), Hattersley & Roksandic (1983), Hattersley & Stone (1986), and Hattersley (chapter 2, this volume).

The subtribe Setariinae (66 genera) is the main group of the Paniceae. All the genera have a characteristically hardened upper lemma. It is centred on the genus *Panicum*, which, with 470 species, is second only to *Poa* in size. The diversity of *Panicum* is such that all the genera in the subtribe, those of two other subtribes and of three minor tribes can be related to it as direct or indirect descendants with some degree of credibility. In view of the fact that the inflorescence and spikelet structure are remarkably uniform, the key to its success would seem to lie in the fact that it has retained the C_3 pathway and evolved all three variants of the C_4 pathway.

The result of this photosynthetic diversity has been the exploitation of the full range of habitats which the tropics has to offer and even an extension into the temperate zone in North America. The fact that it has retained the C_3 pathway may be seen as a primitive trait but if this is the case it is one which has proved remarkably profitable, contributing to its habitat diversity by allowing expansion into mesic situations in the tropics where there is no serious competition from any other grass group. In addition to successfully exploiting mesic tropical habitats, the C_3 sector of *Panicum* appears to have been the source of a number of specialized aquatic or shade-loving genera.

The selection pressures which have influenced the C_4 variants in this genus are less obvious. An understanding of the adaptive significance of the three variants of the C_4 pathway still eludes us and will continue to do so until a comprehensive survey of all C_4 genera is complete, although Hattersley (1988) has addressed this problem.

The majority of C_4 *Panicum* species appear to have evolved with the PCK pathway, but on that line have produced relatively few genera (9). In contrast, the NADP-ME pathway, which has evolved in fewer species, has led to the evolution of many more genera (28) (Renvoize,

1987). One interpretation of this phenomenon is that the group with the NADP-ME pathway has had limited success and adaptive combinations of characters other than the C$_4$ photosynthetic pathway have proved more effective, leading to the evolution of numerous satellite genera. The result is fewer genera in the Paniceae with the PCK pathway and more genera with the NADP-ME pathway. Only two genera, *Yakirra* and *Yvesia*, are considered to be of the NAD-ME type, based on anatomical analysis (Prendergast & Hattersley, 1987; Renvoize, 1987).

The genus *Panicum* offers a fascinating array of photosynthetic variations which, combined with a wide variety of habit forms, appear to hold the key to its remarkable success. In contrast, its spikelet morphology is extremely uniform. *Panicum* species are found in every kind of habitat in the tropics and subtropics and for future enquiry into the adaptive significance of the various C$_4$ pathways provide immense scope.

The seeming increased complexity of the morphology in the tribe is supported by evidence from the photosynthetic pathways, which show a progression from C$_3$ to C$_4$. However, with *Panicum* taken as the central genus, evolution appears to have taken two principal directions, one into the major genera *Paspalum*, *Axonopus* and *Setaria* and numerous smaller genera, all of which are of the C$_4$ NADP-ME type, and the other into the genus *Brachiaria* and smaller genera, which are mostly of the C$_4$ PCK type. A third group of genera, principally C$_3$ and centred on *Acroceras*, make the transition to C$_4$ through the inclusion of the NADP-ME type genus *Echinochloa*.

Two other instances of transition are found in *Chaetium* and in *Alloteropsis*, the latter presenting several atypical suites of characters which pose many questions for C$_3$/C$_4$ evolution. The transition from C$_3$ to C$_4$ appears to have been made five times within the subtribe, seven times within the tribe and eight times in the subfamily. The only other instances of the C$_4$ pathway are in the tribe Aristideae (*Aristida* and *Stipagrostis*), the tribe Arundineae (*Centropodia*) and of course the subfamily Chloridoideae, almost in its entirety (but see *Eragrostis walteri* (Ellis, 1984)). The subtribe Melinidinae (two genera), identified by its distinctive spikelet features, is linked to *Panicum* through *Tricholaena*, which resembles *Panicum* sect. *Hiantes* (a possible link being *P. dregeanum*; they also share a similar C$_4$ PCK photosynthetic pathway).

The subtribe Digitariinae (10 genera) appears to be an early offshoot from the ancestral C$_3$ Paniceae line. It progresses from genera which are

C_3 to several genera which are of the C_4 mestome sheath (MS) type, including the principal genus *Digitaria* (230 species), which leads on to the similar C_4 NADP-ME type subtribe Cenchrinae. The subtribe Cenchrinae (13 genera) is characterized by the bristles which subtend the spikelets, often forming involucres or conspicuous burs and probably evolved as an adaptation to protection or dispersal by animals: a similar feature has evolved independently in some members of the Setariinae. The soft upper lemma in the Cenchrinae indicates a possible derivation from advanced C_4 Digitariinae, in contrast to the *Setaria* group where the upper lemma is crustaceous; the central genera are *Pennisetum* and *Cenchrus*.

The subtribe Arthropogoninae (two genera) has a hyaline upper lemma, indicating a possible relationship with *Leptocoryphium* in the subtribe Digitariinae, which shares the same feature. The leaf-blade anatomy of *Arthropogon* is of the intermediate C_3/C_4 MS type and, although *Leptocoryphium* is of the C_4 MS type, an origin for *Arthropogon* from some C_3 ancestor of *Leptocoryphium* would seem a possibility. *Reynaudia*, the other genus in the subtribe, which is apparently more advanced, has evolved the complete C_4 NADP-ME type of anatomy.

The subtribe Spinificinae (three genera) is a highly specialized assemblage which has adapted to the rigours of arid environments in Australia, India and China. The genera are united by their compact, compound inflorescences and strongly xerophytic habit. Two of the genera are dioecious, which, coupled with having the C_4 photosynthetic pathway and a similarity to *Uranthecium* (an advanced member of the *Setariinae*), suggests that they are an advanced group. The NADP-ME pathway is recorded on the basis of leaf-blade anatomy, although *Spinifex* appears to be unconventional in some aspects as a partial mestome sheath is sometimes present and, in some species, the extent of the photosynthetic carbon reduction sheath is restricted. Hattersley & Browning (1981) also recorded variation from the typical arrangement in the chloroplasts of *Spinifex hirsutus*.

The other half of the subfamily, the tribes Andropogoneae and Arundinelleae, has followed a course of evolution which to a large extent has responded to the challenge of the savanna environment. In the Andropogoneae especially, many species are characteristic of tropical grassland habitats.

ARUNDINELLEAE
The evolutionary line starts with the tribe Arundinelleae (12 genera) where a parallel progression from a panicle to triads of spikelets has been made along several separate lines. As with other members of the Panicoideae, the spikelets are 2-flowered with the lower floret male or barren and the upper floret bisexual. Unlike other panicoids, however, the glumes are persistent; only the florets are deciduous.

Jansenella has the typical panicoid arrangement of the lower floret being male or barren and the upper floret bisexual; it also has the C_3 pathway (Renvoize, 1985) which, in a tribe that is predominantly of the C_4 NADP-ME type, indicates that this is a primitive member.

The anomalous 2-flowered genus *Chadrasekharania* is included in the Arundinelleae on the basis of its papery brown spikelets. However, with both florets fertile, the lemmas bearing straight awns and the C_3 pathway, it appears to be even more primitive than *Jansenella* and indicates a possible link with *Thysanolaena* and the tribe Arundineae.

Two genera, *Garnotia* and *Arundinella*, form an independent line of development from the rest. *Arundinella* is another example of a genus which does not conform to the usual suite of characters associated with the NADP-ME pathway (Renvoize, 1982*a,b*; Hattersley, 1988). It also shares with *Garnotia* an auriculate palea and auxillary vascular bundles in the leaf blades. The 1-flowered spikelets of *Garnotia* appear to be a derived condition from the 2-flowered spikelets of *Arundinella*.

The remaining eight genera constitute the main line of evolution, the success of which must in some measure be due to the evolution of the C_4 pathway. The dimorphic florets and geniculate awn in the *Loudetia* group are typical of the Andropogoneae, for which *Loudetia* is a suggested origin.

ANDROPOGONEAE
Evolution in the tribe Andropogoneae has followed a course of increasing morphological complexity which can be summarized as follows:

1 Modification of the panicle: reduction of a large terminal inflorescence to a short single raceme, followed by expansion into a compound panicle.

2 Fragile racemes bearing paired spikelets: initially the spikelets in each pair were similar, becoming dissimilar, with the pedicelled spikelet reduced, in the more advanced genera.

3 Emphasis on internode and pedicel evolution: the internodes vary from simple and slender in the primitive genera to shortened, swollen and fused with the pedicel in the advanced genera.

Some of the most complex inflorescence structures in the family are found in the tribe Andropogoneae.

As already discussed, the origins of the tribe probably lie with the Arundinelleae, members of which have similarly 2-flowered spikelets, in which the florets are dimorphic and the lemmas bear a geniculate awn. The advanced genera exhibit the C_4 photosynthetic pathway. There is a strong similarity between the panicles of *Arundinella* and those of members of the Saccharinae, the most primitive subtribe, although *Arundinella* is not on the main line of evolution for the Arundinelleae. At the opposite end of the tribe, the Tripsacinae, Chionachninae and Coicinae are the most advanced subtribes, with unisexual spikelets occurring in separate inflorescences or in separate zones of the same inflorescence. The spikelets vary in complexity and may be in single racemes or in racemes which are gathered into compound panicles.

The tribe may be conveniently segregated into 11 subtribes, distributed throughout the tropics and warm temperate regions of the world; all genera have the typical C_4 anatomical character suite associated with the NADP-ME pathway.

The Saccharinae (13 genera) is the most primitive subtribe, with rachis internodes unspecialized and both spikelets in each pair similar and fertile. The genera can be divided into two natural groups, one centred on *Saccharum*, in which the inflorescence is a panicle with the rachis fragile or tough, the other centred on *Eulalia*, and characterized by its digitate inflorescence.

Closely related to the Saccharinae is the subtribe Germainiinae (three genera), which appears to have evolved from *Eulalia*. The spikelets are dissimilar and the sessile spikelet may be reduced and barren, but this is a state which is derived from a male or bisexual spikelet and in this instance is not interpreted as primitive.

The Sorghinae (14 genera) is similar to the Saccharinae in the paniculate or digitate inflorescence, but differs in its male or sterile pedicelled spikelets. Two groups are recognized: *Sorghum* is at the centre of the first group, its more primitive relative *Hemisorghum* linking it with the subtribe Saccharinae; and *Dichanthium* is at the centre of the second group, with *Pseudosorghum* a possible link with *Sorghum*.

In the subtribe Ischaeminae (eight genera), several primitive features persist, such as the male lower floret and the rare occurrence of a bisexual pedicelled spikelet, which suggests a derivation from *Eulalia* in the Saccharinae. On the other hand, various advanced features, such as the strongly 2-keeled glumes, the thickened internodes, the development of spatheoles and the reduction of the inflorescences to one or two racemes, suggest that this subtribe is at a point of divergence where the advanced features have developed further in the descendant subtribes.

In the subtribe Dimeriinae (one genus *Dimeria*, 35 species), the spikelets are unusual as they are single and pedicelled, but in this case the absence of a sessile spikelet appears to be derived from the orthodox arrangement of paired spikelets with one sessile and the other pedicelled. Some species have distant spikelets and resemble *Pogonachne* in Ischaeminae, thus suggesting a possible link with that subtribe. The genus is probably derived from Ischaeminae through suppression of the sessile spikelet.

It is unusual in the tribe Andropogoneae for a genus which has found a niche in more mesic environments to evolve numerous species; the larger genera are usually associated with savanna habitats. In *Dimeria* the annual habit is prevalent, which is probably an adaptation to the seasonally mesic environments on thin or sandy soils in open situations; some species, however, favour marshy habitats.

The subtribe Andropogoninae (six genera) is scarcely distinguishable from the subtribe Ischaeminae. It includes *Andropogon*, *Cymbopogon* and *Schizachyrium*, all important constituents of savannas.

Diheteropogon, an advanced, closely related offshoot from *Andropogon*, resembles *Parahyparrhenia* in the Anthistriinae and provides a possible link between the two subtribes.

The subtribe Anthistiriinae (12 genera) includes the genus *Hyparrhenia*, which, although the largest in the subtribe, is on a line divergent from the other members. The characteristic pointed callus, placed obliquely on the internode tip, has similarities with *Cymbopogon* and *Diheteropogon* and suggests a possible link with the subtribe Andropogoninae.

The genera of this subtribe are typical of savanna or dry open woodland, and most species are important as a source of grazing. The subtribe Rottboelliinae (21 genera) appears to have followed several lines of evolution, favouring a diverse range of habitats from swamps to savanna, dry woodland and disturbed ground.

The least specialized member of the subtribe is *Phacelurus*, with digitate racemes and unspecialized internodes. *Urelytrum*, *Loxodera* and *Elionurus* form an intermediate group which, although they have advanced features, such as thickened internodes and an awnless upper glume, also have primitive features which link the subtribe to the Ischaeminae and Andropogoninae.

The final three subtribes have unisexual spikelets and thus represent the most advanced evolutionary stage of spikelet sexual development in the tribe.

The subtribes Tripsacinae and Chionachninae are sidelines from the subtribe Rottboelliinae, whereas the subtribe Coicinae appears to be an advancement from *Apluda* in the Ischaeminae. All are tropical and subtropical and have favoured mesic situations, which is not the type of habitat usually associated with a tribe where the largest genera are associated with the savanna environment. The apparently poor adaptation to mesic habitats is reflected by the small number of species in the genera of these three subtribes: the maximum is 13.

The subtribe Tripsacinae (two genera, *Tripsacum* and *Zea*) forms a divergent line which in some ways is scarcely distinct from Rottboelliinae, but advanced features such as unisexual spikelets borne on separate parts of the same inflorescence or on different inflorescences set this group apart.

In the subtribe Chionachninae (four genera), the pedicel is fused to the internode, forming a distinctive slender composite unit which bears the spikelet in a lateral position.

The subtribe is linked to the Rottboelliinae by the peg and socket callus joint of the sessile spikelet present in *Chionachne*, but the unisexual spikelets, bisexual racemes and compound inflorescence are sufficiently different to accord the group separate status.

A link to *Heteropholis* in the subtribe Rottboelliinae is indicated by *C. javanica*, which has a relatively broad internode.

In the monotypic subtribe Coicinae, the highly modified inflorescence is composed of paired unisexual racemes, the female enclosed by a bony utricle, the male borne on a long peduncle and exserted. In spite of the unisexual spikelets and racemes it appears to be an advanced genus related to *Apluda* in the Ischaeminae, and therefore on a separate evolutionary line to the other two subtribes with unisexual spikelets.

Conclusion

The picture which we have of the evolution of the grasses is of character combinations which result primarily in the exploitation of open environments. The divergence at an early stage of the bambusoids

36   S. A. Renvoize and W. D. Clayton

into a forest environment is one which has met with only limited success. Although they represent a significant component of many tropical and temperate mesic ecosystems, the bambusoids have not overcome the dominance of the woody dicotyledons in such habitats. On the other hand, the mainstream of grass evolution, which has led the family into the open habitats of savanna and steppe, has achieved a measure of success which is unrivalled in the flowering plant kingdom.

Through the combination of efficient seed production and dissemination and the evolution of an additional and more efficient biochemical pathway for photosynthetic carbon reduction, the grasses have been able to exploit open environments in a wide range of climatic types. The result is a plant family which is far reaching in its ecological adaptations and efficient and diverse in its reproductive mechanisms.

References

Andersson, L. (1976). The synflorescence of Marantaceae. *Botaniska notiser*, **129**, 39–48.

Clayton, W. D. (1990). The spikelet. In *Reproductive versatility in the grasses*, ed. G. P. Chapman, pp. 32–51. Cambridge: Cambridge University Press.

Clayton, W. D. & Renvoize, S. A. (1986). *Genera Graminum: grasses of the world*. Kew Bulletin Additional Series 13. London: Royal Botanic Gardens, Kew.

Cronquist, A. (1987). A botanical critique of cladism. *Botanical Reviews*, **53**, 1–52.

Davidse, G. & Ellis, R. P. (1984). *Steyermarkochloa unifolia*, a new genus from Venezuela and Colombia (Poaceae: Arundinoideae: Steyermarkochloeae). *Annals of the Missouri Botanical Garden*, **71**, 994–1012.

Davidse, G. & Ellis, R. P. (1987). *Arundoclaytonia*, a new genus of the Steyermarkochloeae (Poaceae: Arundinoideae) from Brazil. *Annals of the Missouri Botanical Garden*, **74**, 479–490.

Ellis, R. P. (1984). *Eragrostis walteri* – a first record of non-Kranz leaf anatomy in the sub-family Chloridoideae (Poaceae). *South African Journal of Botany*, **3**, 380–386.

Hattersley, P. W. (1988). Variations in photosynthetic pathway. In *Grass systematics and evolution*, ed. T. R. Soderstrom, K. W. Hilu, C. S. Campbell & M. E. Barkworth, pp. 49–64. Washington DC: Smithsonian Institution Press.

Hattersley, P. W. (1992). Significance of intra C_4 photosynthetic pathway variation in grasses of arid and semi-arid regions. In *Desertified grasslands, their biology and management*, ed. G. P. Chapman. Academic Press (in press).

Hattersley, P. W. & Browning, A. J. (1981). Occurrence of the suberized lamella in leaves of grasses of different photosynthetic types. 1. In parenchymatous bundle sheaths and PCR (Kranz) sheaths. *Protoplasma*, **109**, 371–401.

Hattersley, P. W. & Roksandic, Z. (1983). 13C values of C$_3$ and C$_4$ species of Australian *Neurachne* and its allies (Poaceae). *Australian Journal of Botany*, **31**, 317–321.

Hattersley, P. W. & Stone, N. E. (1986). Photosynthetic enzyme activities in the C$_3$–C$_4$ intermediate *Neurachne minor* S. T. Blake (Poaceae). *Australian Journal of Plant Physiology*, **13**, 399–408.

Kellogg, E. A. & Campbell, C. S. (1987). Phylogenetic analyses of the Gramineae. In *Grass systematics and evolution*, ed. T. R. Soderstrom, K. W. Hilu, C. S. Campbell & M. E. Barkworth, pp. 310–322. Washington DC: Smithsonian Institution Press.

Nadgauda, R. S., Parasharmi, V. A. & Mascarenhas, A. F. (1990). Precocious flowering and seeding behaviour in tissue-cultured bamboos. *Nature*, **344**, 335–336.

Prendergast, H. D. V. & Hattersley, P. W. (1985). Distribution and cytology of Australian *Neurachne* and its allies (Poaceae), a group containing C$_3$, C$_4$ and C$_3$–C$_4$ intermediate species. *Australian Journal of Botany*, **33**, 317–336.

Prendergast, H. D. V. & Hattersley, P. W. (1987). Australian C$_4$ grasses (Poaceae): leaf blade anatomical features in relation to C$_4$ acid decarboxylation types. *Australian Journal of Botany*, **35**, 355–382.

Renvoize, S. A. (1982a). A survey of leaf-blade anatomy in grasses. (II) Arundinelleae. *Kew Bulletin*, **37**, 489–495.

Renvoize, S. A. (1982b). A survey of leaf-blade anatomy in grasses. (III) Garnotieae. *Kew Bulletin*, **37**, 497–500.

Renvoize, S. A. (1985). A note on *Jansenella* (Gramineae). *Kew Bulletin*, **40**, 470.

Renvoize, S. A. (1987). A survey of leaf-blade anatomy in grasses. (XI) Paniceae. *Kew Bulletin*, **42**, 739–768.

Soderstrom, T. R. (1987). Bamboo systematics: yesterday, today and tomorrow. *Journal of the American Bamboo Society*, **6**, 4–16.

Soderstrom, T. R. & Zuolaga, F. O. (1988). A revision of the genus *Olyra* and the new segregate genus *Parodiolyra* (Poaceae: Bambusoideae: Olyreae). *Smithsonian Contributions to Botany*, **69**.

Note added in proof: The term 'pseudospikelet' is sometimes applied to the spikelet constituents of the bamboo inflorescence. For a detailed discussion of this topic see Clayton (1990) cited above [G.P.C.].

2

Diversification of photosynthesis
P. W. Hattersley and L. Watson

Introduction

Photosynthetic pathway variation in the Poaceae is more comprehensively understood than in any other angiosperm family. There are several reasons for this. C_4 photosynthesis, which was the third major variant of photosynthetic carbon metabolism to be discovered in higher plants (after C_3 photosynthesis and Crassulacean Acid Metabolism), was first detected in this family. This initial incentive to work on grasses in C_4 photosynthesis research was later reinforced, not only because of the paramount economic importance of the Poaceae, but also by virtue of the early discovery of interesting C_4 diversity (e.g. in *Aristida* and *Panicum*: Bisalputra, Downton & Tregunna, 1969; Laetsch, 1971). Furthermore, grasses proved to be excellent experimental subjects (cf. C_4 dicotyledonous species) regarding applicability of tissue separation techniques (which were essential for demonstrating differential compartmentation of photosynthetic enzymes), ready availability of seed and relative ease of culture.

A multidisciplinary approach to the phenomenon of C_4 photosynthesis has resulted in an extraordinarily detailed appreciation of the relationships of its biochemistry, physiology, leaf structure, phytogeography and ecology (e.g. Hattersley, 1987; Hatch, 1988). Together with taxonomic knowledge of the Poaceae and increasing inquiry concerning the molecular biology of C_4 photosynthesis, this appreciation forms the basis not only for a fully integrated understanding of C_4 biology in the family but also for insights into the evolutionary relationships of its component taxa. Less esoteric is the concomitant likelihood that photosynthetic pathway will one day be engineered, such that corn (*Zea mays* L., C_4) can be grown in temperate climates and wheat (*Triticum aestivum* L., C_3) can be grown in the tropics. Also, the ecological impacts of global climate change will need to be countered and

38

managed, and the known differential climatic responses not only of C_3 and C_4 grasses but also of different C_4 types, will be of value at least in predicting species compositional changes in savannas and other grasslands, and perhaps even in active management strategies (Hattersley, 1992).

In this chapter we describe photosynthetic pathway diversity and its taxonomic occurrence in grasses, outline knowledge of, and speculate on, the diversification of photosynthesis (in the context of taxonomic and evolutionary concepts), and suggest how our comprehension of this diversity might usefully be exploited in relation to genetic engineering, domestication and the 'greenhouse' effect. Clues to the evolution of C_4 photosynthesis in the grasses seem to lie in the study of C_3–C_4 intermediacy and in molecular systematics using DNA sequences of chloroplast and nuclear genes for photosynthetic enzymes. If an understanding of the diversification of photosynthesis in the Poaceae can be gained, much more will be appreciated about the adaptive radiation of the grasses. Such a synthesis would be part of knowing grasses, and would therefore contribute to the undoubted continuing value of their domestication, which has yet to take cognizance of photosynthetic pathway variation, let alone to exploit and manipulate it.

Diversity of photosynthetic pathways

BIOCHEMICAL VARIANTS AND THEIR RELATIONSHIPS

Grass species are either C_3 (fixing CO_2 via the Calvin–Benson cycle only, in mesophyll cells), or C_4 (initially fixing CO_2 by mesophyll cell primary carboxylation in the C_4 acid cycle, with subsequent C_4 acid decarboxylation and Calvin–Benson cycle activity in another type of chlorenchymatous cell) or, rarely, C_3–C_4 intermediates; no Crassulacean Acid Metabolism species are known. In C_4 plants, photosynthetic carbon reduction (PCR) cycle (Calvin–Benson cycle) activity is restricted to this second type of chlorenchyma in the leaf, which has therefore been termed PCR tissue (Hattersley, Watson & Osmond, 1977). Usually PCR tissue comprises cells in the parenchyma bundle sheath position, but it can occur in other ground tissue sites or in the mestome sheath position and other tissues of procambial origin (Brown, 1977; Dengler, Dengler & Hattersley, 1985). PCR tissue is more usually referred to as 'Kranz' tissue, a term originally coined by Haberlandt (1884), though he used it to refer to the mesophyll in C_4 plant leaves.

NAD-ME TYPE

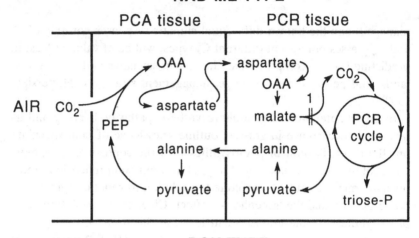

PCK TYPE

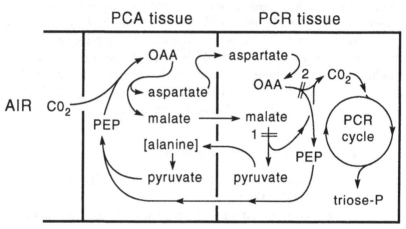

NADP-ME TYPE

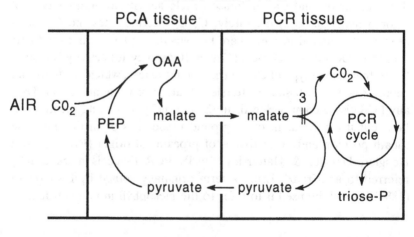

The latter is now known as 'C_4 mesophyll' or 'primary carbon assimilation' (PCA) tissue, to distinguish it from the mesophyll of C_3 plants (Hattersley *et al.*, 1977). Use of the PCR and PCA terminology is further advantageous because it unambiguously refers to C_4 leaf tissues on a functional basis, thus being applicable to all C_4 leaf-structure variants.

The three known biochemical variants of the C_4 acid cycle are each characterized by confinement of the PCR cycle to PCR cells and the straddling of the C_4 acid cycle between PCA tissue and PCR tissue, with phosphoenolpyruvate (PEP) carboxylation occurring in PCA cells and C_4 acid decarboxylation in PCR cells (Fig. 2.1) (Hatch & Osmond, 1976; Edwards & Walker, 1983; Hatch, 1988). The C_4 acid cycle effectively serves as a biochemical pump to deliver CO_2 to the PCR tissue. The three variants differ in the C_4 acid predominantly transported from PCA to PCR cells, their C_4 acid decarboxylase (NADP malic enzyme, NAD malic enzyme or PEP carboxykinase), intracellular site of C_4 acid decarboxylation and pathway of regeneration of PEP (Fig. 2.1). They are known as the NADP-ME, NAD-ME and PCK types respectively. The pathway of the PCK type has only recently been fully elucidated (for one species: Burnell & Hatch, 1988). Hattersley (1987) provided a checklist of all grasses that have been biochemically typed.

It is plain that plants with C_4 photosynthesis evolved in the first instance(s) from C_3 plants. This is not to say that there have not been reversions, but the fact that C_4 photosynthesis is known only in angio-

---

Fig. 2.1. Simplified diagrams of photosynthetic carbon metabolism in C_4 biochemical types of grasses. Photosynthetic carbon assimilation (PCA; or 'C_4 mesophyll') tissue and photosynthetic carbon reduction (PCR) (or 'Kranz') tissue are represented by two adjacent boxes for each C_4 type. In the NAD–ME type, malate is decarboxylated by NAD malic enzyme (at '1') (actually in mitochondria); in the NADP–ME type, malate is decarboxylated by NADP-malic enzyme (at '3'; actually in chloroplasts); and in the PCK type, oxaloacetate is decarboxylated by PEP carboxykinase (at '2'; actually in the cytosol) and *also* malate is decarboxylated by NAD–ME (at '1'; in mitochondria). Note that in all three C_4 types, the PCR cycle is restricted to PCR tissue; the C_4-acid cycle straddles both PCA and PCR tissues. In the NAD–ME and PCK types, aspartate is the predominant C_4 acid that diffuses from PCA to PCR tissue; in the NADP–ME type, it is malate. CO_2, carbon dioxide; OAA, oxaloacetate; triose-P, triose phosphates; PEP, phosphoenolpyruvate. Based on work reviewed by Edwards & Walker (1983) and Hatch (1988).

sperms, that it occurs in multifarious flowering plant families, and that
the C_4 acid cycle is clearly an 'appendage' to the Calvin–Benson cycle
biochemically, all suggest that C_4 plants evolved from C_3 plants several/
many times, perhaps during the Cretaceous period when the angiosperms originated and radiated.

As with the C_3/C_4 distinction, few generalizations can be made about
the relationships of C_4 types to one another on the basis of the pathways
themselves or of the taxonomic occurrence of C_4 types in angiosperms as
a whole. It does seem reasonable, however, to suggest that the PCK
type evolved from the NAD-ME type. Its C_4 acid cycle could be viewed
as an embellishment of the NAD-ME type C_4 acid cycle (Fig. 2.1; and
see Fig. 3 of Hatch, 1988), and, as the PCK type is at present known
only in grasses, it may be unique to the family and may therefore have
evolved subsequent to the NAD-ME type, which is known in both
monocotyledons and dicotyledons. Alternative hypotheses are that it
exists, but has not yet been discovered elsewhere in the angiosperms, or
that it has become extinct elsewhere in this group. The NADP-ME type
is known in both monocotyledons and dicotyledons. It seems to have
relatively little in common with the other two C_4 types, but perhaps it
too evolved from the NAD-ME type (see later discussion).

At the biochemical level, a search for isozyme patterns (e.g. using
immunoelectrophoretic analyses) and for homology among various C_4
acid cycle enzymes (using amino acid or DNA sequencing techniques)
would help to establish any general relationship(s) between the C_4 types
that may exist. Intraspecific and interspecific isozymic variation of some
C_4 acid cycle enzymes has been demonstrated using electrophoretic
analyses (e.g. for aspartate and alanine aminotransferases, Hatch &
Mau, 1973; for pyruvate P_i dikinase, Sugiyama et al., 1979; for PEP
carboxylase, Adams, Leung & Sun, 1986; and for NAD-ME, Murata et
al., 1989), but such variation has yet to be sampled extensively in a
systematic context. Now that DNA sequencing of genes for several
photosynthetic enzymes becoming routine however (see subsequent section), the likelihood of molecular systematic investigations successfully
addressing issues such as the homology of PEP carboxylases among
different grasses, or the relationship (if any) of NADP-ME to NAD-
ME, is increasing. This is because isozyme variation is being identified
and characterized, and its genetic basis is becoming understood.

LEAF-STRUCTURE VARIANTS AND THEIR RELATIONSHIPS

The significance of C_4 leaf structure has been investigated in detail for over 20 years (e.g. Downton & Tregunna, 1968; Hatch & Osmond, 1976; Hattersley, 1984), though it has been known for much longer (Haberlandt, 1884). A five-point functional definition of it that accommodates all known C_4 leaf-structure variants was given by Hattersley *et al.* (1977): (i) two chlorenchymatous cell types are present (PCA and PCR tissue); (ii) contact between PCR tissue and intercellular-gas-space is limited/minimized; (iii) PCA and PCR tissues are either in direct contact or, unusually, separated only by narrow diameter cells (up to 10 μm); (iv) PCA tissue is always closer to the atmosphere than PCR tissue; and (v) the PCA/PCR tissue ratio is low (usually less than 4.5), i.e. the 'anatomical stoichiometry' of the two tissue types has an upper limit (Hattersley, 1984). These anatomical characteristics are functionally significant and contrast with those of C_3 plants (Fig. 2.2). They are essential to the compartmentation of biochemical steps in C_4 photosynthesis and to the effectiveness of the C_4 acid cycle in concentrating CO_2, at the site at which ribulose-1,5-bisphosphate carboxylase/oxygenase (Rubisco) occurs, in a sufficiently CO_2-tight space (the PCR tissue). Most C_4 plants have a 'standard' type of C_4 leaf anatomy, known usually as 'Kranz' anatomy, and characterized by radiate C_4 mesophyll (PCA) tissue surrounding a parenchymatous bundle sheath (or 'Kranz' sheath) – the PCR tissue e.g. Fig. 2.2*b*. However, many other types of C_4 anatomy are known, in both monocotyledons and dicotyledons (e.g. Fig. 2.2*e,f,g,h*, this chapter; Carolin, Jacobs & Vesk, 1973, 1975; Raghavendra & Das, 1976; Brown, 1977; Bruhl, Stone & Hattersley, 1987).

The three C_4 biochemical types occur in the Poaceae among eight leaf-structural types (Fig. 2.2) which differ in a number of anatomical and ultrastructural characters (Table 2.1). There are three main 'classical' leaf-structural types, one associated with each of the three C_4 biochemical types. These three major biochemical–leaf structural *suites* together characterize the majority of grass species, and have been elucidated by research spanning two decades (Downton, 1970; Gutierrez, Gracen & Edwards, 1974; Hatch, Kagawa & Craig, 1975; Hattersley & Watson, 1976; Brown, 1977; Hattersley & Browning, 1981; Hattersley, 1984, 1987, 1992).

Most 'non-classical' leaf-structural types have also been known for many years, for example: the 'Aristida' type (Bisalputra *et al.*, 1969;

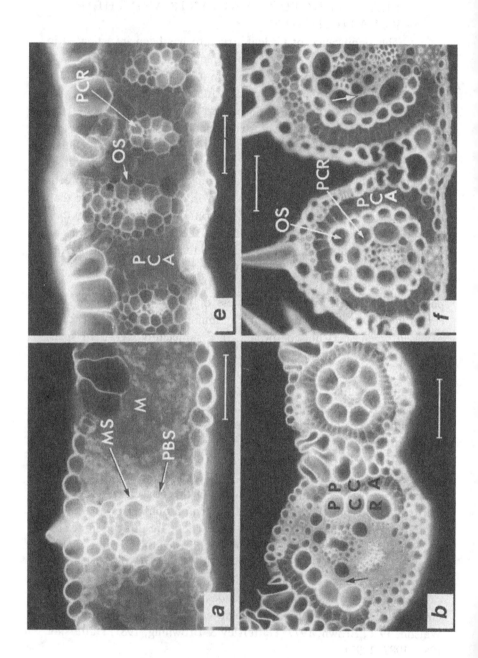

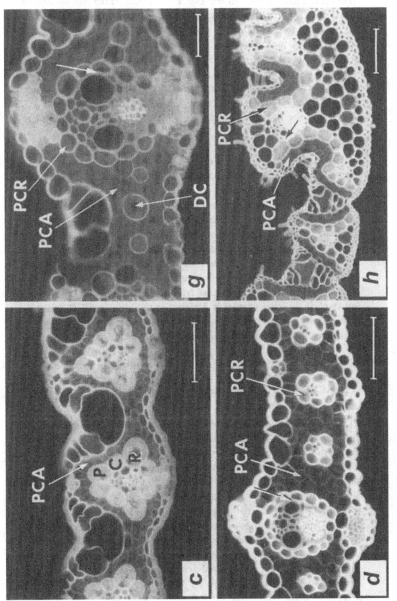

Fig 2.2 For caption see pp. 46–7.

Fig. 2.2. Epifluorescence light micrographs of the transverse sections
of leaf blades of representative species of seven of the C_4 leaf-
structural suites known in the Poaceae. A transverse leaf section of
wheat (C_3) is shown for comparison (*a*). (*a*) *Triticum aestivum* (C_3) –
the primary vascular bundle shown has both a parenchymatous
bundle sheath and a mestome sheath; there is extensive mesophyll
between veins. Bar=50 μm. (*b*) *Buchlöe dactyloides* (C_4) –
'classical NAD–ME' type; there are cells intervening (unlabelled
arrow) between the metaxylem vessel elements and laterally adjacent
'photosynthetic carbon reduction' cells of the primary vascular
bundle (on the left), i.e. this species is XyMS+; chloroplasts in the
PCR cells are found towards their inner tangential walls, i.e. they are
centripetally placed (not seen in this micrograph); PCR cell outlines,
in transverse section, have a relatively 'even' outline (cf. *c*).
Bar=100 μm. (*c*) *Sporobolus fimbriatus* (C_4) – 'classical PCK' type;
leaf is XyMS+ (no primary vein in this transverse section, so XyMS
state cannot be determined from this micrograph); PCR cell
chloroplasts primarily centrifugally positioned within cells, i.e.
towards the outer tangential walls; PCR sheath outlines are 'uneven'
(cf. *b*). Bar=100 μm. (*d*) *Panicum antidotale* (C_4) – 'classical
NADP-ME' type; there are *no* cells intervening (unlabelled arrow)
between the metaxylem vessel elements and laterally adjacent
photosynthetic carbon reduction cells of the primary vascular bundle
(on the left, i.e. the species is designated 'XyMS−'); PCR cell
chloroplasts are centrifugally positioned (but not seen in this
micrograph); PCR sheath outlines are 'uneven'. Bar=100 μm.
(*e*) *Alloteropsis semialata* subsp. *semialata* (C_4) – showing
'Neurachneae' type anatomy; this leaf-structural variant is quite
similar to *d*, except an outer sheath (OS) is present external to the
'Kranz' (PCR) sheath (the OS contains few, or no chloroplasts) (see
also Table 2.1); the species is designated as XyMS−. Bar=100 μm.
(*f*) *Aristida ramosa* (C_4) – showing 'Aristida' type leaf structure; the
outer sheath (functionally equivalent to C_3 mesophyll? – see
Hattersley, 1987) contains centrifugally placed chloroplasts, while the
inner PCR sheath chloroplasts are centripetally oriented (not clearly
shown here); there are no cells between the PCR sheath and laterally
adjacent metaxylem vessel elements in the primary vein shown
(unlabelled arrow), i.e. the species is designated XyMS−.
Bar=50 μm. (*g*) *Arundinella nepalensis* (C_4) – showing the
'Arundinelleae' type leaf structure; there are no cells between the
PCR sheath and laterally adjacent metaxylem vessel elements in the
primary vein shown (unlabelled arrow), i.e. XyMS−; the PCR sheath
and chloroplast position are as for the 'classical NADP-ME' type, but
there are 'distinctive cells' intervening between the veins.
Bar=50 μm. (*h*) *Triodia pungens* (C_4) – exhibiting the 'Triodia' type
of leaf structure; the unlabelled arrow points to cells intervening
between the metaxylem vessel elements and laterally adjacent PCR

Fig. 2.2*f*); the 'Arundinelleae' type (Crookston & Moss, 1973; Fig. 2.2*g*); the 'Triodia' type (McWilliam & Mison, 1974; Fig. 2.2*h*); and the 'Neurachneae' type (=the 'Alloteropsis' type: Ellis, 1974; Brown, 1975; Fig. 2.2*e*). The eighth structural type ('Eriachneae' type) has only recently been characterized (Prendergast & Hattersley, 1987; Prendergast, Hattersley & Stone, 1987).

Recent work (Prendergast *et al.*, 1986, 1987) showed that some of the predictions of C_4 biochemical type for 'non-classical' leaf-structural types, made using anatomical and ultrastructural predictors derived from 'classical' types, are incorrect. Thus, the 'Triodia' type was predicted to be of the PCK biochemical type, but species sampled are of the NAD-ME type; the 'Eriachneae' type was predicted to be of the PCK type (personal pre-1987 observations), yet *Eriachne* and *Pheidochloa* species were later typed as biochemically NADP-ME; and *Alloteropsis semialata* (R. Br.) A. Hitchc. subsp. *semialata* ('Neurachneae' type) was predicted to be of the NADP-ME type, but proved to be of the PCK type (C_4 Neurachneae themselves, however, were typed as NADP-ME as predicted; Hattersley & Stone, 1986).

A further complication was the discovery that not all species with 'classical PCK' type leaf structure are, in fact, of the PCK type biochemically; some are of the NAD-ME type. This has now been shown for some *Panicum* spp. (Ohsugi & Murata, 1980; Ohsugi, Murata & Chonan, 1982; Prendergast *et al.*, 1987), some *Eragrostis* spp. (Prendergast *et al.*, 1986), *Enneapogon* spp., and *Triraphis mollis* R. Br. (Prendergast *et al.*, 1987). Currently, there is no known leaf structure feature which distinguishes so-called PCK 'classical PCK' type species and NAD-ME 'classical PCK' type ('PCK-like NAD-ME') species.

Prendergast *et al.* (1987) recognized nine leaf-structural types. Compared with the treatment presented here: (i) two types were identified within the Eriachneae; (ii) an NAD-ME/PCK intermediate type of

Fig. 2.2 (*cont.*)
cells of the primary vascular bundle, i.e. XyMS+; note that the PCR tissue 'drapes' from vein to vein. Bar=100 μm. (All micrographs were taken on a Zeiss photomicroscope with epifluorescence illumination.) (The Eriachneae C_4 leaf-structural type is not illustrated.)
M=mesophyll tissue; PBS=parenchymatous bundle sheath; MS=mestome sheath; PCR='photosynthetic carbon reduction' tissue (='Kranz' tissue); PCA='primary carbon assimilation' tissue (='C_4 mesophyll' tissue); OS=outer sheath; DC='distinctive cell'.

Table 2.1. Anatomical and ultrastructural characteristics of leaf-structural and biochemical C₄ types in the Poaceae

Three C$_4$ biochemical types (last column) are distributed among eight C$_4$ leaf-structural types (first column). The XyMS character refers to the presence (XyMS+) or absence (XyMS−) of cells in the mestome sheath position between metaxylem vessels and laterally adjacent PCR cells in primary vascular bundles (Hattersley & Watson, 1976). Some XyMS− leaf-structural types are characterized by the presence (+) of an outer non-PCR bundle sheath, external to the PCR sheath (column 3). PCR cell outline refers to the cell outline in leaf transverse section (Prendergast & Hattersley, 1987). Distinctive cells described by Tateoka (1958). For a review on the suberized lamella refer to Hattersley & Browning (1981). See also the general review by Hattersley (1987).

PCR cells, 'photosynthetic carbon reduction' cells ('Kranz' cells). +, present. −, absent. n.a., not applicable. cf, centrifugally oriented. cp, centripetally oriented. NADP-ME, NADP malic enzyme. NAD-ME, NAD malic enzyme. PCK, phosphoenolpyruvate carboxykinase.

C$_4$ leaf structural type	XyMS character	Non-PCR outer sheath	PCR ('Kranz') sheath cells				Mestome sheath suberized lamella	Distinctive cells	PCR sheath contiguous between veins	C$_4$ biochemical type
			Chloroplast position	Chloroplast grana	Cell outline	Suberized lamella				
'Classical NADP-ME'	XyMS−	−	cf	−	Uneven	+	n.a.	−	−	NADP-ME
'Classical NAD-ME'	XyMS+	n.a.	cp	+	Even	−	+	−	−	NAD-ME
'Classical PCK'	XyMS+	n.a.	cf/even	+	Uneven	+	+	−	−	PCK or NAD-ME
Arundinelleae	XyMS−	−	cf	−	Uneven	+	n.a.	+	−	NADP-ME
Neurachneae	XyMS−	+	cf	+	Even	+	n.a.	−	−	NADP-ME or PCK
Aristida	XyMS−	+	cf	−	Even	−	n.a.	−	−	NADP-ME
Triodia	XyMS+	n.a.	cf	+	Even	−	+	−	+	NAD-ME
Eriachneae	XyMS+	n.a.	cf or cp	+	(Uneven)	−	−	−	−	NADP-ME

structure for *Enneapogon* spp. and some *Eragrostis* spp. was recognized; and (iii) the 'Arundinelleae' type was regarded as a subset of the 'classical NADP-ME'. Their delimitation of leaf-structural types differs in only minor detail from that shown in Fig. 2.2 and Table 2.1. Superimposing the occurrence of biochemical C_4 type among the eight leaf-structural types, gives 10 biochemical–leaf-structural suites (Table 2.1), which differs only slightly from the 11 suites described by Hattersley (1987, Table 6.5) where two Eriachneae types are shown, differing in their PCR cell chloroplast position and in the evenness of outline of PCR cell walls in transverse section.

Patterns and relationships among the structural types can be discerned, considering only leaf structure. A major distinction among the C_4 structures is the XyMS+/XyMS− 'dichotomy' (Hattersley & Watson, 1976), otherwise known as the MS/PS 'dichotomy' (Brown, 1977). This refers to the presence (XyMS+; Fig. 2.2*b*) or absence (XyMS−; Fig. 2.2*d*) of cells occurring between the metaxylem vessel elements and laterally adjacent PCR cells in primary vascular bundles in leaf transverse sections. Brown (1975) speculated that this esoteric-sounding distinction resulted from different ontogenetic origins of PCR tissue in the two types of leaves. Dengler *et al.* (1985) later showed that PCR tissue is of procambial origin in XyMS− (MS) species and of ground meristem origin in XyMS+ (PS) species.

This distinction provides a unique opportunity to investigate cellular differentiation and regulatory mechanisms. In XyMS+ species the ground meristem of a developing leaf differentiates into PCA and PCR tissue, and the procambium develops into mestome sheath, xylem and phloem; in XyMS− species, the former develops into PCA tissue only (with an hypothesized suppression of outer parenchyma sheath development), and the procambium differentiates into PCR tissue (in the mestome sheath position), xylem and phloem.

In a series of elegant *in situ* mRNA labelling experiments using radio-actively labelled, single-stranded anti-sense RNA probes, Langdale, Rothermel & Nelson (1988*a*) have shown that the mRNAs for PEP carboxylase, NADP-ME and Rubisco accumulate in a cell-specific fashion. Furthermore, they suggest that pattern formation depends upon 'interpretation' of positional information distributed around each vein leading to the correct expression of cell-specific genes (Nelson & Langdale, 1989). However, the evidence for the vascular bundle hypothesis is preliminary, and there are other aspects of PCA/PCR differentiation that require study, in addition to the control of C_4 enzyme expression, such as the developmental control of suberized lamella deposition in PCR cell walls. A satisfactory mechanism for site-specificity determination needs to explain how tissues of different origin become differentiated as the PCR type, and an element of this must involve an actual phenomenon or feature that differs between XyMS+ and XyMS− grasses.

Four of the leaf-structural types are XyMS− (Table 2.1), following strictly the definition originally given by Hattersley & Watson (1976). If one accepts that the 'classical NADP-ME' type structure is an 'advanced' type (or at least a recently derived type), then one could hypothesize that the 'Neurachneae' type represents an intermediary state, as originally proposed by Brown (1977) for *Panicum petersonii* A. Hitchc. & Ekman and *Alloteropsis semialata* subsp. *semialata*; the PCR tissue is fully developed from the inner mestome sheath, but the outer parenchymatous sheath is still present, even if chloroplast-free, i.e. the ground meristem has still differentiated into two tissue types, mesophyll and outer sheath. In the 'Arundinelleae' type, it could be suggested that the so-called 'distinctive cells', which are PCR (Fig. 2.2g; Hattersley *et al.*, 1977), are 'vestiges' or 'extreme reductions' of small vascular bundles, i.e. they are of procambial origin. Indeed, this was originally proposed by Tateoka (1958 – a pre-C_4 date), based on the fact that some *Arundinella* and *Garnotia* species are characterized by incompletely formed vascular bundles, apparently representing a transition state between fully formed veins and solitary, distinctive cell strands as in *A. nepalensis* Trin. (Fig. 2.2g). The 'Arundinelleae' type, then, could be interpreted as derived from the 'classical NADP-ME' type (Fig. 2.2d). It was originally suggested that 'Aristida' type leaf structure exhibited a double 'Kranz' (or PCR) sheath (Carolin *et al.*, 1973; Johnson & Brown, 1973; Fig. 2.2f); this would represent a unique developmental phenomenon in grasses, where cells of both procambial and ground meristem origins were of the PCR type. A more recent hypothesis (Hattersley & Browning, 1981) suggests that only the inner sheath is of the PCR cell type, the mesophyll being of the PCA type, and the outer sheath being functionally like C_3 mesophyll (see Fig. 6.5 of Hattersley, 1987).

Species of the four leaf-structural types just discussed are nearly all of the NADP-ME type biochemically, suggesting further affinity. *Alloteropsis semialata* subsp. *semialata* is the only known exception here, as it is of the PCK type according to Prendergast *et al.* (1987) for an Australian accession (no other PCK type species are XyMS−); Frean *et al.* (1983), however, found a South African accession to be of the NADP-ME type. Three of the four types are also from the subfamily Panicoideae; the exception is the 'Aristida' type, which is variously placed as arundinoid, panicoid, or by itself.

Differences between the structure of 'classical NAD-ME' and 'classical PCK' type species seem less than between either of these and 'classical NADP-ME' species (Table 2.1; cf. Fig. 2.2*b,c,d*). Both are XyMS+, with PCR tissue derived from ground meristem; the inner sheath remains a mestome sheath, like that of C₃ species as far as can be discerned. This seems congruent with the surmise that the NAD-ME and PCK types may have more in common biochemically with each other, than does either with the NADP-ME type, and this notion is strengthened by the fact that some species with 'classical PCK' type leaf structure are, in fact, of the NAD-ME type biochemically (see also the discussion of their taxonomic distribution in the Poaceae).

A third XyMS+ structural type is the 'Triodia' type. This seems related to the 'classical NAD-ME' type, considering structure alone (Table 2.1), despite a difference in PCR cell chloroplast position and the 'draping' of PCR tissue between adjacent vascular bundles (as seen in transversely sectioned leaves: Fig. 2.2*h*); indeed, the Triodieae are also of the NAD-ME type biochemically as far as is known (Prendergast *et al.*, 1987). The fourth XyMS+ structural type ('Eriachneae' type) is more exceptional in that, while very similar to the 'classical PCK' type structurally, it lacks PCR cell *and* mestome sheath suberized lamellae and is of the NADP-ME type biochemically, as far as is known (Prendergast *et al.*, 1987). It is the only structural type, of those having PCR tissue in the outer sheath position, that is NADP-ME biochemically; like the 'Aristida', 'Triodia', and 'Arundinelleae' types, it is taxonomically restricted, being confined to a tribe of the Arundinoideae.

In some respects, the pattern of leaf structure variation found in grasses is broadly congruent with the pattern of occurrence of biochemical variation. It is more complex, however, and leads to the recognition of at least 10 leaf structural–biochemical suites (Hattersley, 1987; Prendergast & Hattersley, 1987; Fig. 2.2 and Table 2.1). Previous authors attempting to understand the diversification of photosynthesis in the Poaceae (Smith & Robbins, 1974; Brown, 1977; Clayton, 1981) were not aware of all of these suites, so their schemes do not accommodate the diversity now known. Furthermore, Prendergast *et al.* (1986) and Prendergast & Hattersley (1987) have extended the taxonomic sampling of leaf structure and biochemical variation by at least 50%, so that the taxonomic distribution of structural and biochemical types is now known, or can be reasonably predicted, for most grass genera of the world (Table 2.2). This taxonomic distribution will be considered after brief reference has been made (next two sections) to physiological and phytogeographical correlates with structural and biochemical C_4 types, and to the molecular biology of C_4 photosynthesis.

PHYSIOLOGY, ECOPHYSIOLOGY AND PHYTOGEOGRAPHY

Variation among C_4 types in physiology, ecophysiology and phytogeography is not as well documented or understood as the biochemical and leaf-structural variation described above. Nevertheless, the possibility that different biochemical–leaf-structural suites confer different physiologies, resulting in differential adaptation to climate, justifies reference to such variation. Indeed, diversification of grasses at subfamily level may have been associated principally with divergence in photosynthetic strategy, driven by the natural selection pressures of changes in CO_2 level, temperature and rainfall. CO_2 and/or temperature seem to have been most important for the C_3/C_4 divergence, while rainfall (in tropical and subtropical climates) may have been most significant for intra-C_4 diversification (Hattersley, 1992).

C_3 and C_4 grasses are fundamentally polarized with respect to their photosynthetic response to changing CO_2 levels, quantum yield, water-use efficiency and maximum growth rates (e.g. Fischer & Turner, 1978; Ehleringer & Pearcy, 1983; Ludlow, 1985). Polarization in these traits is directly attributable to C_3/C_4 biochemical–leaf-structural differences in these traits are ecophysiologically significant. C_4 grasses achieve their maximal net CO_2 assimilation rates at present-day atmospheric CO_2 concentrations. C_4 grasses outcompete C_3 grasses in tropical and subtropical areas where temperatures are >25 °C and where rainfall is predominantly summer-seasonal; in temperate, spring-rainfall areas, C_3 grasses predominate. Temperature correlates very highly with the percentage of C_4 species in the world's grass floras. Correlation coefficients of more than 0.96 have been documented for North America (Teeri & Stowe, 1976), Australia (Hattersley, 1983), Japan (Takeda et al., 1985) and Argentina (Cavagnaro, 1988). The higher water-use efficiencies of C_4 grasses also mean that they are more likely than are C_3 grasses to survive in arid and semi-arid environments that are hot (Archer, 1984); again, this manifests itself in differential phytogeographic distributions (e.g. Hattersley, 1983). Finally, in deep shade in the tropics and subtropics, C_3 grasses seem to prevail, perhaps outcompeting C_4 grasses in these light-limited habitats, even though temperatures are more or less consistently high (e.g. Medina, 1982).

While physiological distinctions between C_4 types have been documented for the O_2-independent CO_2 post-illumination burst (Downton, 1970), $\delta^{13}C$ values (Hattersley, 1982) and quantum yield (Ehleringer & Pearcy, 1983; but see Krall & Edwards, 1990), their functional significance, though speculated upon (Hattersley & Browning, 1981; Hattersley, 1984, 1987), has yet to be demonstrated. Even more obscure is how such differences might translate into ecophysiological distinctions. Nevertheless, such distinctions seem to exist, and they require explanation. In a recent analysis of the relative occurrence of C_4 biochemical types in the C_4 component of subdivisional grass floras of Australia, Hattersley (1992) showed that the percentage of NAD-ME type C_4 grasses was strongly negatively correlated with median annual rainfall ($r = -0.94$). Subdivisional %NADP-ME and %PCK, by contrast, were highly positively correlated with median annual rainfall ($r = 0.92$ and 0.82, respectively). Broadly similar conclusions had previously been reached in a phytogeographical survey of C_4 grass distribution in Namibia in relation to average annual rainfall (Ellis, Vogel & Fuls, 1980); however, no statistical analysis of the results had been made and no clear relationship between %PCK and rainfall had been established. No other studies have been conducted, despite the significance of such findings for land management practices and desertification control (Hattersley, 1992).

The Australian data also showed that the C_4 biochemical type as currently determined for grasses is more closely related to summer rainfall than was the C_4 type determined using criteria applicable at the time of Ellis *et al.*'s (1980) work, i.e. before Prendergast *et al.* (1986, 1987) and Prendergast & Hattersley (1987) had retyped certain key grass genera and tribes. This retyping was based on more extensive biochemical determinations of C_4 type, and on broader taxonomic sampling within the family, as described above. We have estimated that up to 15% of the Namibian grass flora may have been mistyped; for Australia, 22–27% of its grasses would be mistyped using the old criteria for C_4 typing, and correlation coefficients for the relationship between %C_4 type and median annual rainfall would have been -0.85 for the

NAD-ME type, 0.85 for the NADP-ME type, and −0.04 for the PCK type (Hattersley, 1992). The close relationship between relative C_4 type occurrence and rainfall has yet to be explained. Research in this area will not only be of ecophysiological and ecological significance; it will also further the understanding of grass diversification.

MOLECULAR BIOLOGY OF C_4 PHOTOSYNTHESIS

The study of the molecular biology of C_4 photosynthesis is too young to answer questions about the evolutionary relationships of grasses. *Zea mays* has been the almost exclusive subject of these investigations, though *Sorghum bicolor* (L.) Moench and *Pennisetum americanum* (L.) Leeke have occasionally been used. All three are 'classical NADP-ME' type C_4 species of the Panicoideae *sensu lato*. Genomic and/or cDNA clones have now been prepared and characterized for the genes for four C_4 acid cycle enzymes: PEP carboxylase (Harpster & Taylor, 1986; Hudspeth *et al.*, 1986; Izui *et al.*, 1986); pyruvate, P_i dikinase (PPDK; Hague, Uhler & Collins, 1983; Hudspeth *et al.*, 1986; Glackin & Grula, 1990); NADP malate dehydrogenase (NADP-MDH; Crétin *et al.*, 1988); and NADP-ME (Langdale *et al.*, 1988a). Complete cDNA sequences have been obtained for some of these clones, viz. a sorghum PEP carboxylase (Crétin *et al.*, 1990), maize PEP carboxylases (Hudspeth & Grula, 1989; Matsuoka & Minami, 1989; Kamamura *et al.*, 1990), a maize PPDK (Matsuoka & Ibaraki, 1990), and a maize NADP-MDH (Metzler, Rothermel & Nelson, 1989). These are nuclear genes; the enzymes they encode are chloroplastic, however, in the case of PPDK and NADP-MDH. Chloroplast development and the coordination of gene expression between the nuclear and chloroplast genomes have recently been reviewed (Hoober, 1987; Keegstra, Olsen & Theg, 1989; Taylor, 1989).

The common picture emerging for leaf C_4 enzymes is that they are encoded by genes that are members of multigene or small gene families (Harpster & Taylor, 1986; Hudspeth & Grula, 1989; Metzler *et al.*, 1989). The C_4 enzymes are related to isozymes occurring in other organs and in other cellular compartments (see e.g. Aoyagi & Bassham, 1986; Hudspeth *et al.*, 1986; Metzler *et al.*, 1989). 'C_4 enzymes' are not unique in C_4 plants, in the sense that they have their counterparts in C_3 species also. However, the C_4 isoforms are unique, and each is thought to be encoded by a single gene, i.e. a single member of the gene family (Hudspeth & Grula, 1989; Yanagisawa & Izui, 1989; Glackin & Grula, 1990; Kamamura *et al.*, 1990; Matsuoka & Ibaraki, 1990). Intraspecific and interspecific variation in nucleotide sequence for at least one C_4 enzyme (PEP carboxylase) is now sufficiently well known to allow the first attempts at understanding its molecular evolution (Izui *et al.*, 1991). For maize, Sheen (1991) has recently shown that the gene for C_4 chloroplast PPDK differs from a gene for C_4 cytosolic PPDK in having an added exon encoding the chloroplast transit peptide; the latter gene exhibits extensive sequence identity with a second cytosolic PPDK.

Ribulose-1,5-bisphosphate carboxylase/oxygenase (Rubisco), on the other hand, is a photosynthetic enzyme that occurs in both C_3 and C_4 angiosperms (and in all autotrophic plants). In higher plants, the Rubisco holoenzyme comprises eight large subunits (LS) and eight small subunits (SS) (Andrews & Lorimer, 1987). The LS protein is encoded by a single chloroplast gene (*rbc*L), whereas the SS protein is encoded by a multigene family (*rbc*S) in the nuclear genome (Manzara & Gruissem, 1988). The first *rbc*L gene to be cloned (Coen *et al.*, 1977) and sequenced (McIntosh, Poulsen & Bogorad, 1980) was from maize (C_4). The gene has subsequently been cloned from *Aegilops crassa* Boiss., *Hordeum vulgare* L., *Oryza sativa* L., and *Triticum aestivum* (all C_3 grasses), as well as from several C_3 dicotyledons. Hudson *et al.* (1990) have recently compared the *rbc*L nucleotide sequences from three pairs of closely related C_3 and C_4 species, viz. *Atriplex* L. (Chenopodiaceae), *Flaveria* A.L. Juss (Asteraceae), and *Neurachne* (Poaceae) (these genera all contain C_3–C_4 intermediate species also; see below). The kinetic properties of Rubisco differ between C_3 and C_4 species (Yeoh, Badger & Watson, 1980), and this is ascribed to the LS (Hudson *et al.*, 1990). The latter study aimed at identifying (consistent?) predicted amino acid C_3/C_4 differences in LS protein that may be functionally significant. Intrageneric comparisons showed only three to six amino acid differences, of which only one was consistent between pairs (residue 309; Met in C_3 and Ile in C_4). However, in *Zea mays* (the only other C_4 species for which there is currently an *rbc*L sequence), the residue is Met. Hudson *et al.* (1990) concluded that *rbc*L mutations have altered the kinetics, but not the expression of the enzyme. These results may not seem to encourage the use of the *rbc*L sequence for studying evolutionary relationships of C_3 and C_4 grasses. However, the comparisons made were of predicted amino acid residues, and so 'silent' changes were not compared. The promise of *rbc*L sequences for estimating phylogenetic topologies has already been shown (Zurawski & Clegg, 1987), and Gibbs *et al*. (1990) have already made an analysis of the sequence data that makes broad taxonomic sense.

The *rbc*S gene has been cloned in only three grasses: *O. sativa* (C_3), *T. aestivum* (C_3) and *Z. mays* (C_4). In wheat, the hexaploid genome contains more than 10 *rbc*S genes (Smith, Bedbrook & Speirs, 1983). It is not yet known how many genes there are in the maize or rice nuclear genomes (Sheen & Bogorad, 1986; Lebrun, Waksman & Fressinet, 1987; Matsuoka *et al*., 1988). Much more is known about the genes in species of the Solanaceae and there, although the genes within a species may diverge considerably in their sequences (e.g. by 18%), the mature proteins encoded by them are remarkably similar (e.g. Pichersky *et al*., 1986), i.e. most base-pair substitutions are silent. There is much conservation of amino acid sequence of the mature protein therefore, and this is thought to be due to recent gene conversion events rather than to recent gene duplications (Sugita *et al*., 1987). In the Solanaceae, three gene 'subfamilies' can be recognized which are thought to be 'at least as old as the Solanaceae family since the DNA sequences of coding and noncoding regions . . . of genes from different species of the same [gene] subfamily are more similar to each other than to other *rbc*S genes from the same species (but from different subfamilies)' (Dean, Pichersky & Dunsmuir, 1989). The three gene subfamilies are found at distinct genetic loci. Although the data are few (from a taxonomist's perspective), this example from the Solanaceae illustrates the potential of *rbc*S genes, and other multigene families (including those encoding C_4 enzymes), for elucidating relationships below the family level in grasses.

Molecular biological research on C_4 grasses has also focused on the regulation of expression of genes encoding C_4 enzymes, especially in relation to leaf development and cell specific expression in maize (PEP carboxylase, NADP-MDH and PPDK are C_4 mesophyll enzymes, while Rubisco and NADP-ME are PCR cell (bundle sheath) enzymes). Accumulation of the mRNAs for PEP carboxylase, PPDK, NADP-ME, NADP-MDH, Rubisco LS and Rubisco SS is light dependent (Nelson *et al*., 1984; Sheen & Bogorad, 1986, 1987; Cretin *et al*., 1988; Langdale *et al*., 1988*a,b*; Hudspeth & Grula, 1989). Regulation seems to be at the level of transcript accumulation. Accumulation of PEP carboxylase, NADP-MDH and NADP-ME is cell specific, whereas that of PPDK is rarely specific (Langdale, Taylor & Nelson, 1991). The pattern of Rubisco expression is sometimes cell non-specific. Langdale *et al*. (1991) found a correlation between methylation state at a site upstream of the

PEP carboxylase gene and gene expression (but no correlation between methylation state and *rbc*L and *rbc*S expression; cf. Ngernprasirtsiri *et al.*, 1989). This provides some insight into the basis for the developmental patterns of C$_4$ enzyme gene expression in maize (for a review, see Nelson & Langdale, 1989). The differential expression of the plastid-encoded photosystem II genes in mesophyll and bundle sheath cells has also been investigated (Westhoff *et al.*, 1991). Nelson & Langdale (1989) have concluded that 'bundle sheath and mesophyll cells must interpret positional information distributed around each vein to express correctly cell-specific genes. Light plays a crucial role in the generation or interpretation of this positional information'. The understanding of leaf development in C$_4$ grasses may soon constitute one of the best understood examples of cellular differentiation in plants. Although comparative investigations with respect to the molecular biology of different C$_4$ anatomical and biochemical types have yet to be done, the exciting pioneering work already undertaken clearly shows that they are possible. Appreciation of C$_4$ grass diversity in this new 'molecular dimension' will pave the way for understanding the processes that generated diversity, and, therefore, for knowing how to alter components of the 'C$_4$ syndrome' first described only 25 years ago.

Taxonomic occurrence

SUBFAMILIES AND TRIBES
The preceding sections have described the diversity of the biochemistry, leaf structure, physiology, ecophysiology and phytogeography of photosynthesis in the grasses. Patterns in this diversity have been identified and relationships suggested (e.g. homologies of PCR tissue in the various leaf-structural types), and these clearly constitute some foundations for an understanding of the pathways and processes by which such diversity arose. Diversification would have been complex and intricate, and it comprises the evolutionary history of the family. The current classification of the family can be scrutinized in this context.

The taxonomic occurrence of photosynthetic pathway variants in the family may yield clues about homologies and the comparative biology of photosynthetic variants that are currently unknown. If current grass classification is 'any good' (i.e. it is a 'natural' classification), then it may say more about the evolution of photosynthesis than does photosynthesis about classification. To date, the former has tended to be the case, but modern approaches of DNA systematics using sequence data for photosynthetic enzyme genes may turn the tables, especially at the tribal level and below.

The taxonomic occurrence of photosynthetic variants as presently perceived definitely has pattern, a pattern that has been consolidated as a result of the retyping by Prendergast *et al.* (1986, 1987) and Prendergast & Hattersley (1987). All Pooideae and Bambusoideae are C_3, as are most Arundinoideae. Arundinoids that are C_4 (Table 2.2) are the exception rather than the rule, and they are mostly characterized by exceptional C_4 biochemical–leaf-structural suites. Thus, the Eriachneae exhibit a unique suite that superficially resembles that of 'classical PCK' type species (see Fig. 1H of Hattersley, 1992; Table 2.1). However, not all species in the tribe possess centrifugal PCR cell chloroplasts or have an 'uneven' PCR sheath outline, and all probably lack a suberized lamella in both PCR and mestome sheath cells, in contrast to the 'classical PCK' type (Prendergast *et al.*, 1987; Hattersley, 1987). More unusually, the six species that have been biochemically typed are of the NADP-ME type, making the Eriachneae the only group of XyMS+ grasses of this C_4 biochemical type. It is a tribe of about 40 species confined to Australasia and parts of the Asian continent.

NADP-ME *Aristida* (Aristideae; also arundinoid) is also characterized by a unique suite (Fig. 2.2f; Table 2.1). Its double chlorenchymatous sheath may comprise a double PCR sheath (e.g. Brown, 1977), or an inner PCR sheath and an outer C_3 sheath (Hattersley & Browning, 1981). Again, we have a discrete taxonomic group that is characterized by a unique biochemical–leaf-structural suite, suggesting that it is a lineage that is not part of mainstream evolution in the family. *Aristida* is even more intriguing, in that it is placed in the same tribe (Aristideae) as C_3 *Sartidia* and C_4 *Stipagrostis*. Present information suggests that *Stipagrostis* is likely to be of the C_4 'classical NAD-ME' type (Table 2.2), though it has yet to be biochemically typed. The relationships of *Sartidia* need investigating.

Centropodia (=*Asthenatherum* Nevski) is the only other known C_4 arundinoid genus, and the only C_4 genus in the Danthonieae. Not all four species of *Centropodia* have been examined for leaf structure, and none has been biochemically typed. The genus is predicted to be of the 'classical NAD-ME' type (Ellis, 1984a). All XyMS+ grass species with centripetal PCR cell chloroplasts that have been typed are of the NAD-ME type biochemically (Hattersley, 1992), but *Centropodia* may prove to be exceptional, as other C_4 arundinoids are atypical in some respect.

The Chloridoideae have been extensively researched with respect to

Table 2.2. *Leaf-structural types for grass genera of the world*

Genera of subfamily Pooideae and subfamily Bambusoideae are consistently C_3 as far as is known, and are not listed. Most genera of subfamily Arundinoideae are C_3 as far as is known, and not listed; known C_4 genera of the Arundinoideae are listed (and marked with a *) viz. *Aristida* and *Stipagrostis* (Aristideae), *Eriachne* and *Pheidochloa* (Eriachneae) and *Centropodia* (Danthonieae); *Nematopoa* is an arundinoid genus of unknown photosynthetic pathway. All genera of the Chloridoideae and the Panicoideae are cited, with their leaf-structural type if known; or with '?', where unknown if C_3 or C_4; 'C_4:?', where the genus is known to be C_4 but leaf-structural type has not been determined; or 'C_3–C_4', where the genus contains one or more species which are C_3–C_4 intermediates. **, the genus has one or more 'XyMS variable' species (see Hattersley & Watson, 1976). ***, *Bouteloua curtipendula* may be a C_4 NAD-ME/PCK intermediate (see Prendergast *et al.*, 1986). Current knowledge suggests that species of most genera are consistently of one leaf-structural type, so that limited sampling accurately predicts type for the whole genus; new variation will be discovered, but much intrageneric variation is already documented, and is indicated here. In the supertribe Andropogonodae, all genera are assumed to be 'C_4: 1', as all species of this supertribe investigated to date are known to be of this type.
Numbers indicate the following C_4 leaf-structural types (biochemical C_4 type known to occur in the leaf-structural type is given in parenthesis): 1, 'classical NADP-ME' type (NADP-ME); 2, 'classical NAD-ME' type (NAD-ME); 3, 'classical PCK type' (PCK or NAD-ME); 4, 'Arundinelleae type' (NADP-ME); 5, 'Neurachneae type' (NADP-ME or PCK); 6, 'Aristida type' (NADP-ME); 7, 'Triodia type' (NAD-ME); 8, 'Eriachneae type' (NADP-ME). Refer to Fig. 2.2.
Data mostly from the current version of L. Watson's grass genera databank (Research School of Biological Sciences, The Australian National University; see also Watson *et al.*, 1986; Watson & Dallwitz, 1988) but with some interpretation, based on, in part, copious additional information therein. There are some discrepancies with data from Brown (1977), viz. *Chaetium* C_4: 3; *Holcolemma* C_4: 1; *Leptocoryphium* C_4: 1; *Odontelytrum* C_4: 1; *Oryzidium* C_4: 3. There are also discrepancies with data from Renvoize (1987), viz. *Dissochondrus* C_4: 1; *Holcolemma* C_4: 1; *Hylebates* C_3; *Leptocoryphium* C_4: 1; *Odontelytrum* C_4: 1; *Oryzidium* C_4: 3; *Tatianyx* C_4: 1.

Genus	Type
Acamptoclados Nash	C_4: 2 or 3
Achlaena Griseb.	C_4: 4
Acostia Swallen	?
Acrachne Chiov.	C_4: 2
Acritochaete Pilger	C_3
Acroceras Stapf	C_3
Aegopogon Willd.	C_4: 2 or 3
Aeluropus Trin.	C_4: 2
Afrotrichloris Chiov.	C_4: 2 or 3
Agenium Nees	C_4: 1

Table 2.2. (*cont.*)

Genus	Type
Alexfloydia B.K. Simon	C^4: 1
Allolepis Soderstrom & Decker	C_4: 2 or 3
Alloteropsis Presl	C_3, C_4: 2 & 5. (& C_3–C_4?)
Amphicarpum Kunth	C_3
Anadelphia Hackel	C_4: 1
Ancistrachne S.T.Blake	C_3
Andropogon L.	C_4: 1
Andropterum Stapf	C_4: 1
Anthaenantia P. Beauv.	C_4: 1
Anthaenantiopsis Pilger	C_4: 1
Anthephora Schreber	C_4: 1 (& 4?)
Apluda L.	C_4: 1
Apochiton C.E.Hubb.	C_4: 2 or 3
Apocopis Nees	C_4: 1
*Aristida L.	C_4: 6
Arthragrostis Lazarides	C_4: 2 or 3
Arthraxon P. Beauv.	C_4: 1
Arthropogon Nees	C_4: 4, & 1, 2 or 3
Arundinella Raddi	C_4: 1 & 4
Asthenochloa Buese	C_4: 1
Astrebla F.Muell.	C_4: 2
Austrochloris Lazarides	C_4: 2
Axonopus P.Beauv.	C_4: 1
Baptorhachis W. Clayton & Renvoize	C_4: 1
Beckeropsis Figari & de Not.	C_4: 1
Bealia Scribner	C_4: 2
Bewsia Goossens	C_4: 2
Bhidea Bor	C_4: 1
Blepharidachne Hackel	C_4: 2
Blepharoneuron Nash	C_4: 2
Boivinella A.Camus	C_3
Bothriochloa Kuntze	C_4: 1
***Bouteloua Lag.	C_4: 2 & 3
Brachiaria Griseb.	C_4: 3
Brachyachne (Benth.) Stapf	C_4: 2 & 3
Brachychloa Phillips	C_4: 2
Buchloë Engelm.	C_4: 2
Buchlomimus Reeder, Reeder & Rzedowski	C_4: 2 or 3
Calamovilfa Hackel	C_4: 2 or 3
Calyptochloa C.E.Hubb.	C_3
**Camusiella Bosser	C_4: 1
Capillipedium Stapf	C_4: 1
Catalepis Stapf & Stent	C_4: 3
Cathestechum J.Presl	C_4: 2 or 3
Cenchrus L.	C_4: 1
Centrochloa Swallen	C_4: 1
*Centropodia Reichenb.	C_4: ?2 (only?)
Chaetium Nees	C_4: 1, & 2 or 3

Table 2.2. (*cont.*)

Genus	Type
Chaetopoa C.E.Hubb.	C_4: 1
Chaetostichium (Hochst.) C.E.Hubb.	C_4: 2 or 3
Chamaeraphis R.Br.	C_4: 1
Chandrasekharania V.J.Nair, V.S.Ramachandran & P.V.Sreekumar	C_3?
Chasechloa A.Camus	C_3
Chasmopodium Stapf	C_4: 1
Chionachne R.Br.	C_4: 1
Chloachne Stapf	C_3
Chloris O.Swarz	C_4: 3
Chlorocalymma W.Clayton	C_4: 1
Chrysochloa Swallen	C_4: 2 or 3
Chrysopogon Trin.	C_4: 1
Chumsriella Bor	C_4: 1
Cladoraphis Franch.	C_4: 2
Clausospicula Lazarides	C_4: 1
Cleistachne Benth.	C_4: 1
Cleistochloa C.E.Hubb.	C_3
Cliffordochloa B.K.Simon	C_3
Coelachne R.Br.	C_3
Coelachyropsis Bor	C_4: 2 or 3
Coelachyrum Hochst & Nees	C_4: 2 or 3
Coelorhachis Brongn.	C_4: 1
Coix L.	C_4: 1
Commelinidium Stapf	C_3
Cottea Kunth	C_4: 2 or 3
Craspedorhachis Benth.	C_4: 2
Crypsis Aiton	C_4: 2 or 3
Ctenium Panzer	C_4: 2
Cyclostachya J.&C.Reeder	C_4: 2 or 3
Cymbopogon Spreng.	C_4: 1
Cymbosetaria Schweick.	C_4: 1
Cynodon Rich.	C_4: 2
Cyphochlaena Hackel	C_3
Cypholepis Chiov	C_4: 2
Cyrtococcum Stapf	C_3
Dactyloctenium Willd.	C_4: 3
Daknopholis W.Clayton	C_4: 2 or 3
Dallwatsonia B.K.Simon	C_3
Danthoniopsis Stapf	C_4: 1 & 4
Dasyochloa Rydberg	C_4: 2
Decaryella A.Camus	C_4: 2 or 3
Desmostachya P.Beauv.	C_4: 3?
Diandrochloa de Winter	C_4: 3
Diandrostachya Jacq.-Fel.	C_4: 1 (& 4?)
Dichanthelium (A.Hitchc. & Chase) Gould	C_3
Dichanthium Willem.	C_4: 1
Diectomis Kunth	C_4: 1

Table 2.2. (cont.)

Genus	Type
Digastrium (Hackel) A.Camus	C_4: 1
Digitaria Haller	C_4: 1
Digitariopsis C.E.Hubb.	C_4: 1
Dignathia Stapf	C_4: 2 or 3
Diheteropogon Stapf	C_4: 1
Dilophotriche Jacq.-Fel.	C_4: 2 or 3 (or 4?)
Dimeria R.Br.	C_4: 1
Dimorphochloa S.T.Blake	C_3
Dinebra Jacq.	C_4: 3
Diplachne P. Beauv.	C_4: 2
Dissochondrus (Hillebr.) Kuntze	C_4: 4
Distichlis Rafin.	C_4: 2
Drake-Brockmania Stapf	C_4: 2 or 3
Dybowskia Stapf	C_4: 1
Eccoilopus Steud.	C_4: 1
Eccoptocarpha Launert	C_4: 2 or 3
Echinochloa P.Beauv.	C_4: 1
Echinolaena Desv.	C_3
Ectrosia R.Br.	C_4: 2 or 3
Ectrosiopsis (Ohwi) Jansen	C_4: 3?
Eleusine Gaertn.	C_4: 2
Elionurus Humb. & Bonpl.	C_4: 1
Elymandra Stapf	C_4: 1
Enneapogon P.Beauv.	C_4: 3
Enteropogon Nees	C_4: 2
Entolasia Stapf	C_3
Entoplocamia Stapf	C_4: 2
Eragrostiella Bor	C_4: 2 or 3
Eragrostis N.M.Wolf	C_3, C_4: 2 & 3
Eremochloa Buese	C_4: 1
Eremopogon Stapf	C_4: 1
*Eriachne R.Br.	C_4: 8
Erianthus Michx.	C_4: 1
Eriochloa Kunth	C_4: 3
Eriochrysis P.Beauv.	C_4: 1
ErioneuronNash	C_4: 2
Euchlaena Schrad.	C_4: 1
Euclasta Franch.	C_4: 1
Eulalia Kunth	C_4: 1
Eulaliopsis Honda	C_4: 1
Eustachys Desf.	C_4: 2 & 3
Exotheca Anderss.	C_4: 1
Farrago W.Clayton	C_4: 2 or 3
Fingerhuthia Nees	C_4: 2
Garnotia Brongn.	C_4: 1 & 4
Germainia Bal. & Poitr.	C_4: 1
Gilgiochloa Pilger	C_4: 1 (or 4?)
Glyphochloa W.Clayton	C_4: 1

Table 2.2. (*cont.*)

Genus	Type
Gouinia Fourn.	C_4: 2 or 3
Griffithsochloa G.J.Pierce	C_4: 2 or 3
Gymnopogon P.Beauv.	C_4: 2
Hackelochloa Kuntze	C_4: 1
Halopyrum Stapf	C_4: 2 or 3
Harpachne Hochst	C_4: 2 or 3
Harpochloa Kunth	C_4: 2
Hemarthria R.Br.	C_4: 1
Hemisorghum C.E.Hubb.	C_4: 1
Heterachne Benth.	C_4: 2 or 3
Heteranthoecia Stapf	C_3
Heterocarpha Stapf & C.E.Hubb.	C_4: 2 or 3
Heteropholis C.E.Hubb.	C_4: 1
Heteropogon Pers.	C_4: 1
Hilaria Kunth	C_4: 2 or 3?
Holcolemma Stapf & C.E.Hubb.	C_3
Homolepis Chase	C_3
Homopholis C.E.Hubb.	C_3
Homozeugos Stapf	C_4: 1
Hubbardia Bor	C_3
Hubbardochloa Auquier	C_4: 2 or 3
Hydrothauma C.E.Hubb.	C_3
Hygrochloa Lazarides	C_4: 1
Hylebates Chippindall	C_4: 1
Hymenachne P.Beauv.	C_3
Hyparrhenia Anderss.	C_4: 1
Hyperthelia W.Clayton	C_4: 1
Hypogynium Nees	C_4: 1
Ichnanthus P.Beauv.	C_3
Imperata Cirillo	C_4: 1
Indopoa Bor	C_4: 2 or 3
Isachne R.Br.	C_3
Isalus J.Phipps	C_4: 1
Ischaemum L.	C_4: 1
Ischnochloa J.D.Hook.	C_4: 1
Ischnurus Balf.	C_4: 2 or 3
Iseilima Anderss.	C_4: 1
Ixophorus Schlechtd.	C_4: 1
Jansenella Bor	C_3?
Jardinea Steud.	C_4: 1
Jouvea Fourn.	C_4: 2 or 3
Kampochloa W.Clayton	C_4: 2 or 3
Kaokochloa de Winter	C_4: 3
Kengia Packer	C_4: 2 or 3
Kerriochloa C.E.Hubb.	C_4: 1
Lasiacis A.Hitchc.	C_3
Lasiorrachis Stapf	C_4: 1
Lasiurus Boiss.	C_4: 1

Table 2.2. (*cont.*)

Genus	Type
Lecomtella A.Camus	C_3
Lepargochloa Launert	C_4: 1
Leptocarydion Stapf	C_4: 2
Leptochloa P.Beauv.	C_4: 2 & 3
Leptochloopsis Yates	C_4: 2 & 3
Leptocoryphium Nees	C_4: 2 or 3 (4-ish)
Leptoloma Chase	C_4: 1
Leptosaccharum (Hackel) A.Camus	C_4: 1
Leptothrium Kunth	C_4: 2 or 3
Lepturella Stapf	C_4: 2 or 3
Lepturidium Hitchc. & Ekman	?
Lepturopetium Morat	C_4: 2 or 3
Lepturus R.Br.	C_4: 2 and/or 3
Leucophrys Rendle	C_4: 3
Limnopoa C.E.Hubb.	C_3
Lintonia Stapf	C_4: 3
Lophacme Stapf	C_4: 2
Lopholepis Decne.	C_4: 2 or 3
Lophopogon Hackel	C_4: 1
Loudetia Hochst.	C_4: 1 & 4
**Loudetiopsis* Conert	C_4: 1, & 3 or 8
Louisiella C.E.Hubb. & Leonard	C_4: 2 or 3
Loxodera Launert	C_4: 1
Lycurus Kunth	C_4: 2
Manisuris L.	C_4: 1
Megaloprotachne C.E.Hubb.	C_4: 1
Melanocenchris Nees	C_4: 2 or 3
Melinis P.Beauv.	C_4: 3
Mesosetum Steud.	C_4: 1
Microcalamus Franch.	C_3
Microchloa R.Br.	C_4: 2 & 3
Microstegium Nees	C_4: 1
Mildbraediochloa Butzin	C_4: 2 or 3
Miscanthidium Stapf	C_4: 1
Miscanthus Anderss.	C_4: 1
***Mnesithea* Kunth	C_4: 1
Monanthochloe Engelm.	C_4: 2
Monelytrum Hackel	C_4: 2
Monium Stapf	C_4: 1
Monocymbium Stapf	C_4: 1
Monodia S.W.L.Jacobs	C_4: 7
Mosdenia Stent	C_4: 2
Muhlenbergia Schreber	C_4: 2 & 3
Munroa J.Torr.	C_4: 2 or 3
Myriostachya J.D.Hook.	C_4: 2 or 3
Narenga Bor	C_4: 1
Neeragrostis Bush	C_4: 2 or 3

Table 2.2. (*cont.*)

Genus	Type
Neesiochloa Pilger	C_4: 2 or 3
*Nematopoa C.E.Hubb.	?
Neobouteloua Gould	C_4: 2 or 3
Neostapfia Davy	C_4: 2 or 3
Neostapfiella A. Camus	C_4: probably 2 & 3
Neurachne R.Br.	C_3, C_3–C_4, & C_4: 5
Neyraudia Hook.f.	C_4: 2 or 3
Ochthochloa Edgwe.	C_4: 2 or 3
Odontelytrum Hackel	C_4: 3
Odyssea Stapf	C_4: 2
Ophiuros Gaertn.f.	C_4: 1
Opizia J.&C.Presl	C_4: 2 or 3
Oplismenopsis L.Parodi	C_3
Oplismenus P.Beauv.	C_3
Orcuttia Vasey	C_4: 2 or 3
Orinus A.Hitchc.	C_4: 2 or 3
Oropetium Trin.	C_4: 2
Oryzidium C.E.Hubb. & Schweick.	C_4: 2
Otachyrium Nees	C_3
Ottochloa Dandy	C_3
Oxychloris Lazarides	C_4: 2
Oxyrhachis Pilger	C_4: 1
Panicum L.	C_3, & C_4: 1, 2, 3 & 5
	C_4: 2
Pappophorum Schreber	C_4: 1
Paractenium P.Beauv. corr. Stapf	C_4: 1
Parahyparrhenia A.Camus	C_4: 5
Paraneurachne S.T.Blake	C_4: 1
Paratheria Griseb.	C_4: 1
Paspalidium Stapf	C_4: 1
Paspalum L.	C_4: 1
Pennisetum Rich.	C_4: 1
Pentarrhaphis Kunth	C_4: 2 or 3
Pereilema J.&C.Presl	C_4: 2 or 3
Perotis Aiton	C_4: 2
Perulifera A.Camus	C_3
Phacelurus Griseb.	C_4: 1
*Pheidochloa S.T.Blake	C_4: 8
Piptophyllum C.E.Hubb.	?
Plagiantha Renvoize	C_3
Plagiosetum Benth.	C_4: 1
Planichloa B.Simon	C_4: 2 or 3
Plectrachne Henrard	C_4: 7
Pleiadelphia Stapf	C_4: 1
Pobeguinea Jacq.-Fel.	C_4: 1
Poecilostachys Hackel	C_3
Pogonachne Bor	C_4: 1

Table 2.2. (cont.)

Genus	Type
Pogonarthria Stapf	C_4: 2
Pogonatherum P.Beauv.	C_4: 1
Pogoneura Napper	C_4: 2 or 3
Pogonochloa C.E.Hubb.	C_4: 2 or 3
Polevansia de Winter	C_4: 2
Polliniopsis Hayata	C_4: 1
Polytoca R.Br.	C_4: 1
Polytrias Hackel	C_4: 1
Pommereulla L.f.	C_4: 2 or 3
Pringleochloa Scribner	C_4: 2 or 3
Psammagrostis C.Gardner & C.E.Hubb.	C_4: 2
Pseudanthistiria (Hackel) Hook.f.	C_4: 1
Pseudechinolaena Stapf	C_3
Pseudochaetochloa A.Hitchc.	C_4: 1
Pseudodichanthium Bor	C_4: 1
Pseudopogonatherum A.Camus	C_4: 1
Pseudoraphis Griff.	C_4: 1
Pseudosorghum A.Camus	C_4: 1
Pseudovossia A.Camus	C_4: 1
Pseudozoysia Chiov.	C_4: 2 or 3
Psilolemma Phillips	C_4: 2 or 3
Pterochloris A.Camus	C_4: 2 or 3
Ratzeburgia Kunth	C_4: 1
Redfieldia Vasey	C_4: 2?
Reederochloa Soderstrom & H.F.Decker	C_4: 2 or 3
Reimarochloa A.Hitchc.	C_4: 1
Rendlia Chiov.	C_4: 2
Reynaudia Kunth	C_4: 1
Rhynchelytrum Nees	C_4: 3
Rhytachne Desv.	C_4: 1
Richardsiella Elffers & Kennedy O'Byrne	C_4: 3
Robynsiochloa Jacq.-Fel.	C_4: 1
Rottboellia L.f.	C_4: 1
Saccharum L.	C_4: 1
Sacciolepis Nash	C_3
Saugetia A.Hitchc. & Chase	C_4: 2 or 3
Schaffnerella Nash	C_4: 2 or 3
Schedonnardus Steud.	C_4: 2
Schenckochloa J.J.Ortiíz	C_4: 2 or 3?
Schizachyrium Nees	C_4: 1
Schmidtia Steud.	C_4: 3
Schoenefeldia Kunth	C_4: 2 or 3
Sclerachne R.Br.	C_4: 1
Sclerodactylon Stapf	C_4: 7-like
Scleropogon Phil.	C_4: 2 or 3
Sclerostachya A.Camus	C_4: 1
Scutachne A.Hitchc. & Chase	C_4: 2 or 3

Table 2.2. (*cont.*)

Genus	Type
Sehima Forssk.	C_4: 1
Setaria P.Beauv.	C_4: 1
Setariopsis Millsp.	C_4: 1
Silentvalleya Nair, Sreekumar, Vajravelu & Bhargavan	?
Snowdenia C.E.Hubb.	C_4: 1
Soderstromia Morton	C_4: 2 or 3
Sohnsia Airy Shaw	?
Sorghastrum Nash	C_4: 1
Sorghum Moench	C_4: 1
Spartina Schreber	C_4: 3
Spathia Ewart	C_4: 1
Sphaerocaryum Hook.f.	C_3
Spheneria Kuhlm	C_4: 1
****Spinifex* L.	C_4: 1**
Spodiopogon Trin.	C_4: 1?
Sporobolus R.Br.	C_4: 2 & 3
Steinchisma Raf	C_3–C_4
Steirachne Ekman	C_4: 2 or 3
Stenotaphrum Trin.	C_4: 1
Stereochlaena Hackel	C_4: 1
Stiburus Stapf	C_4: 3?
*Stipagrostis Nees	C_4: 2
Streptolophus Hughes	C_4: 1
Streptostachys Desv.	C_3
Swallenia Soderstrom & Decker	C_4: 2 or 3
Symplectrodia Lazarides	C_4: 7
Tarigidia Stent	C_4: 1
Tatianyx Zuloaga & Soderstrom	C_4: 2 or 3
Tetrachaete Chiov.	C_4: 2 or 3
Tetrachne Nees	C_4: 2
Tetrapogon Desf.	C_4: 2
Thaumastochloa C.E.Hubb.	C_4: 1
Thelepogon Roth.	C_4: 1
Thellungia Stapf	C_4: 2
Themeda Forssk.	C_4: 1
Thrasya Kunth	C_4: 1
Thrasyopsis L. Parodi	C_4: 1
Thuarea Pers.	C_4: 3
Thyridachne C.E.Hubb.	C_3
Thyridolepis S.T.Blake	C_3
Thyrsia Stapf	C_4: 1
**Trachypogon Nees	C_4: 1
Trachys Pers.	C_4: 1
Tragus Haller	C_4: 2
Tricholaena Schrad.	C_4: 3
Trichoneura Anderss.	C_4: 2

Table 2.2. (cont.)

Genus	Type
Trichopteryx Nees	C_4: 1 & 4
Tridens Roem. & Schult.	C_4: 3
Trilobachne Henrard	C_4: 1
Triodia R.Br.	C_4: 7
Triplasis P.Beauv.	C_4: 2
Triplopogon Bor	C_4: 1
Tripogon Roem. & Schult.	C_4: 2
Tripsacum L.	C_4: 1
Triraphis R.Br.	C_4: 3
Triscenia Griseb.	C_3
Tristachya Nees	C_4: 1 (& 4?)
Tuctoria J. Reeder	C_4: 2 or 3
Uniola L.	C_4: 2 or 3
Uranthoecium Stapf	C_4: 1
Urelytrum Hackel	C_4: 1
Urochloa P.Beauv.	C_4: 3
Urochondra C.E.Hubb.	C_4: 2 or 3
Vaseyochloa A.Hitchc.	C_4: 2
Vetiveria Bory	C_4: 1
Viguierella A.Camus	C_4: 2 & 3
Vossia Wall. & Griff.	C_4: 1
Whiteochloa C.E.Hubb.	C_4: 1
Willkommia Hackel.	C_4: 2 or 3?
Xerochloa R.Br.	C_4: 1
Yakirra Lazarides & R.Webster	C_4: 2
Ystia P.Compere	C_4: 1
Yvesia A.Camus	C_4: 2?
Zea L.	C_4: 1
Zonotriche Phipps	C_4: 2 or 3?
Zoysia Willd.	C_4: 3
Zygochloa S.T.Blake	C_4: 1

their photosynthetic pathway and all have been found to be C_4 (Table 2.2) and XyMS+ (Hattersley & Watson, 1976; Brown, 1977; Ellis, 1977), with the remarkable exception of *Eragrostis walteri* Pilger, a perennial C_3 species from arid Namibia (Ellis, 1984b). *E. walteri* may represent an evolutionary reversal, or be a relict C_3 taxon, and is worthy of considerable further study. Mainstream chloridoids (i.e. tribe Chlorideae *sensu lato*; Watson, 1990) are either of the 'classical NAD-ME', or the 'classical PCK' type, the latter including NAD-ME 'classical PCK' type species (refer to Table 2.2 for generic representation). Note the postulate that these suites are closely related to one another on leaf-structure and biochemical grounds (above sections). In the

Chloridoideae, NAD-ME 'classical PCK' type ('PCK-like NAD-ME' type) species are currently known in *Eragrostis* (some species; Chlorideae), *Triraphis* (all species?; Chlorideae) and *Enneapogon* (perhaps all species; Pappophoreae). The C_4 leaf-structure types of two other genera of the Pappophoreae (*Cottea, Kampochloa*) and of the Orcuttieae (*Neostapfia, Orcuttia, Tuctoria*) are not known in detail, but *Pappophorum* has leaf structure of the 'classical NAD-ME' type and *Schmidtia* of the 'classical PCK' type (both Pappophoreae) (Table 2.2).

The NAD-ME 'Triodia' type suite is the only other C_4 suite known in the subfamily; it is restricted to the tribe Triodieae, endemic in Australia. As far as is known, all Triodieae (*Monodia, Plectrachne, Symplectrodia, Triodia*) possess the NAD-ME 'Triodia' type suite. Again, an unusual suite has a restricted taxonomic occurrence (but *Sclerodactylon* leaf structure approaches that of the 'Triodia' type).

The Panicoideae exhibit more photosynthetic pathway variation than either the chloridoids or the arundinoids. Not only are both C_3 and C_4 pathways represented (Table 2.2), but so are five of the eight C_4 leaf-structure types. The four C_3–C_4 intermediate species known in the Poaceae also belong to this subfamily (see below). Nevertheless, even among this variation, at least some clear taxonomic patterns exist. The supertribe Andropogonodae is solidly of the 'classical NADP-ME' type, and in supertribe Panicodae the Isachneae is solidly C_3.

The Arundinelleae may be solidly C_4 (Brown, 1977; Ellis, 1977; Watson *et al.*, 1986; Prendergast *et al.*, 1987), but there are doubts about *Chandrasekharania* and *Jansenella*, and *Isalus* has yet to be examined (Table 2.2). The tribe Neurachneae contains C_3 species (three *Thyridolepis* spp. and four *Neurachne* spp.), C_4 NADP-ME 'Neurachneae' type species (*Neurachne munroi* (F. Muell.) F. Muell. and *Paraneurachne muelleri* (Hackel) S.T. Blake), and the C_3–C_4 intermediate *Neurachne minor* S.T. Blake (Hattersley *et al.*, 1986). The remaining and largest tribe, the Paniceae (*sensu* Watson, 1990), is the most diverse of the Panicodae with C_3 species, C_4 'classical NADP-ME' type species, C_4 'classical NAD-ME' type species, C_4 PCK 'classical PCK' type species, C_4 NAD-ME 'classical PCK' type ('PCK-like NAD-ME') species, C_4 'Neurachneae' type species (*Alloteropsis semialata* subsp. *semialata, Panicum petersonii* and *P. prionitis* Nees only), C_4 'Arundinelleae' type species (see below), and at least three C_3–C_4 intermediate species. At infratribal level there is, nevertheless, taxonomic pattern, and this is described below.

The C_4 NADP-ME 'Arundinelleae type' suite (Fig. 2.2*g*) is found not only in the Arundinelleae but also in the Paniceae (some species of *Arthropogon*, and monotypic *Achlaena* and *Dissochondrus*). Also, not all genera of the Arundinelleae seem to exhibit 'Arundinelleae' type leaf structure (*Diandrostachya, Loudetiopsis, Tristachya*), being of the 'classical NADP-ME' type. Furthermore, none of the genera that do exhibit 'Arundinelleae' type C_4 leaf anatomy is homogeneous in this respect (viz. *Arundinella, Danthoniopsis, Garnotia, Loudetia, Trichopteryx*, where some species are of the 'classical NADP-ME' type) and the leaf structure of others has yet to be adequately determined (*Dilophotriche, Gilgiochloa, Isalus, Jansenella, Leptocoryphium, Zonotriche*). As already noted, on leaf-structure grounds alone it seems reasonable to suggest that the NADP-ME 'classical NADP-ME' type suite and the NADP-ME 'Arundinelleae' type suite are closely related, and the localized taxonomic occurrence of the 'Arundinelleae' type mixed up with taxa of the 'classical NADP-ME' type is consistent with this view.

The taxonomic distribution of C_4 biochemical–leaf-structural suites is not, then, a simple one at subfamily and tribe level (see Table 6.4 of Hattersley, 1987). All three subfamilies that contain C_4 species, also contain C_3 species (though only one in the case of the Chloridoideae). Biochemically, C_4 Arundinoideae are of the NADP-ME type, with the exception that *Stipagrostis* seems likely to be of the NAD-ME type. C_4 Chloridoideae are either of the NAD-ME type or the PCK type. All three C_4 biochemical types occur in the Panicoideae. Three of the eight leaf-structural types (Fig. 2.2*f,h*, and the Eriachneae type) are each restricted to single tribes and for one of these, the 'Aristida' type, to a single genus. For two other types, the 'Arundinelleae' and 'Neurachneae' types (Fig. 2.2*e,g*) representation is not only in the tribes after which they are named, but also in the closely related Paniceae. The 'classical NADP-ME' type leaf structure is confined to the Panicoideae, occurring in both supertribes. The 'classical NAD-ME' and 'classical PCK' leaf-structural types each occur in both the panicoids and chloridoids, and the former is also known in the arundinoids (*Centropodia* and *Stipagrostis*).

If current grass classification (e.g. Watson, 1990) accurately reflects relatedness, then one or more of the C_4 leaf-structural types, and/or one or more of the C_4 biochemical types evolved more than once within the family. Alternatively, the classification may not always reflect phylogenetic relatedness at the subfamily and tribe levels. These issues are addressed in the section entitled 'Diversification of photosynthesis' (see below), but first infratribal variation (mostly among closely related genera) needs describing to complete the picture of taxonomic distribution.

GENERA AND SPECIES

Most grass genera are consistently either C_3 or C_4. There are only four exceptional genera known: *Alloteropsis*, *Eragrostis*, *Neurachne*, and *Panicum* (Table 2.2). Furthermore, most C_4 genera are consistently of one C_4 biochemical type (except *Alloteropsis*, *Bouteloua*, *Leptochloa*, *Panicum*, and *Sporobolus*; and *Muhlenbergia*?) and of one C_4 leaf-structural type (except *Alloteropsis*, *Bouteloua*, *Eragrostis*, *Leptochloa*, *Muhlenbergia*, *Panicum* and *Sporobolus*). These listed exceptions represent an update of the information provided in Fig. 11.9 in Edwards & Walker (1983). Note that some genera are exceptional in more than one way and some of these are very large, viz. *Eragrostis*, with *c*. 350 spp. world-wide, *Muhlenbergia*, 160 spp., *Panicum*, *c*. 370 spp. and *Sporobolus*, 160 spp. Of these genera exhibiting variation in photosynthetic pathway, only *Panicum* is conspicuously taxonomically heterogeneous. Below the species level, there is only one known instance of variation in photosynthetic pathway in grasses (*Alloteropsis semialata*: Ellis, 1974; Brown, 1975). The above genera will now be considered in more detail:

Alloteropsis

This genus is one that has received little attention to date, yet it promises to be especially interesting. According to Butzin (1968) and Clayton & Renvoize (1982) there are two species complexes. The 'Alloteropsis' group comprises *A. semialata*, *A. angusta* Stapf, *A. gwebiensis* Stent & Rattray, and *A. homblei* Robyns, though only the first two taxa are recognized by Clayton & Renvoize. These XyMS− species have the 'Neurachneae' type structure. Only *A. semialata* subsp. *semialata* has been biochemically typed, as of the PCK type by Prendergast *et al.* (1987) using an Australian accession, but as the NADP-ME type for a South African accession (Frean *et al.*, 1983). Enzyme activities reported by the latter, however, were very unusual, being unexpectedly high (47 μmol mg chlor.$^{-1}$ min^{-1}; cf. typical values of up to 25 μmol mg chlor.$^{-1}$ min^{-1}).

The 'Coridochloa' group comprises *A. cimicina* (L.) Stapf, *A. paniculata* (Benth.) Stapf, *A. papillosa* W. D. Clayton, *A. quintasii* (Mez.) Pilger, and *A. latifolia* (Peter) Pilger, though Clayton & Renvoize

(1982) recognize only the first three. In contrast to species in the 'Alloteropsis' group, species of the 'Coridochloa' group are XyMS+ and have centripetal PCR cell chloroplasts (Ellis, 1977, 1981; C. Long & P. W. Hattersley, unpublished data); they are likely to prove to be of the NAD-ME type when biochemically typed. The XyMS− and XyMS+ taxa have been placed in the same genus in recent taxonomic revisions (Butzin, 1968; Clayton & Renvoize, 1982). (*Panicum* and *Neurachne* are the only other grass genera known to contain both XyMS− and XyMS+ species.) Only *A. semialata* and *A. cimicina* occur outside Africa.

Until recently (Ueno *et al.*, 1988), *A. semialata* was the only species in the plant kingdom in which both C_3 and C_4 individuals have been documented. *A. semialata* subsp. *semialata* has been shown to be C_4 on the basis of leaf structure, biochemistry, and physiology, wherever it occurs (Africa, Madagascar, India, SE Asia, Pacific islands, and Australia; Ellis, 1974; Brown, 1975; Hattersley & Browning, 1981; Frean *et al.*, 1983; C. Long & P. W. Hattersley, in preparation). Subsp. *eckloniana* (Nees) Gibbs Russell, by contrast, is C_3 type on the basis of all measures, and is confined to southern Africa (Ellis, 1974; Brown, 1975; Frean *et al.*, 1983; Gibbs Russell, 1983). There is some doubt as to whether the two taxa should be given specific status but a detailed morphological and anatomical comparison led Gibbs Russell (1983) to class them as subspecies.

Interestingly, Ellis (1981) and Gibbs Russell (1983) have located some individuals which are somewhat intermediate morphologically, and one accession of these (Milne-Redhead 3021) was found to have an intermediate $\delta^{13}C$ value (Ellis, 1981). A subsequent study by Long & Hattersley (unpublished data) has located a further two accessions with similar $\delta^{13}C$ values, and seven more accessions whose leaf structure suggests that they too may be C_3–C_4 intermediates (Table 2.3). Unlike other C_3–C_4 intermediates, these individuals are not uniform morphologically, do not comprise a distinct taxon at the species level, intergrade morphologically with the C_3 and C_4 forms of the species, and have a generally intermediate geographical distribution. This situation could be the most interesting yet documented for C_3–C_4 intermediates among angiosperms, with the possibility that natural hybridization between the C_3 and C_4 forms of *A. semialata* is occurring and/or that active evolution of photosynthetic pathway is taking place in Zambia, Zimbabwe and Tanzania. Such introgression would be well worth investigation.

Table 2.3. *Possible C₃–C₄ accessions of* Alloteropsis semialata

The first ten specimens listed are not wholly C_3 anatomically, yet they exhibit intermediate (the first three) or C_3 $\delta^{13}C$ values. (The C_3 range is −22 to −34‰). Bogden & Williams 238 is not wholly C_4 anatomically, yet has a C_4 $\delta^{13}C$ value. (The C_4 range is −10 to−17‰). Leaves of all specimens are XyMS−, except Bogden & Williams 238 which is XyMS+. Morphologically, none of the specimens clearly belongs to *A. semialata* subsp. *eckloniana*, which is the C_3 form of *A. semialata*.

Accession (collector and number)	Country	Holding herbarium	$\delta^{13}C$ value (‰)
Milne-Redhead 3021	Zambia	B	−20.7
Emson 340	Tanzania	EA	−21.4
Stowe 495	Zambia	BOL	−20.8
Brzotowski 26	Tanzania	CANB	−22.6
Robinson 4744	Zambia	EA	−22.7
Proctor 2165	Tanzania	EA	−23.0
Mbano DSM812	Tanzania	EA	−23.9
Proctor 2206	Zambia	EA	−23.9
Simon 932	Zimbabwe	K	−25.3
Tanner 5076	Tanzania	EA	−26.26
Bogden & Williams 238	Kenya	EA	−11.8

Each $\delta^{13}C$ value is the mean of two measurements. For methods see Hattersley & Roksandic (1983). Milne-Redhead 3021 was measured by Ellis (1981) to have a $\delta^{13}C$ value of −20‰.
Data: C. Long & P.W. Hattersley (unpublished).

Bouteloua, Leptochloa and *Sporobolus*

These three chloridoid genera are said to contain both 'classical NAD-ME' type and PCK 'classical PCK' type species (Gutierrez *et al.*, 1974; Hatch *et al.*, 1975; Brown, 1977). However, the only designated PCK 'classical PCK' type species in *Bouteloua* (*B. curtipendula* (Michaux) Torrey) exhibits high NAD-ME activity also (Gutierrez *et al.*, 1974; Prendergast *et al.*, 1987), and so this species could be an NAD-ME–PCK intermediate (Prendergast *et al.*, 1987). As such, it should be included in any study aimed at assessing whether the C_4 PCK type is derived from the C_4 NAD-ME type, as should ×*Cynochloris* (see below).

In *Bouteloua, Leptochloa* and *Sporobolus*, no NAD-ME 'classical PCK' type ('PCK-like NAD-ME') species have been demonstrated, i.e. species that have 'classical PCK' type leaf structure but which exhibit only NAD-ME activity (Prendergast *et al.*, 1987). Such species could represent an early intermediary stage in the evolution of PCK type species from NAD-ME type species, with *B. curtipendula* representing a later stage where actual PCK activity had become expressed. *L. ciliolata* (Jedwabn.) S.T. Blake may be another such species (Prendergast *et al.*, 1987).

Muhlenbergia

This may be another chloridoid genus with both 'classical NAD-ME' type and PCK 'classical PCK' type species. Only *M. schreberi* J. Gmelin has been biochemically typed (Gutierrez *et al.*, 1974), being PCK. Hattersley & Browning (1981), however, found that *M. montana* (Nutt.) A. Hitchc. has the 'classical NAD-ME' type of leaf structure, and also it is quite clear from the micrographs of Peterson, Annable & Franceschi (1989) that more species with such leaf structure occur in the genus.

Eragrostis

This genus was previously suspected of containing both NAD-ME and PCK C_4 species on the basis of leaf structure (Brown, 1977; Ellis, 1977; Hattersley, 1984). In an attempt to demonstrate that *Eragrostis* species with leaf structure intermediate to that of the 'classical NAD-ME' and 'classical PCK' types were biochemically intermediate, Prendergast *et al.* (1986) found instead that all species, including those with the 'classical PCK' type of leaf structure, were biochemically NAD-ME. NAD-ME 'classical PCK' type species (so-called 'PCK-like NAD-ME' species) had, in fact, previously been found in *Panicum* (Ohsugi & Murata, 1980; Ohsugi *et al.*, 1982), and are now known in *Enneapogon* and *Triraphis* also (Prendergast *et al.*, 1987). *Eragrostis* is well worth further study, as some species may be C_4 type intermediates like *B. curtipendula*. Especially noteworthy, too, is the existence of a C_3 *Eragrostis* from Namibia (*E. walteri*) (Ellis, 1984b). There are no taxonomic grounds for excluding *E. walteri* from *Eragrostis*.

Neurachne

This is a small genus of six species endemic in Australia (Blake, 1972) which comprises both C_3 (*N. alopecuroidea* R. Br., *N. lanigera* S.T. Blake, *N. queenslandica* S.T. Blake and *N. tenuifolia* S.T. Blake) and a C_4 species (*N. munroi*) (Hattersley, Watson & Johnston, 1982). Further work showed that *N. minor* is a C_3–C_4 intermediate (Hattersley *et al.*, 1986), the only such species currently known from the Australian flora. There are no taxonomic grounds for excluding C_3 *Neurachne* spp. or C_3–C_4 *N. minor* from *Neurachne*. *Neurachne* was an obscure genus until recently but is now well collected (Prendergast & Hattersley, 1985) and for two of its species the chloroplast gene for the last subunit of Rubisco has been sequenced (Hudson *et al.*, 1990). The related genera *Thyridolepis* (three spp.) and monotypic *Paraneurachne* are C_3 and C_4, respectively (Hattersley & Roksandic, 1983).

Panicum

This genus attracted the attention of C_4 photosynthesis researchers from a very early stage, as it was the first grass genus shown to include both C_3 and C_4 species (Downton, Berry & Tregunna, 1969), and to exhibit more than one C_4 type (Downton, 1971). Modern systematic treatments (e.g. Webster, 1987; Zuloaga, 1987) continue to include some C_3 species and both NADP-ME and NAD-ME type C_4 species in the one genus, though many erstwhile C_3 *Panicum* species have been transferred to *Dichanthelium* (Gould & Clark, 1978). *Panicum* is the only panicoid genus which is known definitely to contain NAD-ME species (for a checklist, see Hattersley, 1987), but three other genera have the 'classical NAD-ME' type of leaf structure (*Oryzidium*, *Yakirra* and *Yvesia*) and so are highly likely to be of the NAD-ME type biochemically. Species of two other genera may have the 'classical NAD-ME' type leaf structure (*Tatianyx* and *Zonotriche*) (Table 2.2). Nevertheless, the representation of C_4 NAD-ME type species in panicoids is very limited, and even less than that of the PCK type. Most C_4 Panicodae genera and all Andropogonodae genera are of the NADP-ME type.

It is not clear whether *Panicum* can still be said to contain C_4 PCK type species. Many of the *Panicum* species that were originally classed as PCK, are now placed in closely related *Brachiaria* or *Urochloa*. Brown (1977) was of the opinion that all PCK *Panicum* species should be placed in other genera, a view shared by Webster (1987), who transferred *P. maximum* Jacq. to *Urochloa*. Watson *et al.* (1986) and Zuloaga (1987) considered that a world-wide treatment of the genus is required before such decisions are made, particularly for the African species with the 'classical PCK' type of leaf structure (e.g. Ellis, 1977). *P. maximum* is currently the only remaining *Panicum* with 'classical PCK' type leaf structure to have been biochemically typed as PCK (for a checklist, see Hattersley, 1987). Other *Panicum* species with 'classical PCK' type structure (*P. dichotomiflorum* Michaux, *P. longijubatum* (Stapf) Stapf, *P. schinzii* Hackel, *P. virgatum* L., and some varieties of *P. coloratum*

L.), though predicted to be biochemically of the PCK type, have recently been found to be of the NAD-ME type (Ohsugi *et al.*, 1982; Prendergast *et al.*, 1987) and are now designated NAD-ME 'classical PCK' type (or 'PCK-like NAD-ME').

One of the diagnostic features of *'Panicum'* species with 'classical PCK' type leaf structure is, according to Brown (1977) and Webster (1987), the possession of a rugose lemma in the upper fertile floret of the spikelet. Indeed, it is the major character used to justify the placement of species such as *P. maximum* in *Urochloa*. It is most interesting, therefore, that the 'PCK-like NAD-ME' *P. shinzii* at least, does *not* have a rugose upper lemma, excluding it from *Urochloa* even though it has the 'classical PCK' type structure. We stress, therefore, that it seems judicious to undertake a revision of *'Panicum'* at world level before making too many taxonomic realignments, especially considering the current concern over the stability of plant names (Hawksworth, 1991).

Panicum is further remarkable in being the first genus in which a C_3–C_4 intermediate species was discovered (*P. hians* Elliott (=*P. milioides* Nees) – Brown & Brown, 1975). This species and relatives were transferred by Brown (1977) to a revalidated *Steinchisma* Raf. on the basis of a number of characters, but Zuloaga (1987) questioned some of these and preferred to treat *Steinchisma* as a subgenus. Two other species of subgenus *Steinchisma* are also known to be C_3–C_4 intermediates (*P. spathellosum* Doell (=*P. schenckii* Hackel) and *P. decipiens* Nees ex Trin.; Brown *et al.*, 1985), and it will be interesting to see if the other three species of this primarily South American group are similar in this respect. No close C_3 or C_4 relatives of *Steinchisma* can be specified. C_3–C_4 intermediacy is discussed more fully later, in the context of the evolution of C_4 photosynthesis.

×*Cynochloris*

×*Cynochloris macivorii* Clifford & Everist and ×*Cynochloris reynoldensis* B.K. Simon are intergeneric C_4 hybrids (Clifford & Everist, 1964) between *Cynodon dactylon* (L.) Pers. (NAD-ME) and two different *Chloris* spp. (both of the PCK type; Prendergast, Stone & Hattersley, 1988). The hybrids exhibit both NAD-ME and PCK C_4 acid decarboxylase activities (Prendergast *et al.*, 1988). Such plants provide a good opportunity to explore the genetic control and expression of C_4 type at both the biochemical and leaf-structural levels. *Lepturopetium* Morat may be another genus of hybrid origin (Fosberg & Sachet, 1982) from parent species of different C_4 type, as *Chloris* is PCK but *Lepturus* is either of the NAD-ME or of the PCK type, depending on the species (Prendergast *et al.*, 1987). If *Lepturus repens* (Forster) R. Br. is the parent species however, as suggested by Fosberg & Sachet (1982), the intergeneric hybridization may simply be between two PCK types, as Prendergast & Hattersley (1987) reported *L. repens* to have 'uneven' PCR sheath cell outlines.

C_3–C_4 intermediacy

Panicum hians (=*P. milioides*; =*Steinchisma hians*) was the first species shown to be neither wholly C_3 nor wholly C_4 (Brown & Brown, 1975). In the Poaceae, two other *Panicum* species (Morgan & Brown, 1979) and *Neurachne minor* (Hattersley & Roksandic, 1983) are also now known to be 'C_3–C_4 intermediates'. No intraspecific variation is known in these taxa, unlike the situation with *Alloteropsis semialata*, where most accessions are C_4, some are C_3 and a few seem likely to be C_3–C_4 (Table 2.3). C_3–C_4 plants have been discovered in other angiosperm families (for a review, see Edwards & Ku, 1987), and suspected intermediates are known from the Cyperaceae as well (Bruhl *et al.*, 1987).

How can a plant be neither C_3 nor C_4, and yet still function effectively? This is especially enigmatic as it first seems that any intermediary stage would involve having competing carboxylases (Rubisco and PEP carboxylase), with different kinetic properties, in the same cells – the mesophyll cells. Early interpretations of C_3–C_4 intermediacy revolved around the concepts of recycling of photorespiratory CO_2 (Brown & Brown, 1975) and limited C_4 acid cycle activity (Hattersley *et al.*, 1977; see also Fig. 6.1 in Hattersley, 1987). At one time, it seemed that the former could not represent an intermediary state, however, as it did not invoke the existence of (any part of) a C_4 acid cycle, thus introducing the notion that 'C_3–C_4 intermediates' may not really be what their name implies. The 'limited C_4 acid cycle concept', on the other hand, explained the intermediate physiology of C_3–C_4 species and how two carboxylases could occur in the same cell (see pp. 162–5 in Hattersley, 1976). In the case though of the species for which the hypothesis was originally invoked (*P. milioides*), enzyme assay studies and $^{14}CO_2$–$^{12}CO_2$ pulse-chase labelling work did not show the presence of such a limited cycle (Bauwe & Apel, 1979; Morgan, Brown & Reger, 1980; Holaday *et al.*, 1981; Edwards, Ku & Hatch, 1982).

One characteristic physiological indicator of C_3–C_4 intermediacy is an

Table 2.4. *Activities of photosynthetic enzymes in leaves of Neurachne minor (C_3–C_4) and C_4 relatives and controls*

Enzyme activities are given for whole leaf extracts (WL), and for isolated mesophyll protoplast (MP) and bundle sheath strand (BSS) fractions. NADP-MDH, NADP malate dehydrogenase; PEPC, phosphoenolpyruvate carboxylase; PPDK, pyruvate, P_i dikinase; NADP-ME, NADP malic enzyme; Rubisco, ribulose-1,5-bisphosphate carboxylase/oxygenase. In C_4 plants, NADP-MDH, PEPC and PPDK are C_4 mesophyll (PCA tissue) enzymes, and NADP-ME and Rubisco are PCR cell (Kranz) enzymes.

| Species | Photosynthetic type | Tissue | Enzyme activity (μmol min^{-1} mg^{-1} chlorophyll) | | | | |
			NADP-MDH	PEPC	PPDK	NADP-ME	Rubisco
Neurachne minor	C_3–C_4	WL	6.76	2.38	0.60	1.95	4.27
		MP	2.55	0.44	0.09	0.00	1.59
		BSS	0.13	0.47	0.10	1.90	0.67
Neurachne munroi	C_4	WL	24.85	24.64	1.84	9.22	1.67
		MP	15.74	35.40	1.48	0.09	0.01
		BSS	0.43	0.32	0.05	18.37	3.12
Paraneurachne muelleri	C_4	WL	14.46	18.71	3.81	5.89	0.71
		MP	11.80	23.36	2.76	0.12	0.07
		BSS	0.33	0.28	0.12	9.28	2.19
Zea mays	C_4	WL	12.53	14.46	6.09	11.90	2.80
Panicum bulbosum	C_4	WL	–	27.17	4.05	14.72	1.52

Methods followed are those described in Hattersley & Stone (1986).
Data of P.W. Hattersley & N.E. Stone (unpublished).

intermediate CO_2 compensation point (Γ) (10 to 25 $\mu l\ l^{-1}$; cf. $>30\ \mu l\ l^{-1}$ for C_3 plants, and zero for C_4 plants). Given that, as far as is known, the kinetic properties of Rubisco in C_3–C_4 species are no different from those of the enzyme in C_3 species (Keck & Ogren, 1976; Holbrook, Jordan & Chollett, 1985), such low Γ values can only be effected by increasing CO_2 at a tissue site which provides at least some of the Rubisco of the leaf with a 'CO_2-tight' space (Hattersley, 1976, 1987). There has to be a 'biochemical carbon pump' of some sort (Hattersley, 1976) pumping CO_2 from one tissue type to another, between which the leaf's Rubisco is distributed. This pump could be a limited C_4 acid cycle or a glycolate shuttle, but not mere recycling of photorespiratory CO_2 within just one tissue type (Hattersley, 1987). In a plant with a limited C_4 acid cycle, at Γ, there would be net CO_2 production from the cooperating photosynthetic carbon reduction cycle and the photo-respiratory carbon oxidation cycle in mesophyll cells. PEP carboxyl-ation and C_4 acid cycle activity would effectively shuttle this CO_2 into the bundle sheath, where there would be a net fixation of CO_2 at a rate equal to the net mesophyll cell CO_2 loss via the PCO cycle (see Fig. 6.1 in Hattersley, 1987). Monson, Edwards & Ku's (1984) 'glycolate shut-tle' scheme provides an alternative mechanism for concentrating CO_2 in bundle sheath cells. It involves glycolate from the PCO cycle in mesophyll cells being moved to bundle sheath cells where, therefore, photorespiratory CO_2 is produced preferentially. At Γ, there would be a net loss of carbon from mesophyll cells (as glycolate) and a net gain of carbon in bundle sheath cells as CO_2 released from glycine during decar-boxylation in PCO cycle activity was refixed in RuBP carboxylation.

It now seems clear that intermediates with a limited C_4 acid cycle activity and with a glycolate shuttle both occur. The C_3–C_4 *Panicum* (*Steinchisma*) and *Moricandia* A. DC. species (Brassicaceae) have the glycolate shuttle (e.g. Hunt, Smith & Woolhouse, 1987; Rawsthorne *et al.*, 1988; von Caemmerer & Hubick, 1989), but many *Flaveria* species (Asteraceae), and perhaps *Neurachne minor*, have a limited C_4 acid cycle, as assessed principally by measurement of C_4 acid cycle enzyme activities in both whole leaf extracts, and in mesophyll protoplast and bundle sheath strand preparations (Ku *et al.*, 1983; Hattersley & Stone, 1986; Monson *et al.*, 1986; Moore *et al.*, 1988). For *N. minor*, while enzyme activities in whole leaf extracts (Hattersley & Stone, 1986) and in cell separation work (P. W. Hattersley & N. E. Stone, unpublished data, see Table 2.4) suggest there is a limited C_4 acid cycle, $^{14}CO_2$-$^{12}CO_2$

pulse-chase experiments to date have not substantiated this (Moore & Edwards, 1989). However, plants for these latter experiments were grown at only one sixth full sunlight; this is below the photon flux density required either fully to activate pyruvate, P_i dikinase or to induce sufficient PPDK levels in the leaf to sustain C_4 acid cycle activity (Usuda, Ku & Edwards, 1984; Edwards et al., 1985). It is possible, then, that PPDK activities in *N. minor* were not optimized and, in a species with limited C_4 cycle activity anyway, significant incorporation of ^{14}C label into C_4 acids should not perhaps be expected under the growth conditions used.

The 'on line' method of von Caemmerer & Hubick (1989) for determining carbon-isotope discrimination associated with CO_2 assimilation, seems an elegant and non-intrusive method for distinguishing the two mechanisms, though few species have so far been investigated using it (*P. milioides*, *Flaveria floridana* Johnson and *F. anomala* B. L. Robinson). C_3–C_4 plants with a glycolate shuttle discriminate *more* against $^{13}CO_2$ when photosynthesizing at low partial pressures of CO_2 than do C_3 plants because there is a double discrimination by Rubisco; firstly in the mesophyll cells, as CO_2 is assimilated via the PCR cycle, and secondly in the bundle sheath cells as photorespired CO_2, imported as glycolate from the mesophyll, is assimilated via the PCR cycle there (von Caemmerer, 1989). This photorespired CO_2 in the bundle sheath probably has a $\delta^{13}C$ value similar to that of leaf carbon in general. Von Caemmerer & Hubick (1989) found that *P. milioides* showed 'on line' ^{13}C discrimination *greater* than predicted for C_3 species at low (subatmospheric) ambient CO_2 partial pressures, whereas the two *Flaveria* spp. did not. All carbon isotope ratios determined, for all three species, were still within the C_3 species range. Von Caemmerer (1989) suspects that even in C_3–C_4 species with limited C_4 cycle activity, such as *F. anomala* and *F. floridana*, there may also be some glycolate shuttling.

Von Caemmerer's (1989) model of carbon isotope discrimination in C_3–C_4 species with a glycolate shuttle in part explains a previously enigmatic feature of C_3–C_4 intermediates, viz. that they consistently exhibited unequivocally C_3-like $\delta^{13}C$ values (Brown & Brown, 1975; Apel & Maass, 1981; Brown et al., 1986; Hattersley et al., 1986). In the case of C_3–C_4 plants with a limited C_4 acid cycle, however, $\delta^{13}C$ values might be expected to be somewhat less negative (i.e. indicative of less discrimination against ^{13}C). This seems likely because the source of the CO_2 that is concentrated in bundle sheath cells is atmospheric CO_2

'fixed' directly via non-discriminating PEP carboxylation, and because at least a portion of the Rubisco is located in the somewhat 'CO_2-tight bundle sheath. However, models of carbon isotope discrimination for C_3–C_4 species with limited C_4 acid activity (Peisker, 1986; Monson *et al.*, 1988) predict that the level of expression of C_4 acid cycle activity can be up to 50% of the full expression in a C_4 species, without C_4-like $\delta^{13}C$ values being evoked. The implication is that only C_3–C_4 plants well on the way to being C_4 biochemically will be identified using $\delta^{13}C$ value determinations; all other C_3–C_4 intermediates would exhibit $\delta^{13}C$ values in the C_3 plant range. In this context, the 'intermediate' $\delta^{13}C$ values of some accessions of *Alloteropsis semialata* subsp. *semialata* become the more interesting (Table 2.3).

Two other physiological traits unique to known C_3–C_4 species are the biphasic response of Γ to oxygen partial pressures (Keck & Ogren, 1976; Apel, Ticha & Peisker, 1978; Holaday, Harrison & Chollett, 1982; Hattersley *et al.*, 1986) and the curvilinear response of Γ to photon flux density (PFD; Brown & Morgan, 1980; Holaday *et al.*, 1982). The former can now be appreciated in the context of both a limited C_4 acid cycle and a glycolate shuttle (Hattersley *et al.*, 1986; von Caemmerer, 1989), but the latter has yet to be satisfactorily documented, let alone explained. The suggestion that declining activities of the light-activated C_4 enzyme PPDK may explain the PFD response of Γ in C_3–C_4 species (Hattersley *et al.*, 1986) has yet to be tested.

The biphasic response of Γ to oxygen partial pressures may be clearly discernible only in quite C_3-like C_3–C_4 intermediates; it is only just detectable in *Neurachne minor* (Hattersley *et al.*, 1986). At the time *N. minor* was being investigated, it was the most C_4-like species that had been designated a C_3–C_4 intermediate, designated even though it exhibited almost zero Γ values and possessed very C_4-like leaf structure. Hattersley *et al.* (1986) suggested that 'C_3–C_4 intermediate species which are more C_4-like than those currently identified presumably will exhibit zero Γ values and a search among species hitherto recognized as C_4 is most likely to reveal such plants'. The first species to be identified as such was *Flaveria brownii* A.M. Powell, which until 1986 was classed as C_4 on the basis of Γ values, $^{14}CO_2$–$^{12}CO_2$ pulse-chase labelling, and $\delta^{13}C$ values (Apel & Maass, 1981; Monson *et al.*, 1986). Unusual cellular compartmentation of photosynthetic enzymes in *F. brownii* had been found, however (Reed & Chollett, 1985), and the species was subsequently reclassed as a C_4-like C_3–C_4 intermediate (Monson, Schuster & Ku, 1987; Holaday *et al.*, 1988; Cheng *et al.*, 1989; Araus *et al.*, 1991).

Several other *Flaveria* species which were previously regarded as C_4 have since been recognized as very C_4-like C_3–C_4 intermediates (Moore, Ku & Edwards, 1989). It is likely that some grasses hitherto classed as C_4 are in fact C_3–C_4 intermediates. Attention has been drawn to *Panicum prionitis* in this regard (Hattersley *et al.*, 1986).

Little is currently known about the adaptive significance of C_3–C_4 intermediacy. The evolution of C_4 photosynthesis, at least in its earliest stages, may have involved structural and biochemical changes that were pre-adaptive (Hattersley, 1987), conferring advantages unrelated to those now associated with being C_4. For example, a development of C_4-like leaf structure that may have preceded any metabolic changes (Brown & Hattersley, 1989) could, in the case of the bundle sheath, be functionally important in leaf water relations. Later stages of C_3–C_4 intermediacy could function to refix photorespiratory CO_2 (Brown, 1980), conferring photosynthetic advantages at warm leaf temperatures (in *Flaveria* at least; Schuster & Monson, 1990).

The C_3–C_4 species now known represent a wide range of intermediacy states (especially in *Flaveria*). Doubtless this range will be extended and the understanding of the nature of C_3–C_4 intermediates improved. In turn, this knowledge will provide the basis for an appreciation of how C_4 photosynthesis evolved. We know that 'the evolution of C_4 from C_3 plants required the development of C_4 acid cycle enzyme activity (in the appropriate cellular compartments), the coupling of C_4 acid cycle elements, an increase in the proportion of total leaf Rubisco located in incipient PCR tissue, the ultimate phasing-out of mesophyll Rubisco activity, the development of as CO_2-tight a PCR system as possible, and the reduction of the C_4 mesophyll/incipient PCR tissue ratio' (not necessarily in that order; Hattersley, 1987). What we do not know is the evolutionary origin and status of any single C_3–C_4 intermediate species that has been described. Possibilities include: (i) a hybrid origin (from C_3 and C_4 parent species); (ii) a descendant from a C_3 ancestor, and representing an incipient C_4 lineage (iii) a descendant from a C_4 ancestor, and representing a 'reversionary' stage towards C_3 photosynthesis; or (least likely) (iv) a syndrome unrelated to C_4 photosynthesis. Known C_3–C_4 intermediates probably exemplify more than one of these options. Or, even if they did all represent plants evolving into C_4 taxa (Monson & Moore, 1989), it may be that they are not all following the same course and sequence of change (Hattersley, 1987). The diversity of C_3–C_4 intermediates is probably worth considering with this in mind.

Fossil record

Although the known fossil record of grasses is not extensive, it suggests that taxonomic diversity was well established at least by the Miocene period (*c.* 23.7 million years ago) (for a review, see Thomasson, 1987). Older grass fossils are known, suggesting a time of origin at least as early as the Upper Cretaceous period (66 million years ago), but how quickly the family diversified is not clear. The subfamilial afffinities of these earliest fossils have not been ascertained (Thomason, 1987; Crepet & Feldman, 1991).

Some leaf macrofossils are sufficiently well preserved to demonstrate C_4 leaf anatomy, the first known of which was a Miocene leaf from California (originally reported as Pliocene; Nambudiri *et al.*, 1978). Tidwell & Nambudiri (1989) have subsequently found a C_4 grass from the Upper Miocene that resembles extant *Paspalum commersonii*, a C_4 'classical NADP-ME' type species. Electron microscopic examination of well-preserved Miocene leaf fragments from Minium's Dead Cow Quarry (Kansas) has clearly shown C_4 leaf anatomy in a chloridoid species of the 'classical NAD-ME' type or 'classical PCK' type (Thomason, Nelson & Zakrzewski, 1986). Use of L. Watson's Grass Genera database (Watson, 1987) has proved valuable in identifying taxonomic affinities of fossil taxa (J. R. Thomason, personal communication).

The few reports cited above indicate that C_4 photosynthesis in grasses had definitely evolved by the Miocene period. The demonstrated capability to determine photosynthetic pathway in fossil material shows that the fossil record has much potential for dating and describing the evolutionary history of grasses. The acquisition of more fossil evidence will be essential for corroborating proposed phylogenies based upon molecular or other data.

Phylogeny of the Poaceae in relation to C_4 photosynthesis

PHENETICS AND PHYLOGENETICS

Concepts of taxa within the Poaceae have been divided into 'genetic' (involving 'neural', phenetic, experimental and cytogenetic analyses) and 'phylogenetic' (involving 'traditional', experimental, and cladistic analyses; Estes & Tyrl, 1987). Such a distinction, though useful, implies that only the phylogenetic concepts of taxa recognize that it is evolution by descent that has produced discontinuous variability among organisms. A classification based on overall similarity, rather than 'actual' (usually speculative) phylogeny, must also be 'phylogenetic' by virtue of the fact that 'genetic' similarities and discontinuities exist because of phylogeny. When 'actual' phylogeny is not known, as in most angiosperms (and certainly in grasses), phenetic approaches to classification may often produce as plausible a 'phylogenetic' classification as any, and one that consistently has excellent predictive value. This seems to be the case for the classification of the Poaceae (e.g. Watson, 1990), in as much as any hierarchical classification can possibly reflect phylogeny. Hierarchies are most suitable when evolution has been, or seems to have been, non-reticulate; monophyletic groups can be validly sought. In a group where reticulate evolution has been extensively demonstrated (such as the Poaceae: Stebbins, 1956, 1987), it seems unwise to set too much store by techniques or concepts that demand that 'the taxa in the system must be monophyletic'.

The foregoing is not to suggest that cladistics has no role in grass systematics. If plesiomorphies and synapomorphies can be accurately identified, then classification will improve (particularly at infrafamilial levels) and possible homoplasies will be recognized. The empirical means for demonstrating homoplasy in photosynthetic pathway characters now exist, and this is important in a group where it seems likely that there has been significant parallel and convergent evolution with respect to C_4 photosynthesis. Furthermore, there may be some refinements in assigning taxonomic level, e.g. the Andropogonodae may be better regarded as a sister group of a tribe of the Panicodae, rather than as a sister group of the Panicodae.

A notable feature of a recent symposium on the systematics and evolution of grasses (Soderstrom et al., 1987) was the consensus on grass classification with respect to circumscription of subfamilies and, in most cases, tribes, as exemplified by the essentially phenetic classification of Watson (1987, 1990). There are, nevertheless, several areas of debate, e.g. the Stipeae has been variously regarded as a pooid or an arundinoid tribe; the Aristideae and Eriachneae are not always placed in the Arundinoideae; and the Bambusoideae are sometimes regarded as primitive grasses (Clayton & Renvoize, 1986) but more usually as advanced. The Panicoideae (Panicodae and Andropogonodae) and Chloridoideae are generally regarded as closely related (e.g. Watson, 1987) and, together with the Pooideae, as 'young' groups, while the Arundinoideae is mostly recognized as the 'basal' and/or polyphyletic assemblage (Tateoka, 1957; Kellogg & Campbell, 1987; Watson, 1987).

Cladistic analyses of the grasses are just beginning. Linder &

Ferguson (1985) and Campbell & Kellogg (1987) concluded that the grass family is linked with six small families of the Southern Hemisphere (Anarthriaceae, Centrolepidaceae, Ecdeiocoleaceae, Flagellariaceae, Joinvillaeceae and Restionaceae). None of these is known to contain C_4 species, though they have yet to be sampled adequately. In fact, the only other monocotyledon family known to contain C_4 species (the Cyperaceae), is now generally placed in an Order different from that of the grasses (Dahlgren, Clifford & Yeo, 1985).

Cladistic analyses of the Poaceae claim much more than do any phenetic analyses. Kellogg & Campbell (1987) state that the monophyly of each of the Pooideae, Bambusoideae, and Panicoideae has been established, and that the Chloridoideae may also be monophyletic. They further conclude that the arundinoids are clearly polyphyletic. Amongst the characters used in the analyses, photosynthetic pathway (C_3 or C_4) was included. Kellogg & Campbell's (1987) published phylogenetic hypothesis for the grasses involves the evolution of C_4 photosynthesis six times (yet C_4 type was concluded to have only 'one parallelism'). 'Small' changes in some assumptions led to large changes in tree structure; regarding the two-flowered spikelet of the Isachneae as a 'reversal' within the Paniceae or not, led to the Isachneae (C_3) being taken as basal to the Paniceae, or to the panicoids being 'closely related to the bambusoids'. For both cases though, the Panicoideae had been assigned as C_3, as it was assumed that C_4 photosynthesis was the derived condition in the group (Kellogg & Campbell, 1987). There is no definitive evidence for this, however.

At present, then, it seems that neither phenetic nor phylogenetic analyses have answered any basic questions about the evolution of C_4 photosynthesis within the Poaceae. Such analyses will answer questions only if the empirical data which they manipulate are relevant, and we suggest that a firm set of molecular data pertinent to C_4 photosynthesis is required before any conclusions can be credible. Such data are now being acquired in a few laboratories.

MOLECULAR APPROACHES TO PHYLOGENY

It is too early for molecular approaches to have contributed much to an understanding of grass phylogeny in general. Nevertheless, some interesting results have been obtained, sometimes challenging previous taxonomic concepts in 'difficult' groups.

Hilu (1987) has reviewed the success of work on chloroplast DNA (cpDNA) in contributing to phylogenetic studies on the Triticeae (all

C_3), including studies of wheat phylogeny, and on the origin of maize (C_4). Zimmer, Jupe & Walbot (1988) have studied the restriction endo-nuclease mapping of nuclear rRNA genes, confirming the close relation-ship of maize to Mexican teosintes (all C_4). An evolutionary tree based on nucleotide sequence data for different regions of the 18S and 26S rRNA molecules (Hamby & Zimmer, 1988) was consistent with Wat-son, Clifford & Dallwitz's (1985) phenetic classification, but only nine species were sampled, the four C_4 species all being Andropogonodae (i.e. not representative as a C_4 sample).

Restriction site analysis has also been applied to cpDNA to inves-tigate the subgeneric classification of *Poa* (C_3), a 'reticulating group' (Soreng, 1990), and to subfamily Pooideae (all C_3; Soreng, Davis & Doyle, 1990). In *Poa*, the data suggested relationships where none had been previously detectable, and also allowed an assessment of bio-geographical events associated with colonizing *Poa* groups. The Pooideae molecular study is the first for grasses to include a broad sample for the subfamily, and it illustrates the usefulness of such data for reassessing traditional taxonomic concepts. It also, however, shows how detailed and complex the hypotheses concerning 'problem' groups can become (i.e. where data of different kinds are not congruent!). It will be interesting to see what relationships such cpDNA studies suggest for groups that are variable for photosynthetic pathway.

Gene sequencing investigations have included sequencing of the spacer region between nuclear 18 S and 26 S nuclear rRNA genes for a few grasses (Appels, Scoles & Chapman, 1987) and sequencing of chloroplast Rubisco LS genes (*rbc*L) from maize, barley, rice, wheat and *Neurachne* (Zurawski, Clegg & Brown, 1984; Hudson *et al.*, 1990). The number of species sampled is too small as yet to be taxonomically useful within the Poaceae. rRNA and *rbc*L genes are good candidates for molecular systematic studies in which congruence can be assessed and comparisons made with non-nucleotide sequence data such as those from cpDNA restriction site analyses and even from cpDNA reassoci-ation (Hilu & Johnson, 1991) and cell DNA thermal denaturation data (King & Ingrouille, 1987).

Non-DNA 'molecular' data that have previously been collected for systematic purposes in the grasses, include protein amino acid profiles (Yeoh & Watson, 1987), various isozyme polymorphisms (especially for *Avena*, *Hordeum*, *Triticum* and *Zea*; Kahler & Price, 1987), and flavonoid patterns (Harborne & Williams, 1987). As with DNA studies, this work essentially corroborates broad taxonomic affinities already established for the family, and has not contributed to an understanding of the diversification of photosynthesis. More recently, however, endosperm prolamin size heterogeneity and immunological cross-reactivies have been studied (Esen & Hilu, 1989). The Arundinoideae are judged to be monophyletic using prolamin data (Hilu & Esen, 1990; in contrast to Kellogg & Campbell, 1987). Attention has focused on the phylogenetic affinities of the Aristideae (Esen & Hilu, 1991), now included in the arundinoids (Watson, 1990). The Aristideae comprises *Aristida* (C_4 NADP-ME 'Aristida' type), *Stipagrostis* (C_4, probably 'classical NAD-ME' type) and *Sartidia* (C_3). *Sartidia* was not included in Esen & Hilu's (1991) sample, but the data suggested that the tribe is an artificial assemblage, and that *Stipagrostis* has high taxonomic affinity with the chloridoids, whereas *Aristida* should not be placed in either the chloridoids or the arundinoids. Such a taxonomic rearrangement would make the Aristideae much less heterogeneous, indeed perhaps homogeneous with respect to phytosynthetic pathway variation, than is currently circumscribed.

The use of molecular methods (DNA studies) in the biosystematics of grasses is in its infancy, but has already included work on cpDNA (restriction and sequence analyses) and the nuclear-encoded ribosomal RNA gene family. Genes that encode small nuclear gene families have yet to be used for grass systematics, but the obvious candidates are those that encode 'C_4 enzymes'. Several of these have now been characterized and sequenced in at least a few species (see section above on the molecular biology of C_4 photosynthesis). The enzymes have both C_3 and C_4 isoforms. Nucleotide sequence analysis of these isoforms (cf. Ritland & Clegg, 1987) in a taxonomically representative sample of C_3, C_4 and C_3–C_4 grasses, from an evolutionary viewpoint, should enable much to be learned about the diversification of photosynthesis in the grasses. This becomes the more exciting when coupled with the prospect of sequencing 'fossil' DNA. Golenberg *et al.* (1990) have recently sequenced *rbc*L from a leaf compression fossil of *Magnolia* from the Miocene period (17–20 million years ago).

Diversification of photosynthesis

SOME CONCEPTS

Grasses are generally believed to have originated in the Late Cretaceous period, more than 66 million years ago (Brown, 1977; Campbell, 1985; Clayton & Renvoize, 1986; Stebbins, 1987). The family is thought to have a tropical origin, perhaps in forests (Bews, 1929), or in the forest–savanna ecotone (Clayton, 1981), or in open habitats (Stebbins, 1987). There is seldom conjecture about when the major grass groups differentiated, but it may have been before the break up of the continents, as the latter each have 'a full array of suprageneric taxa' (Clayton, 1981). Stebbins (1987) suggested that the bambusoids, pooids, arundinoids, and chloridoids (=eragrostoids) differentiated from a now extinct ancestral group. He further conjectured that the stipoids are derived from primitive pooids, and 'the panicoids and the andropogonoids both from the arundinoids via *Arundinella* and related genera'.

The time of emergence of savannas or extensive grasslands is variously dated as Eocene (in South America and/or Africa: Stebbins, 1987) or middle Oligocene (Clayton, 1981), though the main adaptive radiation of major grass groups (including tribes) is often taken to be in the Miocene period (Campbell, 1985). Perhaps radiations that post-dated breakup of the continents led to the evolution of those genera which tend to be restricted to individual continental blocks (Clayton & Renvoize, 1986); diversification is also associated with the coevolution of ungulates (Stebbins, 1981). In Australia, grass pollen is abundant in mid-Miocene deposits, but the main expansion of grassland and open woodland is put at 10–2.5 million years ago (Martin, 1981); this is tentative, as palynological sampling has not been evenly spread over the continent (Galloway & Kemp, 1981).

The time of origin of grasses, and of their phases of adaptive radiation, probably coincided with global climatic shifts. Some of these climatic shifts may have been caused or initiated by catastrophic asteroid or comet impacts, such as that/those at the Cretaceous/Tertiary boundary (e.g. Florentin, Maurrasse & Sen, 1991). Others are the indirect result of continental drift, e.g. the global aridization of the climate in the Late Oligocene period (c. 24 million years ago) following the drop in sea surface temperatures owing to the development of a strong Antarctic Circumpolar Current (Zubakov & Borzenkova, 1990). During the Cretaceous period, there was a sustained decline in atmospheric CO_2 concentration, reaching a minimum level during the Palaeocene period (Lasaga *et al.*, cited by Ehleringer *et al.*, 1991). In the Late Ypresian period (52–53 million years ago; Eocene), there may have been a return to a 'greenhouse' climate, with atmospheric CO_2 concentrations 5–10 times the present-day level, warm sea waters, and the 'precipitation being not less than 700–900 mm in the modern arid landscapes' (Zubakov & Borzenkova, 1990). Subsequently, at the Eocene/Oligocene boundary (36 million years ago), there was a major cooling peak, with a drop in sea surface temperatures, sea levels and CO_2 partial pressures; however, global warming, sea temperature rises and increased CO_2 content of the atmosphere characterized a later, lesser warming peak in the Neogene period (17.2–16.5 million years ago, in the Miocene period).

As knowledge of palaeoclimates and grass fossils improves, there is an increasing likelihood of reconstructing the evolutionary history of grasses in relation to the diversification of photosynthesis. For now, one can only wonder about the options. For example, did C_4 photosynthesis originate in grasses in the Cretaceous period, or somewhere between the Palaeocene and Miocene of the Tertiary? Did only one C_4 type first evolve in grasses (the NAD-ME or NADP-ME type?), or did C_4 types evolve independently? Did C_4 photosynthesis evolve primarily in response to temperature (taking advantage of the higher quantum yield that characterizes C_4 plants at more than 28 °C, at current CO_2 levels; Ehleringer, 1978), increasing aridity (Apel, 1988), increasing atmospheric O_2 (Smith, 1976), or decreasing atmospheric CO_2 (Ehleringer *et al.*, 1991), or to some combination of one or more of these variables?

Given that C_4 photosynthesis evolved several times among angiosperms, that angiosperms radiated during the Cretaceous period, that C_4 photosynthesis comprises a system for concentrating CO_2 where Rubisco occurs, and that global CO_2 levels declined dramatically during the Cretaceous period, it is tempting to conclude that C_4 plants evolved primarily in response to the lowering of CO_2 in the atmosphere (Ehleringer *et al.*, 1991). It is likely that this would have been in a high-irradiance, hot habitat because the quantum requirement of C_4 plants is lower than that of C_3 plants only under these conditions. Furthermore, the habitat would probably have needed to be arid, or at least seasonally arid (i.e. 'wet/dry', or monsoonal tropics), because the 'wet' tropics sustain closed forest where C_4 photosynthesis has no advantages. In other words, it may not have been changing CO_2 alone that was the 'catalyst' for the emergence of C_4 plants, but CO_2 and light, temperature, and seasonal aridity. After C_4 grasses originated, there could then have been subsequent specialization and diversification (up to and including the Miocene?) during later cycles of global climatic change and also in the context of regional climatic changes caused by continental movement and separation.

In view of the different climatic preferences of NADP–ME and

NAD-ME grasses (shown for southern Africa and Australia; Ellis *et al.* (1980) and Hattersley (1992), respectively), one could suggest two major lines of specialization of C_4 grasses, with the NAD-ME type having evolved (in the tropical or subtropical zones) primarily in response to aridization of the climate, and the NADP-ME type (in the wet–dry 'monsoon' tropics) primarily in response to decreasing CO_2; the PCK type also seems suited mostly to the wet–dry tropics (Prendergast *et al.*, 1986; Hattersley, 1992), but perhaps to shadier habitats than are the NADP-ME grasses (Ellis, 1988, for *Panicum*).

Interestingly, the C_3–C_4 intermediate *Panicum* (*Steinchisma*) species possess leaf-structural characteristics that suggest an affinity with C_4 NAD-ME type species, whereas features of *Neurachne minor* suggest affinity with C_4 NADP-ME type species (Hattersley *et al.*, 1986), i.e. these two sorts of C_3–C_4 intermediate in grasses are unlikely to represent just different stages in a common, 'standard', linear sequence of changes involved in becoming C_4. Nevertheless, proposed linear sequences (e.g. Monson *et al.*, 1984; Peisker, 1986; Edwards & Ku, 1987) are quite compelling. The glycolate shuttle is seen as preceding development of a limited C_4 acid cycle. A difficulty with this is that the CO_2 concentrating mechanism in C_3–C_4 intermediates with a glycolate shuttle is dependent upon retaining Rubisco in the mesophyll; C_4 plants do not have Rubisco in the mesophyll. Previously, we considered that this would preclude regarding the glycolate shuttle as a forerunner of a plant with a limited C_4 acid cycle (Hattersley, 1987). However, an impetus for the C_4 acid cycle to develop subsequently would be the advantage conferred by recycling CO_2 leaking from the bundle sheath (CO_2 'pumped in' by the glycolate shuttle); the C_4 acid cycle could then itself, in turn, become the CO_2 concentrating mechanism (H. W. Woolhouse, personal communication).

How does photosynthetic pathway variation itself correspond with what little is known or conjectured about the emergence and divergence of grasses? Various phylogenetic schemes for the family have been constructed. We note, as an example of a scheme that precedes the discovery of C_4 photosynthesis, that the phylogeny proposed by Tateoka (1957) differs considerably from the scenario of Stebbins (1987, referred to above) with respect to major group relationships, but not with respect to the broad circumscription of the groups themselves (with the exception of the stipoids). These different interpretations reflect improvements in the choice and extent of taxonomic data; for example, photosynthetic pathway was not previously a major factor in assessing or defining relationships. Schemes that have subsequently tried to accommodate photosynthetic pathway data include those of Gutierrez *et al.* (1974), Smith & Robbins (1974), Brown (1977) and Clayton (1981). All of these schemes are 'traditionally' or 'neurally' constructed, and primarily accommodate photosynthesis variation at the 'major group' level. Some take cognizance of the three biochemical C_4 types.

As described above, much more is now known about the diversity of leaf-structural–biochemical suites that occurs in the family, and about their taxonomic occurrence (Hattersley, 1987; Watson & Dallwitz, 1988). Figure 2.3 represents a 'traditional' scheme, constructed by us, that, it could be suggested, accommodates most of the currently known variation in photosynthetic pathway as distributed among major taxonomic groups. We do not, for a moment, believe that the scheme reflects adequately the actual phylogeny of the Poaceae. We doubt that this could ever be done in two dimensions, as it is likely that the 'phylogeny of the entire family is to some extent reticulate' (Stebbins, 1987). The scheme is presented merely to demonstrate that, if the extreme complexity of photosynthetic variation is taken into account, a complex scheme is an inevitable necessity (and one not just addressing variation at the subfamily level). This is because, as described above in detail, C_4 photosynthesis occurs in three subfamilies, no subfamily is uniformly C_3 or C_4, some tribes, some genera, and even a species are variable for pathway, and C_4 biochemical type also is scattered across groups (Fig. 2.3). It is doubtful that any credible scheme can be constructed as a branching phylogeny. We know too much about the grasses! However, we do not yet know them well enough to be able to say, for example, if NADP-ME from an andropogonoid is encoded by the same gene as NAPD-ME from an *Eriachne* species, if the genes that determine the leaf structure of a 'classical NAD-ME' type chloridoid are the same as those that determine the leaf structure of a 'classical NAD-ME' type panicoid, or if a modern C_3 panicoid is descended from a C_3 panicoid ancestor or from a C_4 panicoid ancestor.

Whether or not we mean the same thing genetically when referring to 'C_4' in a panicoid as when referring to 'C_4' in a chloridoid, Fig. 2.3 at least suggests that C_4 photosynthesis evolved more than once (three or four times?). It also illustrates (accepting that the main subfamily groups are probably 'natural') that each biochemical C_4 type probably evolved more than once in the family (e.g. in the mythical phylogeny shown, the NADP-ME type arose three times). Attempting such a scheme has at least one virtue; it engenders many questions. For

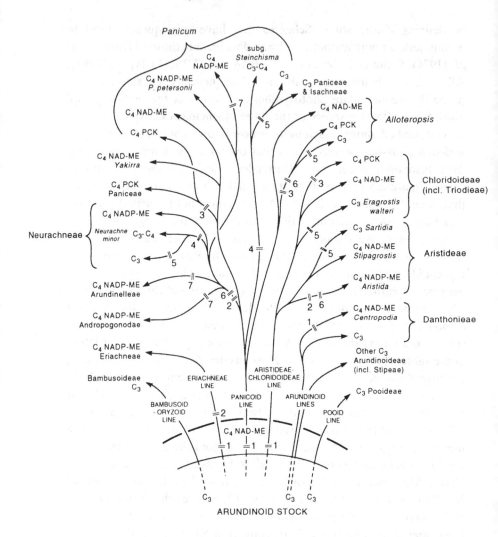

Fig. 2.3. A 'traditional' phylogeny comprising a suite of hypotheses concerning the diversification of photosynthesis in grasses. This scheme is not intended to represent the authors' views on the evolution of C₄ photosynthesis in the Poaceae; these are presented in the text. As evolution has been reticulate in the family (and cannot be shown in two dimensions), and as too much of that which is relevant to diversification in grasses remains unknown, it is premature to draw definite conclusions. This scheme serves (i) to provide a suite of potentially testable hypotheses concerning diversification of photosynthesis within taxonomic groups, e.g. that C₄ NAD–ME *Yakirra* is more closely related to C₄ NAD–ME *Panicum* species than are non-NAD–ME *Panicum* species; and (ii) to highlight the fact that

example, why should the Paniceae be regarded as basically a C_3 tribe (cf. Renvoize, 1987)? It could be that there have been more C_4 to C_3 reversions during the evolutionary history of grasses, than *de novo* C_4 origins (and this premise was accepted in drawing up Fig. 2.3, where five such reversions are indicated); it may even one day be shown to be more parsimonious. Denying reversions in the Paniceae, means that one has to invoke more C_4 origins in the Poaceae as a whole. Is there a 'primitive' biochemical C_4 type? In Fig. 2.3, the NADP-ME and PCK types are each considered to be derived from the NAD-ME type, and to have been able to revert. Is the XyMS+/XYMS− (or PS/MS) distinction a fundamental evolutionary dichotomy in the Panicoideae or not? All XyMS− grasses are panicoid.

SOME CONCLUSIONS

It is not yet possible to be definitive about the diversification of photosynthesis in grasses. However, when consideration is given to all that is so far known in this context (as is described in previous sections of this chapter), the following are testable, and not unreasonable, hypotheses (not necessarily consistent with the scheme presented in Fig. 2.3):

(1) C_4 photosynthesis evolved more than once, perhaps separately in each of the panicoid, Aristideae-chloridoid, danthonioid and Eriachneae lines. These groups differ not only in their C_4 suites, but also in many other characters; a single C_4 ancestral stock seems unlikely, therefore.

(2) There have been several 'reversions' to C_3, e.g. in the Neurachneae, the Paniceae, the Chlorideae and the Aristideae. An alternative hypothesis is that one or more of these groups are

Fig. 2.3. – *cont.*
the construction of a 'phylogeny' that is reasonably compatible with most accepted taxonomic relationships at upper hierarchical levels in the family has to invoke several *de novo* origins of C_4 photosynthesis itself ('1'), as well as multiple origins of the different C_4 types (at '1' for NAD–ME, '2' for NADP–ME, and '3' for PCK) and multiple C_4 to C_3 'reversions' (partial reversions at '4', complete reversions at '5'). The scheme also invokes the transfer of PCR function from the outer bundle sheath to the inner bundle sheath (at '6'), and the loss of the outer bundle sheath in some lines (at '7'). This scheme has presupposed that (i) the NAD–ME type is primitive, and that the PCK and NADP–ME types have evolved from C_4 NAD–ME type species (another testable hypothesis); and (ii) that all extant grasses have evolved from an 'arundinoid' stock.

relicts of the ancestral stock that gave rise to the subfamilies in which these tribes are placed, i.e. the Panicoideae or the Chloridoideae. The Aristideae may be a discrete line.

(3) C_4 types each arose in more than one subfamily. The NAD-ME type occurs in *three* major groups (panicoid, chloridoid and arundinoid) whereas the other two C_4 biochemical types are each known from only two major groups. This suggests that the NAD-ME C_4 acid cycle is perhaps a relatively 'primitive' (or, at least, less complicated) type. Indeed, the 'classical NAD-ME' type leaf structure is the one where ultrastructural and anatomical features need least modification from C_3 leaf structure. There are clear differences between the leaf structures of NADP-ME panicoids, NADP-ME Aristideae, and NADP-ME Eriachneae, not only for characters irrelevant to photosynthetic pathway (e.g. epidermal cell shapes) but also for relevant characters. One could argue that there has been convergent evolution with respect to 'C_4 type'; lack of homology has been shown for the different leaf structures, but not yet for the C_4 acid cycles. Differences between PCK panicoids and PCK chloridoids in their C_4 acid cycles, and in the leaf-structure features functionally important to photosynthesis, have yet to be documented. They may be subtle differences, suggesting a remarkable degree of convergent evolution. The same reasoning can be applied to NAD-ME panicoids *vs* NAD-ME chloridoids *vs* the C_4 (probably NAD-ME) arundinoid *Centropodia*.

(4) Species of the PCK biochemical type evolved from species of the NAD-ME type (as suggested, for example, by Brown, 1977). Circumstantial evidence for this is as follows: biochemically, the C_4 acid cycle pathway of the PCK type seems like an 'embellished' or sophisticated NAD-ME type system (Fig. 2.1); PCK activity is detectable only in PCK species, whereas NAD-ME activity is detectable in both NAD-ME and PCK species (Gutierrez *et al.*, 1974; Prendergast *et al.*, 1987); all PCK type species (except *A. semialata*) have evolved an *extra* suberized lamella (in the walls of cells in the *outer* sheath position; Hattersley & Browning, 1981) – this ultrastructural feature is unique to PCK type grasses; PCK taxa always have closely related NAD-ME taxa (Table 2.2); and NAD-ME 'classical PCK' type ('PCK-like NAD-ME') species could represent intermediate states between the two C_4 types.

(5) NADP-ME type grasses (which, in the panicoids, are all XyMS−) evolved from NAD-ME type grasses. Though no bio-chemical similarity between these two C_4 types is obvious at present, similarities are worth searching for as taxa of the NADP-ME type in the Panicoideae have close relatives of the NAD-ME type. Developmentally, XyMS− species seem dis-tant from XyMS+ species, but to regard the XyMS+/XyMS− distinction as indicative of a single 'deep' evolutionary divergence does not fit current classificatory schemes unless, for example, one regards the XyMS− species of *Panicum*, *Neurachne* and the Arundinelleae as more closely related to each other than to their respective XyMS+ sister species cur-rently placed in each of these genera. Alternatively, and more likely, it may be that species of the NADP-ME type have origi-nated more than once in the subfamily.

Grass classification is so good (or so 'natural') that to date it has contributed considerably to the locating and sampling of photosynthetic pathway variation. This contribution is greater than that made by work on the diversity of photosynthetic pathways to the construction of better classifications. For some groups, this may be about to change (e.g. Aristideae, cf. the work of Esen & Hilu, 1991), but it is more likely that taxonomy will continue to build the foundation for an understanding of the evolutionary events and processes that led to the diversification of photosynthetic pathways in grasses. Experimental work should sample the variation judiciously, by appropriate choice of species, as the 'non-classical' biochemical–leaf-structural suites are primarily discrete taxo-nomic outliers or endemic groups. The apparent complexity of the diversity shown in the Poaceae really comprises a core C_3/C_4 pattern, and a core C_4 type diversity, around which non-mainstream elements are clustered. Trying to understand the processes and patterns that gave rise to this variation in an evolutionary context is what unifies knowl-edge concerning C_4 photosynthesis from diverse disciplines into a syn-thesis of 'C_4 biology' for the Poaceae.

Finally, the idea has occasionally been raised that the capacity for grasses to become C_4 is 'latent' in some way (Watson *et al.*, 1985), and that although the evolution of C_4 photosynthesis must involve much genetic change, the potential to change in that direction resides in angio-sperm groups in general. Similarly, there may be a latent potential to change towards one of only three biochemical C_4 types. In these contexts, the ideas of Cockburn (1985) at least draw attention to biochemical similarities among a 'family' of different kinds of 'photosynthetic acid metabolisms' (including stomatal metabolism). Interspecific and intraspecific DNA sequence analyses of, for example, PEP carboxylase genes should give some insights into these concepts. Relevant here, and not conjectural, are the recent demonstrations that 'syndromes' or developmental processes involving genetic determination and regulation by a large number of genes, can be 'switched on' by a single gene encoding an activator protein (e.g. the floral homeotic genes controlling flower development in angiosperms (Coen, 1991)). Is it possible that an analogous activator sequence is necessary to switch on the 'C$_4$ syndrome'? If this is the case, then only a minor mutation, or repositioning of a 'jumping gene', might result in reversion to C$_3$ photosynthesis.

Domestication, genetic engineering and 'greenhouse' in relation to diversity of photosynthetic pathways

Photosynthetic pathway variation has not been a major consideration in breeding programmes, and has not had much impact on domestication practices for any given cereal group. This is because the major cereals and grass cash crops, though some are C$_3$ and some C$_4$, are each homogeneous for photosynthetic pathway (Table 2.5). There has been recognition that the type of photosynthetic pathway is important in relation to whether a given cereal or crop is suited to tropical, subtropical, or temperate zones; C$_3$ and C$_4$ grasses are typical of temperate and tropical climates, respectively (e.g. see the analysis for Australia by Hattersley (1983)). The useful, empirical designation of grasses as temperate or tropical had preceded an understanding of its basis, however, and attempts to explore the significance of pathway variation in the context of domestication have been limited to the comparison of C$_3$ and C$_4$ physiology in relation to primary productivity, growth rates, net assimilation rates, water-use efficiency etc., as reviewed, for example, by Gifford (1974), Ludlow (1976) and Osmond et al. (1982), the matching of crops and environment (e.g. in the dryland tropics; Sinclair, 1988), and the investigation of the basis of the adaptation of some C$_4$ grasses to cool climates (e.g. Long, 1983).

Much more emphasis has been given to the significance of photosynthetic pathways for (non-domesticated) pasture systems and rangelands (see Hattersley, 1992, and references therein). The significance of C$_3$/C$_4$ and intra-C$_4$ variation for the management of natural grasslands has been emphasized by Hattersley (1992). Commercialized tropical pasture

Table 2.5. *Photosynthetic pathway of the major cereal and cash crops in the grasses*

Biochemical type is also given if taxon is C_4. NADP-ME, NADP malic enzyme; NAD-ME, NAD malic enzyme; PCK, phosphoenolpyruvate carboxykinase. n.a., not applicable.

Genus/species	C_3 or C_4	C_4 type
Avena spp.	All C_3	n.a.
Andropogon spp.	All C_4	NADP-ME
Brachiaria spp.	All C_4	PCK
Echinochloa spp.	All C_4	NADP-ME
Eleusine spp.	All C_4	NAD-ME
Eragrostis tef	C_4	NAD-ME
Euchlaena spp.	All C_4	NADP-ME
Hordeum spp.	All C_3	n.a.
Oryza spp.	All C_3	n.a.
Panicum miliaceum	C_4	NAD-ME
Pennisetum spp.	All C_4	NADP-ME
Saccharum spp.	All C_4	NADP-ME
Secale spp.	All C_3	n.a.
Setaria spp.	All C_4	NADP-ME
Sorghum spp.	All C_4	NADP-ME
Tripsacum spp.	All C_4	NADP-ME
Triticum spp.	All C_3	n.a.
Urochloa spp.	All C_4	PCK
Zea spp.	All C_4	NADP-ME
Zizania spp.	All C_3	n.a.

species are almost exclusively of the NADP-ME or PCK type (see Table 2.6); this is because these types of grasses inhabit those parts of Africa and Asia that were first exploited during colonization (the wetter, more productive savannas). Subsequently, these have been the species that were researched in attempts to evaluate cultivars suitable for semi-arid and arid conditions (e.g. Anning, 1982). As a generalization, however, it is NAD-ME grasses that are best suited to low rainfall, and emphasis might usefully be directed to them, especially for developing sustainable systems and for arresting desertification (Hattersley, 1992).

The diversity of photosynthetic pathways is also relevant to crop and pasture systems in the context of selective herbivory and differential digestibility. Phytophagous insects are often found to prefer C_3 to C_4 grasses if given the choice (e.g. Heidorn & Joern, 1984), and this is probably because the former tends to be characterized by less sclerenchyma and more extensive mesophyll. For the same reasons, C_3 leaves

Table 2.6. *Biochemical type of commercially available C_4 tropical pasture grasses in Australia*

C_4 type has been assigned by us to those C_4 species listed in Humphreys (1980). Individual species are not listed; only the numbers of each C_4 type within major taxonomic groups are given. Supertribes Panicodae and Andropogonodae comprise the subfamily Panicoideae. NADP-ME, NADP malic enzyme; NAD-ME, NAD malic enzyme; PCK, phosphoenolpyruvate carboxykinase.

Biochemical C_4 type	Subfamily or supertribe	No. of species
NADP-ME	Panicodae	15
	Andropogonodae	5
PCK	Panicodae	8
	Chloridoideae	1
NAD-ME	Panicodae	1
	Chloridoideae	3

are generally more digestible for ruminants than are C_4 grass leaves (e.g. Akin & Burdick, 1975; Wilson & Hacker, 1987), and there are also differences in digestibility among the three C_4 types (Wilson & Hattersley, 1983, 1989).

To date then, the impact of knowledge of the diversity of photosynthetic pathways in grasses on domestication has mainly been towards an improved understanding of comparative physiology and ecophysiology. This has led to studies aimed at increasing plant productivity, especially in forage plants (e.g. Ludlow, 1985; Long *et al.*, 1989), and at improving water-use efficiency (e.g. Ludlow & Muchow, 1988). In the latter context a major spin-off from C_4 photosynthesis research for domestication has been the use of measurements of carbon isotopic composition of plant tissues to select for increased water-use efficiency in C_3 plants. This is based on the fact that plants discriminate against $^{13}CO_2$ during photosynthesis (Park & Epstein, 1960). The impetus to research in this area increased dramatically following the discovery that C_3 and C_4 plants discriminate to different degrees (Bender, 1971). Furthermore, among C_4 grasses, the different C_4 types exhibit differential discrimination (Hattersley, 1982). The basis of discrimination has been modelled and reviewed by Farquhar, O'Leary & Berry (1982) and Farquhar (1983). Because the extent to which potential discrimination can actually be realized is dependent upon stomatal conductance, which itself is related to transpiration rate, the carbon isotopic composition of plant tissues is correlated with water-use efficiency in C_3 plants (Farquhar & Richards, 1984). Subsequently, this technique has been exploited to screen for water-use efficient lines within individual species (e.g. peanut (Hubick, Shorter & Farquhar, 1988); and wheat, barley and cotton – see the review by Farquhar, Ehleringer & Hubick, (1989)).

The greatest benefits of an appreciation of the diversification of photosynthesis in grasses, however, are yet to come. These will probably be in two quite distinct areas: genetic engineering and rangeland management. We have described above the current status of molecular biological research on C_4 grasses. It is clear that an understanding of the genetic basis of C_4 determination, expression and regulation is achievable. Coupled with modern approaches to tissue culture and transformation, this knowledge will bring closer the possibility that the photosynthetic pathway of currently domesticated cereals and crops can be manipulated, or even that plants can be transformed to express a different pathway, with the prospect of growing maize in the temperate zone and wheat in the tropics. This knowledge may also further encourage the domestication of currently wild grasses, or at least the transfer of their genes into already domesticated species.

Such high levels of sophistication and technology, however, and the human societies that support them, may be overtaken by the consequences of global climate change (see e.g. Bernard, 1980; Pearman, 1988). While genetic engineering has the potential to contribute to manipulating the adaptation of cereals and crops to changing climates, it seems unrealistic to contemplate that this could ever be done on the scale and in the detail that would be required to maintain diverse, sustainable ecosystems capable of supporting current populations and standards. A doubling of 'greenhouse' gases by 2030–2050 may lead to global temperature rises of up to 4 °C, to a rise in sea level of up to 1.5 m, and to increases in rainfall in many parts of the world (see e.g. Henderson-Sellers & Blong, 1989). Attempts are being made to initiate planning for such change at national (e.g. Australia: Pearman, 1988), regional (e.g. Northern Territory: Moffatt, 1991), and industrial (e.g. pastoralism: McKeon & Howden, 1991) levels.

Grasses are so important, fundamentally as staple food, as well as ecologically and economically, that is just as well we know so much about their basic biology, notwithstanding changing fashions of and priorities given to fundamental research. What we know about the diversity of photosynthesis in grasses gives us the framework, within which to consider their response to changing climate, if not the answers. While net assimilation rates of C_3 grasses increase with increasing CO_2 concentrations, those of C_4 grasses do not. Considering the CO_2 increase alone, therefore, it seems likely that growth rates in predominantly C_3 vegetation will increase overall, whereas that in predominantly C_4 vegetation will not. It is also sometimes suggested or implied that C_3 grasses will increase their ability to compete with C_4 grasses in many, if not all environments where C_3 grasses occur (Henderson-Sellers & Blong, 1989; Ehleringer *et al.*, 1991); but this will depend on the extent to which vapour pressure deficits increase as a result of the overall temperature increase (especially in semi-arid and arid zones), and on the change in availability of water and nutrients in the 'new' rainfall regime of any given area. The obvious advantage in a competitive situation that C_3 plants would have, in a rising CO_2 scenario, is still further eroded by the concomitant temperature rise, as a result of the increased photorespiratory rates that would be sustained by C_3 plants compared with C_4 plants. Furthermore, the adaptation of higher plants to temperature involves many physiological processes, not just photosynthesis (Berry & Björkman, 1980).

In most rangelands, either C_3 or C_4 grasses predominate, so the above considerations are most relevant to the currently mixed C_3/C_4 grasslands, and the geographical relocation of such grasslands. With respect to intra-C_4 variation, solidly C_4 grasslands in a climate-change scenario may be little affected by increased CO_2 and temperature *per se*, depending in large part, perhaps, on the interaction of C_4 grasses with woody shrubs and trees (which are all C_3). However, C_4 species composition of tropical and subtropical savannas would presumably alter with changing *rainfall*, given the differential distributions of C_4 types in relation to aridity (Ellis *et al.*, 1980; Hattersley, 1992). The more economically important NADP-ME and PCK type grasses will become more prevalent in currently arid and semi-arid areas if rainfall increases, at the expense of NAD-ME type grasses, and vice versa.

There has been insufficient research on the response of vegetation to changing climate. That vegetation *will* respond under a 'greenhouse' scenario is certain, and this vegetation includes the grasslands, and the cereals and crops that are so important to civilization. There seems little doubt that the diversification of photosynthesis has been a key process during the evolutionary history and adaptive radiation of grasses. This diversification has occurred in response to natural changes in those same environmental variables that are now being anthropogenically altered:

CO_2, temperature and precipitation. Understanding the past will help us to understand the consequences of the changes that seem imminent. It will also contribute to planning a strategy for managing vegetation change attributable to the 'greenhouse' effect, even if economic rationalism and expediency do not permit funding to implement it.

Acknowledgements
We thank James Whitehead for preparing Figs 2.1 and 2.3

References
Adams, C. A., Leung, F. & Sun, S. S. M. (1986). Molecular properties of phosphoenolpyruvate carboxylase from C_3, C_3–C_4 intermediate, and C_4 *Flaveria* species. *Planta (Berlin)*, **167**, 218–225.
Akin, D. E. & Burdick, D. (1975). Percentage of tissue types in tropical and temperate grass leaf blades and degradation of tissues by microorganisms. *Crop Science*, **15**, 661–668.
Andrews, T. J. & Lorimer, G. H. (1987). Rubisco: structure, mechanisms, and prospects for improvement. In *The biochemistry of plants*, vol. 10, ed. M. D. Hatch & N. K. Boardman, pp. 131–218. San Diego: Academic Press.
Anning, P. (1982). Evaluation of introduced grass species for pastures in the dry tropics of North Queensland. *Tropical Grasslands*, **16**, 136–142.
Aoyagi, K. & Bassham, J. A. (1986). Appearance and accumulation of C_4 carbon pathway enzymes in developing wheat leaves. *Plant Physiology*, **80**, 334–340.
Apel, P. (1988). Some aspects of the evolution of C_4 photosynthesis. *Kultürpflanze*, **36**, 225–236.
Apel, P. & Maass, I. (1981). Photosynthesis in species of *Flaveria*. CO_2 compensation concentration, O_2 influence on photosynthetic gas exchange and $\delta^{13}C$ values in species of *Flaveria* (Asteraceae). *Biochemie und Physiologie der Pflanzen*, **176**, 396–399.
Apel, P., Ticha, I. & Peisker, M. (1978). CO_2-Kompensationspunkt von *Moricandia arvensis* (L.) D.C. bei Blättern unterschiedlicher Insertionshöhe und bei verschiedenen O_2-Konzentrationen. *Biochemie und Physiologie der Pflanzen*, **172**, 547–552.
Appels, R., Scoles, G. & Chapman, C. G. D. (1987). The nature of change in nuclear DNA in the evolution of the grasses. In *Grass systematics and evolution*, ed. T. R. Soderstrom, K. W. Hilu, C. S. Campbell & M. E. Barkworth, pp. 73–87. Washington DC: Smithsonian Institution Press.
Araus, J. L., Brown, R. H., Byrd, G. T. & Serret, M. D. (1991). Comparative effects of growth irradiance on photosynthesis and leaf anatomy of *Flaveria brownii* (C_4-like), *Flaveria linearis* (C_3–C_4), and their F1 hybrid. *Planta (Berlin)*, **183**, 497–504.
Archer, S. (1984). The distribution of photosynthetic pathway types on a mixed-grass Prairie hillside. *American Midland Naturalist*, **111**, 138–142.
Bauwe, H. & Apel, P. (1979). Biochemical characterization of *Moricandia arvensis* (L.) D.C., a species with features intermediate between C_3 and C_4

photosynthesis, in comparison with the C_3 species *Moricandia foetida* Bourg. *Biochemie und Physiologie der Pflanzen*, **174**, 251–254.

Bender, M. M. (1971). Variations in the $^{13}C/^{12}C$ ratios of plants in relation to the pathway of photosynthetic carbon dioxide fixation. *Phytochemistry*, **10**, 1239–1244.

Bernard, H. W. (1980). *The greenhouse effect*. Cambridge, Massachusetts: Ballinger Publ. Co.

Berry, J. A. & Björkman, O. (1980). Photosynthetic response and adaptation to temperature in higher plants. *Annual Review of Plant Physiology*, **31**, 491–543.

Bews, J. W. (1929). *The world's grasses; their differentiation, distribution, economics and ecology*. London: Longmans Green & Co.

Bisalputra, T., Downton, W. J. S. & Tregunna, E. B. (1969). The distribution and ultrastructure of chloroplasts in leaves differing in photosynthetic carbon metabolism. I. Wheat, *Sorghum*, and *Aristida* (Gramineae). *Canadian Journal of Botany*, **47**, 15–21.

Blake, S. T. (1972). *Neurachne* and its allies (Gramineae). *Contributions from the Queensland Herbarium*, **13**, 1–53.

Brown, R. H. (1980). Photosynthesis of grass species differing in carbon dioxide fixation pathways IV. Analysis of reduced oxygen response in *Panicum milioides* and *Panicum schenckii*. *Plant Physiology*, **65**, 346–349.

Brown, R. H., Bassett, C. L., Cameron, R. G., Evans, P. T., Bouton, J. H., Black, C. C., Sternberg, L. O. & DeNiro, M. J. (1986). Photosynthesis of F1 hybrids between C_4 and C_3–C_4 species of *Flaveria*. *Plant Physiology*, **82**, 211–217.

Brown, R. H., Bouton, J. H., Evans, P. T., Malter, H. E. & Rigsby, L. L. (1985). Photosynthesis, morphology, leaf anatomy, and cytogenetics of hybrids between C_3 and C_3/C_4 *Panicum* species. *Plant Physiology*, **77**, 653–658.

Brown, R. H. & Brown, W. V. (1975). Photosynthetic characteristics of *Panicum milioides*, a species with reduced photorespiration. *Crop Science*, **15**, 681–685.

Brown, R. H. & Hattersley, P. W. (1989). Leaf anatomy of C_3–C_4 species as related to evolution of C_4 photosynthesis. *Plant Physiology*, **91**, 1543–1550.

Brown, R. H. & Morgan, J. A. (1980). Photosynthesis of grass species differing in carbon dioxide fixation pathways VI. Differential effects of temperature and light intensity on photorespiration in C_3, C_4 and intermediate species. *Plant Physiology*, **66**, 541–544.

Brown, W. V. (1975). Variations in anatomy, associations, and origins of Kranz tissue. *American Journal of Botany*, **62**, 395–402.

Brown, W. V. (1977). The Kranz syndrome and its subtypes in grass systematics. *Memoirs of the Torrey Botanical Club*, **23**, 1–97.

Bruhl, J. J., Stone, N. E. & Hattersley, P. W. (1987). C_4 acid decarboxylation enzymes and anatomy in sedges (Cyperaceae): first record of NAD-malic enzyme species. *Australian Journal of Plant Physiology*, **14**, 719–728.

Burnell, J. N. & Hatch, M. D. (1988). Photosynthesis in phosphoenolpyruvate carboxykinase-type C_4 plants: pathways of C_4 acid decarboxylation in bundle sheath cells of *Urochloa panicoides*. *Archives of Biochemistry and Biophysics*, **260**, 187–199.

Butzin, F. (1968). Bemerkungen zum Umfang und zur Morphologie der Paniceengattung *Alloteropsis*. *Willdenowia*, **5**, 123–143.

Campbell, C. S. (1985). The subfamilies and tribes of Gramineae (Poaceae) in the southeastern United States. *Journal of the Arnold Arboretum*, **66**, 123–199.

Campbell, C. S. & Kellogg, E. A. (1987). Sister group relationships of the Poaceae. In *Grass systematics and evolution*, ed. T. R. Soderstrom, K. W. Hilu, C. S. Campbell & M. E. Barkworth, pp. 217–225. Washington DC: Smithsonian Institution Press.

Carolin, R. C., Jacobs, S. W. L. & Vesk, M. (1973). The structure of the cells of the mesophyll and parenchymatous bundle sheath of the Gramineae. *Botanical Journal of the Linnean Society*, **66**, 259–275.

Carolin, R. C., Jacobs, S. W. L. & Vesk, M. (1975). Leaf structure in Chenopodiaceae. *Botanische Jahrbücher für Systematik Pflanzengeschichte und Pflanzengeographie*, **95**, 226–255.

Cavagnaro, J. B. (1988). Distribution of C_3 and C_4 grasses at different altitudes in a temperate arid region of Argentina. *Oecologia (Berlin)*, **76**, 273–277.

Cheng, S.-H., Moore, B.d., Wu, J., Edwards, G. E. & Ku, M. S. B. (1989). Photosynthetic plasticity in *Flaveria brownii*. Growth irradiance and the expression of C_4 photosynthesis. *Plant Physiology*, **89**, 1129–1135.

Clayton, D. (1981). Evolution and distribution of grasses. *Annals of the Missouri Botanical Gardens*, **68**, 5–14.

Clayton, D. & Renvoize, S. A. (1982). *Flora of tropical East Africa: Gramineae*. Part 3. Rotterdam: A. A. Balkema.

Clayton, D. & Renvoize, S. A. (1986). *Genera Graminum. Grasses of the world*. Kew Bulletin Additional Series XIII. London: HMSO.

Clifford, H. T. & Everist, S. L. (1964). ×*Cynochloris macivorii* gen. et sp. nov., a suspected spontaneous hybrid between *Cynodon dactylon* (L.) Pers. and *Chloris divaricata* R. Br. *Proceedings of the Royal Society of Queensland*, **75**, 45–49.

Cockburn, W. (1985). Variation in photosynthetic acid metabolism in vascular plants: CAM and related phenomena. *New Phytologist*, **101**, 3–24.

Coen, D. M., Bedbrook, J. R., Bogorad, L. & Rich, A. (1977). Maize chloroplast DNA fragment encoding the large subunit of ribulose-bisphosphate carboxylase. *Proceedings of the National Academy of Sciences USA*, **74**, 5487–5491.

Coen, E. S. (1991). The role of homeotic genes in flower development and evolution. *Annual Review of Plant Physiology and Plant Molecular Biology*, **42**, 241–279.

Crepet, W. L. & Feldman, G. D. (1991). The earliest remains of grasses in the fossil record. *American Journal of Botany*, **78**, 1010–1014.

Crétin, C., Keryer, E., Tagu, D., Lepiniec, L., Vidal, P. & Gadal, P. (1990). Complete cDNA sequence of sorghum phosphoenolpyruvate carboxylase involved in C_4 photosynthesis. *Nucleic Acids Research*, **18**, 658.

Crétin, C., Luchetta, P., Joly, C., Miginiac-Maslow, M., Decottignies, P.,

Jacquot, J., Vidal, J. & Gadal, P. (1988). Identification of a cDNA clone for sorghum leaf malate dehydrogenase (NADP). Light-dependent mRNA accumulation. *European Journal of Biochemistry*, **174**, 497–501.

Crookston, R. K. & Moss, D. N. (1973). A variation of C_4 leaf anatomy in *Arundinella hirta* (Gramineae). *Plant Physiology*, **52**, 397–402.

Dahlgren, R. M. T., Clifford, H. T. & Yeo, P. F. (1985). *The families of the monocotyledons. Structure, evolution and taxonomy.* Berlin: Springer-Verlag.

Dean, C., Pichersky, E. & Dunsmuir, P. (1989). Structure, evolution, and regulation of *RbcS* genes in higher plants. *Annual Review of Plant Physiology and Plant Molecular Biology*, **40**, 415–439.

Dengler, N. G., Dengler, R. E. & Hattersley, P. W. (1985). Differing ontogenetic origins of PCR ('Kranz') sheaths in leaf blades of C_4 grasses (Poaceae). *American Journal of Botany*, **72**, 284–302.

Downton, W. J. S. (1970). Preferential C_4-dicarboxylic acid synthesis, the post-illumination CO_2 burst, carboxyl transfer step, and grana configurations in plants with C_4 photosynthesis. *Canadian Journal of Botany*, **48**, 1795–1800.

Downton, W. J. S. (1971). Further evidence for two modes of carboxyl transfer in plants with C_4 photosynthesis. *Canadian Journal of Botany*, **49**, 1439–1442.

Downton, W. J. S., Berry, J. & Tregunna, E. B. (1969). Photosynthesis: temperate and tropical characteristics within a single grass genus. *Science*, **163**, 78–79.

Downton, W. J. S. & Tregunna, E. B. (1968). Carbon dioxide compensation – its relation to photosynthetic carboxylation reactions, systematics of the Gramineae, and leaf anatomy. *Canadian Journal of Botany*, **46**, 207–215.

Edwards, G. E. & Ku, M. S. B. (1987). Biochemistry of C_3–C_4 intermediates. In *The Biochemistry of Plants*, vol. 10, ed. M. D. Hatch & N. K. Boardman, pp. 275–325. San Diego: Academic Press.

Edwards, G. E., Ku, M. S. B. & Hatch, M. D. (1982). Photosynthesis in *Panicum milioides*, a species with reduced photorespiration. *Plant and Cell Physiology*, **23**, 1185–1195.

Edwards, G. E., Nakomoto, H., Burnell, J. N. & Hatch, M. D. (1985). Pyruvate, P_i dikinase and NADP-malate dehydrogenase in C_4 photosynthesis: properties and mechanisms of light/dark regulation. *Annual Review of Plant Physiology*, **36**, 255–286.

Edwards, G. E. & Walker, D. A. (1983). C_3, C_4: *mechanisms, and cellular and environmental regulation of photosynthesis.* Oxford: Blackwell Scientific.

Ehleringer, J. R. (1978). Implications of quantum yield differences on the distributions of C_3 and C_4 grasses. *Oecologia (Berlin)*, **31**, 255–267.

Ehleringer, J. R. & Pearcy, R. W. (1983). Variation in quantum yield for CO_2 uptake among C_3 and C_4 plants. *Plant Physiology*, **73**, 555–559.

Ehleringer, J. R., Sage, S. F., Flanagan, L. B. & Pearcy, R. W. (1991). Climate change and the evolution of C_4 photosynthesis. *Trends in Ecology and Evolution*, **6**, 95–99.

Ellis, R. P. (1974). The significance of the occurrence of both Kranz and non-Kranz leaf anatomy in the grass species *Alloteropsis semialata*. *South African Journal of Science*, **70**, 169–173.

Ellis, R. P. (1977). Distribution of the Kranz syndrome in the Southern African Eragrostoideae and Panicoideae according to bundle sheath anatomy and cytology. *Agroplantae*, **9**, 73–110.

Ellis, R. P. (1981). *Relevance of comparative leaf blade anatomy in taxonomic and functional research on the South African Poaceae.* Ph.D. thesis: University of Pretoria, Pretoria.

Ellis, R. P. (1984a). Leaf anatomy of the South African Danthonieae (Poaceae). IX. *Asthenatherum glaucum. Bothalia*, **15**, 153–159.

Ellis, R. P. (1984b). *Eragrostis walteri* – a first record of non-Kranz leaf anatomy in the sub-family Chloridoideae (Poaceae). *South African Journal of Botany*, **3**, 380–386.

Ellis, R. P. (1988). Leaf anatomy and systematics of *Panicum* (Poaceae: Panicoideae) in southern Africa. *Monographs in Systematic Botany, Missouri Botanic Gardens*, **25**, 129–156.

Ellis, R. P., Vogel, J. C. & Fuls, A. (1980). Photosynthetic pathways and the geographical distribution of grasses in South West Africa/Namibia. *South African Journal of Science*, **76**, 307–314.

Esen, A. & Hilu, K. W. (1989). Immunological affinities among subfamilies of the Poaceae. *American Journal of Botany*, **76**, 196–203.

Esen, A. & Hilu, K. W. (1991). Electrophoretic and immunological studies of prolamins in the Poaceae: II. Phylogenetic affinities of the Aristideae. *Taxon*, **40**, 5–17.

Estes, J. R. & Tyrl, R. J. (1987). Concepts of taxa in the Poaceae. In *Grass systematics and evolution*, ed. T. R. Soderstrom, K. W. Hilu, C. S. Campbell & M. E. Barkworth, pp. 325–333. Washington DC: Smithsonian Institution Press.

Farquhar, G. D. (1983). On the nature of carbon isotope discrimination in C_4 species. *Australian Journal of Plant Physiology*, **10**, 205–226.

Farquhar, G. D., Ehleringer, J. R. & Hubick, K. T. (1989). Carbon isotope discrimination and photosynthesis. *Annual Review of Plant Physiology and Plant Molecular Biology*, **40**, 503–537.

Farquhar, G. D., O'Leary, M. H. & Berry, J. A. (1982). On the relationship between carbon-isotope discrimination and the intercellular carbon dioxide concentration in leaves. *Australian Journal of Plant Physiology*, **9**, 121–137.

Farquhar, G. D. & Richards, R. A. (1984). Isotopic composition of plant carbon correlates with water-use efficiency of wheat genotypes. *Australian Journal of Plant Physiology*, **11**, 539–552.

Fischer, R. A. & Turner, N. C. (1978). Plant productivity in the arid and semiarid zones. *Annual Review of Plant Physiology*, **29**, 277–317.

Florentin, J.-M., Maurrasse, R. & Sen, G. (1991). Impacts, Tsunamis, and the Haitian Cretaceous–Tertiary boundary layer. *Science*, **252**, 1690–1693.

Fosberg, F. R. & Sachet, M.-H. (1982). Micronesian Poaceae: critical and distributional notes. *Micronesica*, **18**, 71–74.

Frean, M. L., Barrett, D. R., Ariovich, D., Wolfson, M. & Cresswell, C. F. (1983). Intraspecific variability in *Alloteropsis semialata* (R.Br.) Hitchc. *Bothalia*, **14**, 901–913.

Galloway, R. W. & Kemp, E. M. (1981). Late Cainozoic environments in

Australia. In *Ecological biogeography of Australia*, 2nd edn, ed. A. Keast, pp. 51–80. The Hague: W. Junk.

Gibbs, A. J., Ding, S., Howe, J., Keese, P., Mackenzie, A., Skotnicki, M., Srifah, P. & Torronen, M. (1990). Old *versus* new characters for systematics: cautionary tales from virology. *Australian Systematic Botany*, **3**, 159–163.

Gibbs Russell, G. E. (1983). The taxonomic position of C_3 and C_4 *Alloteropsis semialata* (Poaceae) in southern Africa. *Bothalia*, **14**, 205–213.

Gifford, R. M. (1974). A comparison of potential photosynthesis, productivity and yield of plant species with differing photosynthetic metabolism. *Australian Journal of Plant Physiology*, **1**, 107–118.

Glackin, C. A. & Grula, J. W. (1990). Organ-specific transcripts of different size and abundance derive from the same pyruvate orthophosphate dikinase gene in maize. *Proceedings of the National Academy of Sciences USA*, **86**, 3004–3008.

Golenberg, E. M., Giannasi, D. E., Clegg, M. T., Smiley, C. J., Durbin, M., Henderson, D. & Zurawski, G. (1990). Chloroplast DNA sequence from a Miocene *Magnolia* species. *Nature*, **344**, 656–658.

Gould, F. W. & Clark, C. A. (1978). *Dichanthelium* (Poaceae) in the United States and Canada. *Annals of the Missouri Botanical Garden*, **65**, 1088–1132.

Gutierrez, M., Gracen, V. E. & Edwards, G. E. (1974). Biochemical and cytological relationships in C_4 plants. *Planta (Berlin)*, **119**, 279–300.

Haberlandt, G. (884). *Physiologische Pflanzenanatomie*. Leipzig: Engelmann.

Hague, D. R., Uhler, M. & Collins, D. (1983). Cloning of cDNA for pyruvate, Pi dikinase from maize leaves. *Nucleic Acids Research*, **11**, 4853–4865.

Hamby, R. K. & Zimmer, E. A. (1988). Ribosomal RNA sequences for inferring phylogeny within the grass family (Poaceae). *Plant Systematics and Evolution*, **160**, 29–37.

Harborne, J. B. & Williams, C. A. (1987). Flavonoid patterns of grasses. In *Grass systematics and evolution*, ed. T. R. Soderstrom, K. W. Hilu, C. S. Campbell & M. E. Barkworth, pp. 107–113. Washington DC: Smithsonian Institution Press.

Harpster, M. H. & Taylor, W. C. (1986). Maize phosphoenolpyruvate carboxylase. Cloning and characterization of mRNAs encoding isozymic forms. *Journal of Biological Chemistry*, **261**, 6132–6136.

Hatch, M. D. (1978). Regulation of enzymes in C_4 photosynthesis. In *Current topics in cellular regulation*, vol. 14, ed. B. L. Horecker & E. R. Stadtman, pp. 1–27. New York: Academic Press.

Hatch, M. D. (1988). C_4 photosynthesis: a unique blend of modified biochemistry, anatomy and ultrastructure. *Biochimica et Biophysica Acta*, **895**, 81–106.

Hatch, M. D., Kagawa, T. & Craig, S. (1975). Subdivision of C_4-pathway species based on differing C_4 acid decarboxylating systems and ultrastructural features. *Australian Journal of Plant Physiology*, **2**, 111–128.

Hatch, M. D. & Mau, S.-L. (1973). Activity, location and role of aspartate aminotransferase and alanine aminotransferase isoenzymes in leaves with C_4 pathway photosynthesis. *Archives of Biochemistry and Biophysics*, **156**, 195–206.

Hatch, M. D. & Osmond, C. B. (1976). Compartmentation and transport in C_4 photosynthesis. In *Transport in plants. III. Intracellular interactions and transport processes.* Encyclopedia of Plant Physiology (New Series), vol. 3, ed. C. R. Stocking & U. Heber, pp. 144–184. Berlin: Springer-Verlag.

Hattersley, P. W. (1976). *Specification and functional significance of the leaf anatomy of C_4 plants.* Ph.D. thesis: Australian National University, Canberra.

Hattersley, P. W. (1982). $\delta^{13}C$ values of C_4 types in grasses. *Australian Journal of Plant Physiology*, **9**, 139–154.

Hattersley, P. W. (1983). The distribution of C_3 and C_4 grasses in Australia in relation to climate. *Oecologia (Berlin)*, **57**, 113–128.

Hattersley, P. W. (1984). Characterization of C_4 type leaf anatomy in grasses (Poaceae). Mesophyll: bundle sheath area ratios. *Annals of Botany*, **53**, 163–179.

Hattersley, P. W. (1987). Variations in photosynthetic pathway. In *Grass systematics and evolution*, ed. T. R. Soderstrom, K. W. Hilu, C. S. Campbell & M. E. Barkworth, pp. 49–64. Washington DC: Smithsonian Institution Press.

Hattersley, P. W. (1992). C_4 photosynthetic pathway variation in grasses (Poaceae): its significance for arid and semi-arid lands. In *Desertified grasslands – their biology and management*, ed. G. P. Chapman, pp. 181–212. London: Academic Press.

Hattersley, P. W. & Browning, A. J. (1981). Occurrence of the suberized lamella in leaves of grasses of different photosynthetic types. I. In parenchymatous bundle sheaths and PCR ('Kranz') sheaths. *Protoplasma*, **109**, 371–401.

Hattersley, P. W. & Roksandic, Z. (1983). $\delta^{13}C$ values of C_3 and C_4 species of Australian *Neurachne* and its allies (Poaceae). *Australian Journal of Botany*, **31**, 317–321.

Hattersley, P. W. & Stone, N. E. (1986). Photosynthetic enzyme activities in the C_3–C_4 intermediate *Neurachne minor* S. T. Blake (Poaceae). *Australian Journal of Plant Physiology*, **13**, 399–408.

Hattersley, P. W. & Watson, L. (1975). Anatomical parameters for predicting photosynthetic pathways of grass leaves: the maximum lateral cell count and the maximum cells distant count. *Phytomorphology*, **25**, 325–333.

Hattersley, P. W. & Watson, L. (1976). C_4 grasses: an anatomical criterion for distinguishing between NADP-malic enzyme species and PCK or NAD-malic enzyme species. *Australian Journal of Botany*, **24**, 297–308.

Hattersley, P. W., Watson, L. & Johnston, C. R. (1982). Remarkable leaf anatomical variations in *Neurachne* and its allies (Poaceae) in relation to C_3 and C_4 photosynthesis. *Botanical Journal of the Linnean Society*, **84**, 265–272.

Hattersley, P. W., Watson, L. & Osmond, C. B. (1977). *In situ* immunofluorescent labelling of ribulose-1,5-bisphosphate carboxylase in leaves of C₃ and C₄ plants. *Australian Journal of Plant Physiology*, **4**, 523–539.

Hattersley, P. W., Wong, S. C., Perry, S. & Roksandic, Z. (1986). Comparative ultrastructure and gas exchange characteristics of the C₃–C₄ intermediate *Neurachne minor* S.T. Blake (Poaceae). *Plant, Cell and Environment*, **9**, 217–233.

Hawksworth, D. L. (ed.) (1991). *Improving the stability of names: needs and options*. Taunus, Germany: Koeltz Scientific.

Heidorn, T. & Joern, A. (1984). Differential herbivory on C₃ versus C₄ grasses by the grasshopper *Ageneotettix deorum* (Orthoptera: Acrididae). *Oecologia*, **65**, 19–25.

Henderson-Sellers, A. & Blong, R. (1989). *The greenhouse effect: living in a warmer Australia*. Kensington, NSW: University of New South Wales Press.

Hilu, K. W. (1987). Chloroplast DNA in the systematics and evolution of the Poaceae. In *Grass systematics and evolution*, ed. T. R. Soderstrom, K. W. Hilu, C. S. Campbell & M. E. Barkworth, pp. 65–72. Washington DC: Smithsonian Institution Press.

Hilu, K. W. & Esen, A. (1990). Prolamins in systematics of Poaceae subfam. Arundinoideae. *Plant Systematics and Evolution*, **173**, 57–70.

Hilu, K. W. & Johnson, J. L. (1991). Chloroplast DNA reassociation and grass phylogeny. *Plant Systematics and Evolution*, **176**, 21–31.

Holaday, A. S., Brown, R. H., Barlett, J. M., Sandlin, E. A. & Jackson, R. C. (1988). Enzymic and photosynthetic characteristics of reciprocal F₁ hybrids of *Flaveria pringlei* (C₃) and *Flaveria brownii* (C₄-like species). *Plant Physiology*, **87**, 484–490.

Holaday, A. S., Harrison, A. T. & Chollett, R. (1982). Photosynthetic/photorespiratory CO₂ exchange characteristics of the C₃–C₄ intermediate species *Moricandia arvensis*. *Plant Science Letters*, **27**, 181–189.

Holaday, A. S., Shieh, Y.-J., Lee, K. W. & Chollett, R. (1981). Anatomical, ultrastructural and enzymic studies of leaves of *Moricandia arvensis*, a C₃–C₄ intermediate species. *Biochemica et Biophysica Acta*, **637**, 334–341.

Holbrook, G. P., Jordan, D. B. & Chollett, R. (1985). Reduced apparent photorespiration by the C₃–C₄ intermediate species, *Moricandia arvensis* and *Panicum milioides*. *Plant Physiology*, **77**, 578–583.

Hoober, J. K. (1987). The molecular basis of chloroplast development. In *The biochemistry of plants*, vol. 10, ed. M. D. Hatch & N. K. Boardman, pp. 1–74. San Diego: Academic Press.

Hubick, K. T., Shorter, R. & Farquhar, G. D. (1988). Heritability and genotype×environment interactions of carbon-isotope discrimination and transpiration efficiency in peanut (*Arachis hypogaea* L.). *Australian Journal of Plant Physiology*, **15**, 799–813.

Hudson, G. S., Mahon, J. D., Anderson, P. A., Gibbs, M. J., Badger, M. R., Andrews, T. J. & Whitfield, P. R. (1990). Comparisons of *rbc*L genes for the large subunit of ribulose-bisphosphate carboxylase from closely related C₃ and C₄ plant species. *Journal of Biological Chemistry*, **265**, 808–814.

Hudspeth, R. L., Glackin, C. A., Bonner, J. & Grula, J. W. (1986). Genomic and cDNA clones for maize phosphoenolpyruvate carboxylase and pyruvate, orthophosphate dikinase: expression of different gene-family members in leaves and roots. *Proceedings of the National Academy of Sciences USA*, **83**, 2884–2888.

Hudspeth, R. L. & Grula, J. W. (1989). Structure and expression of the maize gene encoding the phosphoenolpyruvate carboxylase isozyme involved in C_4 photosynthesis. *Plant Molecular Biology*, **12**, 579–589.

Humphreys, L. R. (1980) *A guide to better pastures for the tropics and subtropics*, 4th edn. Silverwater, NSW: Wright-Stevenson & Co.

Hunt, S., Smith, A. & Woolhouse, H. W. (1987). Evidence for a light-dependent system for reassimilation of photorespiratory CO_2, which does not include a C_4 cycle, in the C_3–C_4 intermediate species *Moricandia arvensis*. *Planta (Berlin)*, **171**, 227–234.

Izui, K., Ishijima, S., Yamaguchi, Y., Katagiri, F., Murata, T., Shigesada, K., Sugiyama, T. & Katsuki, H. (1986). Cloning and sequence analysis of cDNA encoding active phosphoenolpyruvate carboxylase of the C_4 pathway from maize. *Nucleic Acids Research*, **14**, 1615–1628.

Izui, K., Kawamura, T., Okumura, S., Yanagisawa, S. & Toh, H. (1991). Molecular evolution of phosphoenolpyruvate carboxylase gene involved in C_4 photosynthesis in maize. *Journal of Cell Biochemistry*, Suppl., 15A.

Johnson, Sister C. & Brown, W. V. (1973). Grass leaf ultrastructural variations. *American Journal of Botany*, **60**, 727–735.

Kahler, A. L. & Price, S. C. (1987). Isozymes in population genetics, systematics, and evolution of grasses. In *Grass systematics and evolution*, ed. T. R. Soderstrom, K. W. Hilu, C. S. Campbell & M. E. Barkworth, pp. 97–106. Washington DC: Smithsonian Institution Press.

Kamamura, T., Shigesada, K., Yanagisawa, S. & Izui, K. (1990). Phosphoenolpyruvate carboxylase prevalent in maize roots. Isolation of a complementary DNA clone and its use for analyses of the gene and gene expression. *Journal of Biochemistry (Tokyo)*, **107**, 165–168.

Keck, R. W. & Ogren, W. L. (1976). Differential oxygen response of photosynthesis in soybean and *Panicum milioides*. *Plant Physiology*, **58**, 552–555.

Keegstra, K., Olsen, L. J. & Theg, S. M. (1989). Chloroplastic precursors and their transport across the envelope membranes. *Annual Review of Plant Physiology and Plant Molecular Biology*, **40**, 471–501.

Kellogg, E. A. & Campbell, C. S. (1987). Phylogenetic analyses of the Gramineae. In *Grass systematics and evolution*, ed. T. R. Soderstrom, K. W. Hilu, C. S. Campbell & M. E. Barkworth, pp. 310–322. Washington DC: Smithsonian Institution Press.

King, G. J. & Ingrouille, M. J. (1987). Genome heterogeneity and classification of the Poaceae. *New Phytologist*, **107**, 633–644.

Krall, J. P. & Edwards, G. E. (1990). Quantum yields of photosystem II electron transport and carbon dioxide fixation in C_4 plants. *Australian Journal of Plant Physiology*, **17**, 579–588.

Ku, S. B., Monson, R. K., Littlejohn, R. O., Nakamoto, H., Fisher, D. B. &

Edwards, G. E. (1983). Photosynthetic characteristics of C_3–C_4 intermediate *Flaveria* species. I. Leaf anatomy, photosynthetic responses to O_2 and CO_2, and activities of key enzymes in the C_3 and C_4 pathways. *Plant Physiology*, **71**, 944–948.

Laetsch, W. M. (1971). Chloroplast structural relationships in leaves of C_4 plants. In *Photosynthesis and photorespiration*, ed. M. D. Hatch, C. B. Osmond & R. O. Slatyer, pp. 323–349. New York: Wiley-Interscience.

Langdale, J. A., Rothermel, B. A. & Nelson, T. (1988*a*). Cellular pattern of photosynthetic gene expression in developing maize leaves. *Genes & Development*, **2**, 106–115.

Langdale, J. A., Taylor, W. C. & Nelson, T. (1991). Cell-specific accumulation of maize phosphoenolpyruvate carboxylase is correlated with demethylation at a specific site >3 kb upstream of the gene. *Molecular and General Genetics*, **225**, 49–55.

Langdale, J. A., Zelitch, I., Miller, E. & Nelson, T. (1988*b*). Cell position and light influence C_4 versus C_3 patterns of photosynthetic gene expression in maize. *European Molecular Biology Organisation Journal*, **7**, 3643–3652.

Lebrun, M., Waksman, G. & Fressinet, G. (1987). Nucleotide sequence of a gene encoding corn ribulose-1,5-bisphosphate carboxylase/oxygenase small subunit (*rbc*S). *Nucleic Acids Research*, **15**, 4360–4364.

Linder, H. P. & Ferguson, I. K. (1985). On the pollen morphology and phylogeny of the Restionales and Poales. *Grana*, **24**, 65–76.

Long, S. P. (1983). C_4 photosynthesis at low temperatures. *Plant, Cell and Environment*, **6**, 345–363.

Long, S. P., Garcia Moya, E., Imbamba, S. K., Kamnalrut, A., Piedade, M. T. F., Scurlock, J. M. O., Shen, Y. K. & Hall, D. O. (1989). Primary productivity of natural grass ecosystems of the tropics: a reappraisal. *Plant and Soil*, **115**, 155–166.

Ludlow, M. M. (1976). Ecophysiology of C_4 grasses. In *Water and plant life. Problems and modern approaches*, ed. O. L. Lange, L. Kappen & E. D. Schulze, pp. 364–386. Berlin: Springer-Verlag.

Ludlow, M. M. (1985). Photosynthesis and dry matter production in C_3 and C_4 pasture plants, with special emphasis on tropical C_3 legumes and C_4 grasses. *Australian Journal of Plant Physiology*, **12**, 557–572.

Ludlow, M. M. & Muchow, R. C. (1988). Critical evaluation of the possibilities for modifying crops for high production per unit of precipitation. In *Drought research priorities for the dryland tropics*, ed. F. R. Bidinger & C. Johansen, pp. 179–211. Patancheru, Andhra Pradesh: International Crops Research Institute for the Semi-Arid Tropics.

McIntosh, L., Poulsen, C. & Bogorad, L. (1980). Chloroplast gene sequence for the large subunit of ribulose bisphosphate carboxylase of maize. *Nature*, **288**, 556–560.

McKeon, G. M. & Howden, S. M. (1991). Adapting northern Australian grazing systems to climatic change. *Climate Change Newsletter*, **3**, 5–8.

McWilliam, J. R. & Mison, K. (1974). Significance of the C_4 pathway in *Triodia irritans* (spinifex), a grass adapted to arid environments. *Australian Journal of Plant Physiology*, **1**, 171–175.

Manzara, T. & Gruissem, W. (1988). Organization and expression of the genes encoding ribulose-1,5-bisphosphate carboxylase in higher plants. *Photosynthesis Research*, **16**, 117–139.

Martin, H. A. (1981). The Tertiary flora. In *Ecological biogeography of Australia*, 2nd edn, ed. A. Keast, pp. 391–406. The Hague: W. Junk.

Matsuoka, M. & Ibaraki, T. (1990). Structure, genetic mapping and expression of the gene for pyruvate orthophosphate dikinase from maize. *Journal of Biological Chemistry*, **265**, 16772–16777.

Matsuoka, M., Kano-Murakami, Y., Tanaka, Y., Ozeki, Y. & Yamamoto, N. (1988). Classification and nucleotide sequence of cDNA encoding the small subunit of ribulose-1,5-bisphosphate carboxylase from rice. *Plant and Cell Physiology*, **29**, 1015–1022.

Matsuoka, M. & Minami, E.-I. (1989). Complete structure of the gene for phosphoenolpyruvate carboxylase from maize. *European Journal of Biochemistry*, **181**, 593–598.

Medina, E. (1982). Physiological ecology of neotropical savanna plants. In *Ecology of tropical savannas*, Ecological Studies No. 42, ed. B. J. Huntley & B. H. Walker, pp. 308–335. Berlin: Springer-Verlag.

Metzler, M. C., Rothermel, B. A. & Nelson, T. (1989). Maize NADP-malate dehydrogenase: cDNA cloning, sequence, and mRNA characterization. *Plant Molecular Biology*, **12**, 713–722.

Moffatt, I. (1991). Possible ecological impacts of the greenhouse effect on the Northern Territory, Australia. *Global Ecology and Biogeography Letters*, **1**, 102–107.

Monson, R. K., Edwards, G. E. & Ku, M. S. B. (1984). C_3–C_4 intermediate photosynthesis in plants. *BioScience*, **34**, 563–574.

Monson, R. K. & Moore, B.d. (1989). On the significance of C_3–C_4 intermediate photosynthesis to the evolution of C_4 photosynthesis. *Plant, Cell and Environment*, **12**, 689–699.

Monson, R. K., Moore, B.d., Ku, M. S. B. & Edwards, G. E. (1986). Cofunction of C_3, C_4 and C_3–C_4 intermediate *Flaveria* species. *Planta (Berlin)*, **168**, 493–502.

Monson, R. K., Schuster, W. S. & Ku, M. S. B. (1987). Photosynthesis in *Flaveria brownii* A. M. Powell, a C_4-like C_3–C_4 intermediate. *Plant Physiology*, **85**, 1063–1067.

Monson, R. K., Teeri, J. A., Ku, M. S. B., Gurevitch, J., Mets, L. J. & Dudley, S. (1988). Carbon-isotope discrimination by leaves of *Flaveria* species exhibiting different amounts of C_3- and C_4-cycle co-function. *Planta (Berlin)*, **174**, 145–151.

Moore, B.d. & Edwards, G. E. (1989). Metabolism of $^{14}CO_2$ by leaves of different photosynthetic types of *Neurachne* species. *Plant Science*, **60**, 155–161.

Moore, B.d., Ku, M. S. B. & Edwards, G. E. (1989). Expression of C_4-like photosynthesis in several species of *Flaveria*. *Plant, Cell and Environment*, **12**, 541–549.

Moore, B.d., Monson, R. K., Ku, M. S. B. & Edwards, G. E. (1988). Activities of principal photosynthetic and photorespiratory enzymes in leaf mesophyll and bundle sheath protoplasts from the C_3–C_4 intermediate *Flaveria ramosissima*. *Plant and Cell Physiology*, **29**, 999–1006.

Morgan, J. A. & Brown, R. H. (1979). Photosynthesis in grass species differing in carbon dioxide fixation pathways. II. A search for species with intermediate gas exchange and anatomical characteristics. *Plant Physiology*, **64**, 257–262.

Morgan, J. A., Brown, R. H. & Reger, B. J. (1980). Photosynthesis in grass species differing in carbon dioxide fixation pathways. III. Oxygen response and enzyme activities of species in the Laxa group of *Panicum*. *Plant Physiology*, **65**, 156–159.

Murata, T., Ohsugi, R., Matsuoka, M. & Nakamoto, H. (1989). Purification and characterization of NAD malic enzyme from leaves of *Eleusine coracana* and *Panicum dichotomiflorum*. *Plant Physiology*, **89**, 319–324.

Nambudiri, E. M. V., Tidwell, W. D., Smith, B. N. & Herbert, N. P. (1978). A C_4 plant from the Pliocene. *Nature*, **276**, 816–817.

Nelson, T., Harpster, M. H., Mayfield, S. P. & Taylor, W. C. (1984). Light-regulated gene expression during maize leaf development. *Journal of Cell Biology*, **98**, 558–564.

Nelson, T. & Langdale, J. A. (1989). Patterns of leaf development in C_4 plants. *Plant Cell*, **1**, 3–13.

Ngernprasirtsiri, J., Chollet, R., Kobayashi, H., Sugiyama, T. & Akazawa, T. (1989). DNA methylation and the differential expression of C_4 photosynthesis genes in mesophyll and bundle sheath cells of greening maize leaves. *Journal of Biological Chemistry*, **264**, 8241–8248.

Ohsugi, R. & Murata, T. (1980). Leaf anatomy, post-illumination CO_2 burst and NAD-malic enzyme activity of *Panicum dichotomiflorum*. *Plant and Cell Physiology*, **21**, 1329–1333.

Ohsugi, R., Murata, T. & Chonan, N. (1982). C_4 syndrome of the species in the *Dichotomiflora* group of the genus *Panicum* (Gramineae). *Botanical Magazine (Tokyo)*, **95**, 339–347.

Osmond, C. B., Winter, K. & Ziegler, H. (1982). Functional significance of different pathways of CO_2 fixation in photosynthesis. In *Physiological plant ecology. III. Water relations and carbon assimilation*. Encyclopedia of Plant Physiology (New Series), vol. 12B, ed. O. L. Lange, P. S. Nobel, C. B. Osmond & H. Ziegler, pp. 479–547. Berlin: Springer-Verlag.

Park, R. & Epstein, S. (1960). Carbon isotope fractionation during photosynthesis. *Geochimica et Cosmochimica Acta*, **21**, 110–126.

Pearman, G. I. (ed.) (1988). *Greenhouse: planning for climate change*. Australia: CSIRO.

Peisker, M. (1986). Models of carbon metabolism in C_3–C_4 intermediate plants as applied to the evolution of C_4 photosynthesis. *Plant, Cell and Environment*, **9**, 627–635.

Peterson, P. M., Annable, C. R. & Franceschi, V. R. (1989). Comparative leaf anatomy of the annual *Muhlenbergia* (Poaceae). *Nordic Journal of Botany*, **8**, 575–583.

Pichersky, E., Bernatzky, R., Tanksley, S. D. & Cashmore, A. R. (1986).

Evidence for selection as a mechanism in the concerted evolution of *Lycopersicon esculentum* (tomato) genes encoding the small subunit of ribulose-1,5-bisphosphate carboxylase/oxygenase. *Proceedings of the National Academy of Sciences USA*, **83**, 3880–3884.

Prendergast, H. D. V. & Hattersley, P. W. (1985). Distribution and cytology of Australian *Neurachne* and its allies (Poaceae), a group containing C_3, C_4 and C_3–C_4 intermediate species. *Australian Journal of Botany*, **33**, 317–336.

Prendergast, H. D. V. & Hattersley, P. W. (1987). Australian C_4 grasses (Poaceae): leaf blade anatomical features in relation to C_4 acid decarboxylation types. *Australian Journal of Botany*, **35**, 355–382.

Prendergast, H. D. V., Hattersley, P. W. & Stone, N. E. (1987). New structural/biochemical associations in leaf blades of C_4 grasses (Poaceae). *Australian Journal of Plant Physiology*, **14**, 403–420.

Prendergast, H. D. V., Hattersley, P. W., Stone, N. E. & Lazarides, M. (1986). C_4 acid decarboxylation type in *Eragrostis* (Poaceae): patterns of variation in chloroplast position, ultrastructure and geographical distribution. *Plant, Cell and Environment*, **9**, 333–344.

Prendergast, H. D. V., Stone, N. E. & Hattersley, P. W. (1988). Leaf blade structure and C_4 acid decarboxylation enzymes in ×*Cynochloris* spp. (Poaceae), intergeneric hybrids between species of different C_4 type. *Botanical Journal of the Linnean Society*, **96**, 381–389.

Raghavendra, A. S. & Das, V. S. R. (1976). Diversity in the biochemical and biophysical characteristics of C_4 dicotyledonous plants. *Indian Journal of Plant Physiology*, **19**, 101–112.

Rawsthorne, S., Hylton, C. M., Smith A. & Woolhouse, H. W. (1988). Photorespiratory metabolism and immunogold localization of photorespiratory enzymes in leaves of C_3 and C_3–C_4 intermediate species of *Moricandia*. *Planta (Berlin)*, **173**, 298–308.

Reed, J. E. & Chollet, R. (1985). Immunofluorescent localisation of phosphoenolpyruvate carboxylase and ribulose, 1,5-bisphosphate carboxylase/oxygenase proteins in leaves of C_3,C_4 and C_3–C_4 intermediate *Flaveria* species. *Planta (Berlin)*, **165**, 439–445.

Renvoize, S. A. (1987). A survey of leaf-blade anatomy in grasses XI. *Paniceae*. *Kew Bulletin*, **42**, 739–768.

Ritland, K. & Clegg, M. T. (1987). Evolutionary analysis of plant DNA sequences. *American Naturalist*, **130**, S74–S100.

Schuster, W. S. & Monson, R. K. (1990). An examination of the advantages of C_3–C_4 intermediate photosynthesis in warm environments. *Plant, Cell and Environment*, **13**, 903–912.

Sheen, J.-Y. (1991). Molecular mechanisms underlying the differential expression of maize pyruvate, orthophosphate dikinase genes. *Plant Cell*, **3**, 225–245.

Sheen, J.-Y. & Bogorad, L. (1986). Expression of the ribulose-1,5-bisphosphate carboxylase large subunit gene and three small subunit genes in two cell types of maize leaves. *EMBO Journal*, **5**, 3417–3422.

Sheen, J.-Y. & Bogorad, L. (1987). Differential expression of C_4 pathway genes

114   P. W. Hattersley and L. Watson

in mesophyll and bundle sheath cells of greening maize leaves. *Journal of Biological Chemistry*, **262**, 11726–11730.

Sinclair, T. R. (1988). Selecting crops and cropping systems for water-limited environments. In *Drought research priorities for the dryland tropics*, ed. F. R. Bidinger & C. Johansen, pp. 87–94. Andhra Pradesh, India: ICRISAT.

Smith, B. N. (1976). Evolution of C_4 photosynthesis in response to changes in carbon and oxygen concentrations in the atmosphere through time. *Bio-Systems*, **8**, 24–32.

Smith, B. N. & Robbins, M. J. (1974). Evolution of C_4 photosynthesis: an assessment based on $^{13}C/^{12}C$ ratios and Kranz anatomy. In *Proceedings of the third international congress on photosynthesis*, ed. M. Avron, pp. 1579–1587. Amsterdam: Elsevier.

Smith, S. M., Bedbrook, J. & Speirs, J. (1983). Characterization of three cDNA clones encoding different mRNAs for the precursor to the small subunit of wheat ribulose bisphosphate carboxylase. *Nucleic Acids Research*, **11**, 8719–8734.

Soderstrom, T. R., Hilu, K. W., Campbell, C. S. & Barkworth, M. E. (1987). *Grass systematics and evolution*. Washington DC: Smithsonian Institution Press.

Soreng, R. J. (1990). Chloroplast-DNA phylogenetics and biogeography in a reticulating group: study in *Poa* (Poaceae). *American Journal of Botany*, **77**, 1383–1400.

Soreng, R. J., Davis, J. I. & Doyle, J. J. (1990). A phylogenetic analysis of chloroplast DNA restriction site variation in Poaceae subfam. Pooideae. *Plant Systematics and Evolution*, **172**, 83–97.

Stebbins, G. L. (1956). Cytogenetics and evolution of the grass family. *American Journal of Botany*, **43**, 890–905.

Stebbins, G. L. (1981). Coevolution of grasses and herbivores. *Annals of the Missouri Botanic Gardens*, **68**, 75–86.

Stebbins, G. L. (1987). Grass systematics and evolution: past, present and future. In *Grass systematics and evolution*, ed. T. R. Soderstrom, K. W. Hilu, C. S. Campbell & M. E. Barkworth, pp. 359–367. Washington DC: Smithsonian Institution Press.

Sugita, M., Manzara, T., Pichersky, E., Cashmore, A. & Gruissem, W. (1987). Genomic organization, sequence analysis and expression of all five genes encoding the small subunit of ribulose-1,5-bisphosphate carboxylase/oxygenase from tomato. *Molecular and General Genetics*, **209**, 247–256.

Sugiyama, T., Schmitt, M. R., Ku, S. B. & Edwards, G. E. (1979). Differences in cold lability of pyruvate, P_i dikinase among C_4 species. *Plant and Cell Physiology*, **20**, 965–971.

Takeda, T., Tanikawa, T., Agata, W. & Hakoyama, S. (1985). Studies on the ecology and geographical distribution of C_3 and C_4 grasses. I. Taxonomic and geographical distribution of C_3 and C_4 grasses in Japan with special reference to climatic conditions. *Japanese Journal of Crop Science*, **54**, 54–64. [In Japanese]

Tateoka, T. (1957). Miscellaneous papers on the phylogeny of Poaceae (10).

Proposition of a new phylogenetic system of Poaceae. *Journal of Japanese Botany*, **32**, 275–288.

Tateoka, T. (1958). Notes on some grasses. VIII. On the leaf structure of *Arundinella* and *Garnotia*. *Botanical Gazette*, **120**, 101–109.

Taylor, W. C. (1989). Regulatory interactions between nuclear and plastid genomes. *Annual Review of Plant Physiology and Plant Molecular Biology*, **40**, 211–233.

Teeri, J. A. & Stowe, L. G. (1976). Climatic patterns and the distribution of C_4 grasses in North America. *Oecologia (Berlin)*, **23**, 1–12.

Thomason, J. R. (1987). Fossil grasses: 1820–1986 and beyond. In *Grass systematics and evolution*, ed. T. R. Soderstrom, K. W. Hilu, C. S. Campbell & M. E. Barkworth, pp. 159–167. Washington DC: Smithsonian Institution Press.

Thomason, J. R., Nelson, M. E. & Zakrzewski, R. J. (1986). A fossil grass (Gramineae: Chloridoideae) from the Miocene with Kranz anatomy. *Science*, **233**, 876–878.

Tidwell, W. D. & Nambudiri, E. M. V. (1989). *Tomlinsonia thomassonii, gen. et sp. nov.*, a permineralized grass from the upper Miocene Ricardo formation, California. *Review of Palaeobotany and Palynology*, **60**, 165–177.

Ueno, O., Samejima, M., Muto, S. & Miyachi, S. (1988). Photosynthetic characteristics of an amphibious plant, *Eleocharis vivipara*. Expression of C_4 and C_3 pathway modes in contrasting environments. *Proceedings of the National Academy of Sciences USA*, **85**, 6733–6737.

Usuda, H., Ku, M. S. B. & Edwards, G. E. (1984). Rates of photosynthesis relative to activity of photosynthetic enzymes, chlorophyll and soluble protein content among ten C_4 species. *Australian Journal of Plant Physiology*, **11**, 509–517.

von Caemmerer, S. (1989). A model of photosynthetic CO_2 assimilation and carbon-isotope discrimination in leaves of certain C_3–C_4 intermediates. *Planta (Berlin)*, **178**, 463–474.

von Caemmerer, S. & Hubick, K. T. (1989). Short-term carbon-isotope discrimination in C_3–C_4 intermediate species. *Planta (Berlin)*, **178**, 475–481.

Watson, L. (1987). Automated descriptions of grass genera. In *Grass systematics and evolution*, ed. T. R. Soderstrom, K. W. Hilu, C. S. Campbell & M. E. Barkworth, pp. 343–351. Washington DC: Smithsonian Institution Press.

Watson, L. (1990). The grass family, Poaceae. In *Reproductive versatility in the grasses*, ed. G. P. Chapman, pp. 1–31. Cambridge: Cambridge University Press.

Watson, L., Clifford, H. T. & Dallwitz, M. J. (1985). The classification of Poaceae: subfamilies and supertribes. *Australian Journal of Botany*, **33**, 433–484.

Watson, L. & Dallwitz, M. J. (1988). *Grass genera of the world: illustrations of characters, descriptions, classification, interactive identification, information retrieval.* Canberra: Research School of Biological Sciences, Australian National University.

116   P. W. Hattersley and L. Watson

116   *P. W. Hattersley and L. Watson*

Watson, L., Dallwitz, M. J. & Johnston, C. R. (1986). Grass genera of the world: 728 detailed descriptions from an automated database. *Australian Journal of Botany*, **34**, 223–230 (with 3 microfiches).

Webster, R. D. (1987). *The Australian Paniceae*. Berlin & Stuttgart: J. Kramer.

Westhoff, P., Offermann-Steinhard, K., Höfer, M., Eskins, K., Oswald, A. & Streubel, M. (1991). Differential accumulation of plastid transcripts encoding photosystem II components in the mesophyll and bundle-sheath cells of monocotyledonous NADP-malic enzymic-type C_4 plants. *Planta (Berlin)*, **184**, 377–388.

Wilson, J. R. & Hacker, J. B. (1987). Comparative digestibility and anatomy of some sympatric C_3 and C_4 arid zone grasses. *Australian Journal of Agricultural Research*, **38**, 287–295.

Wilson, J. R. & Hattersley, P. W. (1983). *In vitro* digestion of bundle sheath cells in rumen fluid and its relation to the suberized lamella and C_4 photosynthetic type in *Panicum* species. *Grass Forage Science*, **38**, 219–223.

Wilson, J. R. & Hattersley, P. W. (1989). Anatomical characteristics and digestibility of leaves of *Panicum* and other grass genera with C_3 and different types of C_4 photosynthetic pathway. *Australian Journal of Agricultural Research*, **40**, 125–136.

Yanagisawa, S. & Izui, K. (1989). Maize phosphoenolpyruvate carboxylase involved in C_4 photosynthesis. Nucleotide sequence analysis of the 5' flanking region of the gene. *Journal of Biochemistry (Tokyo)*, **106**, 982–987.

Yeoh, H.-H., Badger, M. R. & Watson, L. (1990). Variations in K_m (CO_2) of ribulose-1,5-bisphosphate carboxylase among grasses. *Plant Physiology*, **66**, 1110–1112.

Yeoh, H.-H. & Watson, L. (1987). Taxonomic patterns in protein amino acid profiles of grass leaves and caryopses. In *Grass systematics and evolution*, ed. T. R. Soderstrom, K. W. Hilu, C. S. Campbell & M. E. Barkworth, pp. 88–96. Washington DC: Smithsonian Institution Press.

Zimmer, E. A., Jupe, E. R. & Walbot, V. W. (1988). Ribosomal gene structure, variation and inheritance in maize and its ancestors. *Genetics*, **120**, 1125–1136.

Zubakov, V. A. & Borzenkova, I. I. (1990). *Global palaeoclimate of the late Cenozoic*. Developments in Palaeontology and Stratigraphy, No. 12. Amsterdam: Elsevier.

Zuloaga, F. O. (1987). Systematics of New World species of *Panicum* (Poaceae: Paniceae). In *Grass systematics and evolution*, ed. T. R. Soderstrom, K. W. Hilu, C. S. Campbell & M. E. Barkworth, pp. 287–306. Washington DC: Smithsonian Institution Press.

Zurawski, G. & Clegg, M. T. (1987). Evolution of higher-plant chloroplast DNA-encoded genes: implications for structure-function and phylogenetic studies. *Annual Review of Plant Physiology*, **38**, 391–418.

Zurawski, G., Clegg, M. T. & Brown, A. H. D. (1984). The nature of nucleotide sequence divergence between barley and maize chloroplast DNA. *Genetics*, **106**, 735–749.

3

The S–Z incompatibility system

D. L. Hayman

Introduction

Self-incompatibility is a genetically determined outbreeding system found in flowering plants and fungi. In flowering plants with perfect flowers a genetically specified functional pollen grain is recognized as compatible or incompatible by a genetically specified style. If it is compatible the pollen grain is accepted as capable of effecting fertilization; if it is incompatible it is rejected and fails to effect fertilization. The system operates by selecting as compatible from among the pollen population on a stigma, those pollen grains with an incompatibility phenotype other than that possessed by that stigma. Consequently self-pollination results in no seed or in a very substantial reduction in seed setting. The system was originally called 'self-sterility', but the term self-incompatibility is now generally used as it more accurately reflects the basis for reduced seed setting. There are two major self-incompatibility systems not associated with floral heteromorphisms which have been described following studies in many dicotyledons. In these systems genetic control of incompatibility is by one gene with many alleles. The two systems differ: in one the specification of the pollen grain is gametophytic and the genes act independently in the style, in the other the pollen is specified by the sporophyte's genotype and dominance relationships may occur in both the pollen and the style. The subject has been reviewed most recently by de Nettancourt (1977).

The S–Z system and its detection

Self-incompatibility has been identified in many grasses, usually, but not exclusively, perennial species. However, an understanding of its genetic basis was much later in coming than was a knowledge of the systems found in many dicotyledons. The first formal genetic

117

Table 3.1. *Studies on the genetics of incompatibility in grasses*

Species	Author	Tribe
(a) Species in which the *S–Z* system has been identified by studies using parental genotypes of known relationship		
Secale cereale	Lundqvist (1954)	Triticae
Festuca pratensis	Lundqvist (1955)	Poeae
Phalaris coerulescens	Hayman (1956)	Avenae
Hordeum bulbosum	Lundqvist (1965)	Triticae
Dactylis aschersoniana	Lundqvist (1965)	Poeae
Briza media	Murray (1974)	Poeae
Lolium perenne	Cornish *et al.* (1979)	Poeae
Lolium multiflorum	Fearon *et al.* (1983)	Poeae
(b) Species in which a gametophytic system has been identified and there are differences between reciprocal crosses		
Alopecurus myosuroides	Leach & Hayman (1987)	Avenae
Cynosurus cristatus	Weimarck (1968)	Poeae
Holcus lanatus	Weimarck (1968)	Avenae
Alopecurus pratensis	Weimarck (1968)	Avenae
Arrhenatherum elatius	Weimarck (1968)	Avenae
Festuca rubra	Weimarck (1968)	Poeae
Deschampsia flexuosa	Weimarck (1968)	Avenae
Phalaris arundinacea	Weimarck (1968)	Avenae

analyses were performed in *Secale cereale* (Lundqvist, 1954) and *Phalaris coerulescens* (Hayman, 1956) and a novel and identical interpretation was offered to explain the results. This has been further confirmed by studies of other grasses (see Table 3.1).

Self-incompatibility in grasses is determined by the action of two independently segregating genes, given the symbols *S* and *Z*, each with a series of multiple alleles. A pollen grain is specified gametophytically by the complementary interaction of its *S* and *Z* genes. A pollen grain is incompatible when both the *S* and *Z* genes it carries are present in the stigma on which it lands.

A pollen grain with only an *S* or *Z* gene in common with the stigma suffers no impairment in its compatibility and is fully functional. Fertilization by such a pollen grain may result in plants homozygous at one of the two incompatibility loci. Thus a considerable range of incompatible genotypes is possible with only a few alleles at each locus.

As a consequence of the complementary interaction determining specificity, there are a number of features of the *S–Z* incompatibility system which are absent from single-locus systems with gametophytic control or with sporophytic control. These features include differences between reciprocal crosses in compatibility itself and differences in the degree of compatibility (as measured by the percentage of compatible pollen) between compatible crosses. Depending upon the genotypes of the plants involved in a pollination, the expected ratio of compatible to incompatible pollen can be 0, 50, 75 or 100%. Finally, the number of intra-incompatible, inter-compatible groups in the progeny obtained after compatible intercrosses may vary from 16 to 3. These features are illustrated in Table 3.2.

Genetic analyses necessary to identify the precise system of control which is operating are most easily performed using the progeny from intercrosses of genotypes with shared incompatibility factors. Such crosses reduce the number of intra-incompatible inter-compatible groups within any progeny. Thus inbred progeny from selfing and the progeny from two plants which are compatible only in one direction would give the simplest progeny for genetic analysis. Lundqvist (1975) cautioned that there is a risk in using such restricted material. If there are more than two loci concerned in the incompatibility system, a close relationship between the original materials may restrict the chance of recognizing them, owing to fixation at one of the loci. Lundqvist argued that, preferably analyses should be based on segregation patterns from crosses among materials of remote provenance.

Genetic analyses of the *S–Z* incompatibility system have employed several different techniques, both in the criteria adopted to identify a compatible pollination and in the material studied. While seed setting has been used as an index of a compatible pollination, it is not always reliable because of the reduction in seed setting that may occur under bagged conditions and because it is a slow procedure, since the results of a cross are not known for some time.

The most informative index of compatibility is provided by the behaviour of the pollen grains. A detailed account of this behaviour is given by Mogensen (1990) in the companion volume to this book. Compatible and incompatible pollen grains germinate equally well, within minutes of pollination, and the pollen tubes of both penetrate the stigma surface. Compatible pollen tubes grow freely into the style and down towards the ovary. Callose deposition is present along the pollen tube. The contents of the pollen grain enter the pollen tube and the grain is emptied. The incompatible pollen tube does not penetrate far into the stigma and may fail to reach or only just reach the style. The pollen tube is short and the pollen grain does not release its contents. Thus inspec-

Table 3.2. *The results of crossing a variety of S–Z genotypes between which there are different degrees of genetic identity. The features shown are those which identify the S–Z system*

♀ Parental genotype	♂ Parental genotype	% compatible pollen	No. of incompatible genotypes in the progeny	♂ Parental genotype present	% compatible pollen in reciprocal cross
$S_{34}Z_{34}$	$S_{12}Z_{12}$	100	16	No	100
$S_{13}Z_{34}$	$S_{12}Z_{12}$	100	16	No	100
$S_{12}Z_{34}$	$S_{12}Z_{12}$	100	16	No	100
$S_{13}Z_{13}$	$S_{12}Z_{12}$	75	12	Yes	75
$S_{12}Z_{13}$	$S_{12}Z_{12}$	50	6	Yes	50
$S_{11}Z_{13}$	$S_{12}Z_{12}$	75	6	Yes	50
$S_{11}Z_{12}$	$S_{12}Z_{12}$	50	3	Yes	0

tion of a pollination, usually after 24 hours, detects these differences. Staining with cotton blue in lactophenol (Watkins, 1925) and observation with the light microscope or using decolorized aniline blue and ultraviolet fluorescence microscopy (Martin, 1959) enables the two classes of pollen to be recognized. Lundqvist's (1961) technique of planting large numbers of stigmas in 2% agar so that stigmas with different genotypes may have pollen applied from a common source is commonly used.

The advantages of using the appearance of the pollen as an index of incompatibility are that the answer is obtained quickly and that uncertain results may be checked at once. Further, a distinction may be made between the degrees of compatibility, i.e. 0, 50, 75 and 100%. The presence of both compatible and incompatible pollen in a pollination is indicative of gametophytic control. The interpretation of the degree of compatibility is not easy and most authors have indicated that not all pollinations can be reliably scored. The pollen must be of high quality so that it all germinates and crowding of pollen grains should be avoided. Most workers would agree with Lundqvist's comments that 'the proportions of emptied pollen grains are almost constantly below the values expected' (Lundqvist, 1961). The initial differences in the interpretation of the genetic control of incompatibility in *Lolium* species were almost certainly the result of problems in classifying the partly compatible pollinations (Cornish, Hayward & Lawrence, 1979; Fearon, Hayward & Lawrence, 1983); McGraw & Spoor, 1983*a*,*b*; Spoor & McGraw, 1984).

The ideal experiment involves the intercrossing of a group of progeny with their parents in a diallel. This enables all the features of the *S–Z* system to be identified. However, the two significant features of the presence of both compatible and incompatible pollen in a cross (which indicates gametophytic determination), and of differences in compatibility between reciprocal crosses (which indicates the involvement of more than one locus) may be demonstrated in material from other sources (Weimarck, 1968).

All the species listed in Table 3.1 have a basic chromosome number of $x=7$. The species in group (a) are diploid, as are the first three species in group (b); the remainder are polyploid.

Evidence of a close similarity in genetic organization between these species comes from a study of the effects of linkage between an incompatibility locus and a segregating locus. In crosses between appropriate genotypes which share *S* and *Z* alleles and in which, therefore, a proportion of the pollen is incompatible, the operation of the

Table 3.3. *Estimates of the recombination frequency between the S locus and GPI-2*

Species	Recombination	S.E.M.
Lolium perenne	0.1538	0.0252
(Cornish *et al.* 1980)		
Lolium multiflorum	0.2414	0.0795
(Fearon *et al.* 1983)		
Alopecurus myusorides	0.1852	0.0449
Phalaris coerulescens	0.1134	0.0312
Festuca pratensis	0.2381	0.0933
Holcus lanatus	0.2987	0.0563
Secale cereale (a)	0.0714	0.0668
(b)	0.1902	

S.E.M., standard error of estimate of mean value.
(a) and (b) are estimates from different sources.
Taken from Leach & Hayman (1987).

incompatibility system restricts the free transmission of genes linked to an incompatible allele. This restriction results in a disturbed segregation for such genes, and the extent of the disturbance provides a measure of the linkage distance between the incompatibility locus and the gene being studied. Appropriate statistical methods are described by Leach (1988).

Leach & Hayman (1987) used this technique to show that the isozyme phosphoglucoisomerase *PGI-2* (E.5.1.3.9.), which has been reported as linked to the *S* locus in *Lolium perenne* by Cornish, Hayward & Lawrence, (1980), is also linked to an incompatibility locus in five of the other grass species listed in Table 3.1 (see Table 3.3). This suggests conservation of a linkage group and provides an objective means of distinguishing between *S* and *Z* loci when comparing different species. Thorogood & Hayward (1992), in a study of a self-fertility mutant at or close to the *Z* locus in *Lolium* spp., reported that segregation for the *GOT/3* isozyme locus was disturbed and estimated the linkage distance as about 27%. Both *GOT/3* and *PGI-2* are known to be on separate chromosomes in *L. perenne* (Lewis, Humphreys & Coton, 1980). Gertz & Wricke (1989) used the conserved linkage group argument to identify the *S* locus in *Secale cereale* as that linked to *PGI-2*, and the *Z* locus as that linked to a β-glucosidase. Independent studies have located this glucosidase to chromosome 2R and *PGI-2* to chromosome 1R in *Secale* (Schmidt, Seliger & Schlegel, 1984; Wricke & Wehring, 1985). There is, thus, direct evidence in two species which shows that the S and Z loci are independently inherited, because they are not on the same chromosome. It is expected that the results in *Lolium* and *Secale* will prove to be typical of those which will be found in other species of the Pooideae with $x=7$.

The general rule in single-locus incompatibility systems is that species from the same taxonomic family share the same incompatibility system. The $S-Z$ incompatibility system is thought to be characteristic of the Poaceae as a whole, but this really remains to be demonstrated and it is not generally realized that the range of species which have been studied carefully is limited when compared to the diversity of species in the family as a whole.

Connor (1979) has provided the most recent compilation of species in which incompatibility has been reported. He lists species from 48 genera and 11 tribes. These genera include representatives of the Panicoideae and Bambusoideae as well as the Pooideae. The absence of genetic data about the incompatibility system present in these first two subfamilies is unfortunate. Many of the species have a basic chromosome number of $x=5$. Should an $S-Z$ incompatibility system be present in these species it would be of interest to see how the linkage relationships identified in the $x=7$ species have been modified.

Lewis & Crowe (1958) reported that in crosses between self-compatible (SC) and self-incompatible (SI) species unilateral incompatibility occurred very frequently. Typically the cross $SI♀ \times SC♂$ was incompatible in species with gametophytic or sporophytic single locus systems. This suggests that there may be a relationship between the incompatibility system and the failure of such a hybridization. Such unilateral incompatibility appears not to be general in grasses. Viable seeds are obtained from the cross *Phalaris coerulescens* (SI) $\times$ *Phalaris minor* (SC) (D. L. Hayman, unpublished data) and *Lolium temulentum* (SC) hybridizes with both *L. perenne* (SI) and *L. multiflorum* (SI) when the cross is made in either direction (Thorogood & Hayward, 1991*b*).

The $S-Z$ system has two other features which distinguish it from the single locus systems. Heslop-Harrison (1975) pointed out that the gametophytic systems are found in species with wet stigma surfaces and that the Poaceae have a dry stigma like those species with single-locus sporophytic systems. Brewbaker (1957) noted that trinucleate pollen was a feature of species with sporophytic incompatibility systems and that the Poaceae did not fit this pattern. Nothing is known of the role of these two features in the operation of incompatibility and of their implications for the particular method of genetic control which is adopted.

The effect of autopolyploidy upon the S–Z system

A fundamental difference between the single-locus gametophytic system of dicotyledons and the S–Z incompatibility system is the effect on it of autopolyploidy. This has been studied in both *Secale cereale* and *Festuca pratensis* by Lundqvist (1957, 1962a) and in *L. perenne* by Lawrence and colleagues (Fearon, Hayward & Lawrence, 1984a,b,c). Unlike many plants with the single-locus system, autopolyploid grasses remain self-incompatible. The S and Z genes continue to act independently in the style and tetraploid stylar genotypes may be elucidated from the behaviour of haploid pollen of known genotype placed upon them. A diploid pollen grain is incompatible if one of the S–Z gene pairs present is matched in the style. The phenomena of dominance and competition between alleles in diploid pollen, found in tetraploids of many species with single-locus gametophytic systems, are both absent. These observations have implications for both the evolution of the S–Z system and the mechanism of incompatibility, as discussed below.

The number of incompatibility alleles

The complementary interaction between the S and Z genes in specifying the S–Z system is a very efficient means of encouraging cross-pollination, since the number of incompatibility specificities present should be given by the product of the number of alleles segregating at the two loci. The efficiency of this system is further maximized by the independent segregation of S and Z and there is a theoretical expectation that such independence would have been favoured by selection.

These very properties, however, make the determination of the number of alleles at each locus and their relative frequency in a population technically difficult. Various systems have been adopted to answer these questions, frequently making use of the fact that enforced self-pollination may result in plants homozygous at both S and Z. Using such genotypes in crosses with unknown genotypes reduces the number of different genotypes in the progeny. These S and Z homozygotes also enable the frequency of specific combinations of S and Z genes to be determined in a population.

Lundqvist (1962a, 1964a) has estimated the number of alleles present in a cultivar of *Festuca pratensis* by a number of approaches and in 1969 determined the number of individual members within the two allelic series as 14 S alleles and 13 Z alleles in one population.

Trang, Wricke & Weber (1982) estimated that in a population from a commercial variety of *Secale cereale* there were 6 to 7 alleles present at one locus and 12 to 13 alleles at the other. Fearon *et al.*(1983) calculated that there may be a total of 40 incompatibility alleles at each locus present in the population of *Lolium multiflorum* from which the plants they studied were drawn. They report that estimates of a similar number have been obtained for *L. perenne*.

Not only is there interest in knowing the number of alleles at each locus which are maintained in a population but it is also important to demonstrate their relative frequency and the relative frequency of singly versus doubly heterozygous plants. The problems involved in predicting these metrics are considerable and have attracted the attention of mathematicians. Charlesworth (1979) believed a general solution for multilocus systems was only possible when the number of S and Z alleles was the same. Weber, Wricke & Trang (1982) have developed an alternative treatment which does not require this restriction and which was used by Trang *et al.* (1982) in their estimate of the number of alleles in a *S. cereale* cultivar.

The experimental testing of the theoretical estimates of these parameters by direct means would appear at present to be a most difficult and time-consuming task. Perhaps, in the future, direct sequencing of the S and Z genes will reveal allele-specific sequences which can be recognized in hybridization with genomic DNA and used as indices. This may be more satisfactory than relying on the phenotypic behaviour of pollen to determine genotypes in the population.

The origin and frequency with which new alleles appear is one of the unsolved problems of all homomorphic incompatibility systems. Nothing is known of the rate at which mutations to new fully functional alleles at S or Z occur. In the only reported large scale attempt to recover such mutants, Hayman & Richter (1992) were unsuccessful. In the system they were using, it can be estimated that they were testing 1.2×10^8 pollen grains for such mutations.

Incompatibility systems in other monocotyledons

The $S–Z$ system has not been found in other groups of monocotyledons in which incompatibility has been studied. The information available at present suggests that there may be more than one system operating in the group as a whole. Incompatibility in *Tradescantia paludosa* (Commelinaceae) is under the control of a single gene with multiple alleles and gametophytic determination of the pollen (Annerstedt & Lundqvist, 1967). A similar interpretation has been offered for the closely related species *Gibasis oaxacana* (Owens, 1977) and for the more distant species *Ananas comosus* (Bromeliaceae; Brewbaker & Gorrez, 1967).

Brandham & Owens (1978) reported that the results they obtained in their study of fertile hybrid progeny obtained by crossing two species of *Gasteria* (Liliaceae) required them to postulate control by more than one gene and a minimum of two independently segregating genes. Eleven F_1 plants tested in a diallel gave no instances of reciprocal incompatibility and three cases of unilateral incompatibility in reciprocal crosses. They suggested that a previous study by Brewbaker & Gorrez (1967), which reported control by one gene with multiple alleles, may be in error.

Such an interpretation would make the results in *Gasteria* consistent with those obtained by Lundqvist (1991) in his study of *Lilium martagon* (Liliaceae). No instances of cross-incompatibility were found amongst 23 siblings in a diallel and there was little unilateral incompatibility. Such a result would require the operation of at least three or four independently segregating genes with complementary interaction and gametophytic pollen control.

Breakdown of the S–Z system – pseudocompatibility

One of the classical approaches used in studying any genetic system involves looking for variants in that system, since these offer the possibility of examining the elements of the system more closely. It is a common observation that in many self-incompatible grass species a few seeds are obtained after self-pollination, i.e. the system is not absolute. Individual plants may differ in their capacity to yield seeds. This partial breakdown of the incompatibility system which results in reduced seed set, i.e. perhaps 10% of the potential seed set, is referred to as 'pseudocompatibility'. The *S* and *Z* components of the incompatibility system are both present and unchanged, but the system is not uniformly efficient. Pseudocompatibility has been investigated most extensively in *Secale cereale* and *Festuca pratensis* by Lundqvist (1958, 1964b), who related the degree of pseudocompatibility present to the S–Z status of the plants. He observed that the amount of seed set decreased with increased homozygosity for *S* or *Z*. Thus the degree of seed setting decreased as the level of inbreeding increased. In terms of genetics, pseudocompatibility was greatest in plants that were heterozygous for both the *S* and *Z* alleles. Such effects had been noted by Myers (1948) in *Dactylis glomerata* and Adams (1953) in *Bromus inermis*. While there was no knowledge of the genetic basis for incompatibility in grasses at that time, Adams believed that such genes existed and recognized that these observations would favour the preservation of incompatibility genes in the heterozygous state.

In his study of *F. pratensis*, Lundqvist followed specific genes in inbred families to determine whether there was any rank order in the individual genes or gene pairs in their capacity to self-pollinate. Consistent results were obtained in that S_1 was more pseudocompatible than was S_2, and Z_3 than Z_4. The order of pseudocompatibility of gene pairs was $S_1Z_4 > S_1Z_3 = S_3Z_4 > S_2Z_3$, which would indicate a simple additive system. In addition to these effects, which are related to specific S and Z gene content and combinations, other non-allelic genes are clearly involved in modifying the degree of self-incompatibility. Plants which give few seeds on selfing yield occasional progeny which give an increased seed set on selfing presumably owing to the segregation of genes other than S and Z. Pseudocompatibility represents a relatively minor perturbation of the incompatibility system. It has, however, considerable implications in breeding programmes for grasses, since inbred lines are likely to favour intercrossing and to produce hybrid progeny.

Breakdown of the *S–Z* system – compatibility

The genetic analysis of mutations which result in self-compatibility has been used in studies of the single locus gametophytic system, most notably by Lewis (for a review see de Nettancourt, 1977). The mutants obtained by Lewis and other investigators have fallen into three classes. Some mutants are complete, that is the mutant fails to specify either the pollen or the stigma, and so is compatible with all other genotypes when used either as a pollen or stigma parent. Other mutants lack pollen specificity only: they are compatible with all plants as pollen parents. However, the stigmas of plants carrying such mutants recognize and reject a pollen grain carrying the unmutated gene. The final class of mutant is one in which the capacity to specify the stigma only has been modified: the pollen carrying such a mutant is properly specified, and is recognized and rejected by a style carrying the normal allele.

This classification provides a basis for the examination of mutants which cause self-compatibility in grasses with an *S–Z* system. Lundqvist (1958, 1962*a*, 1968) carried out a series of studies in *Secale cereale* in which he characterized a number of self-compatible mutants. He was able to demonstrate that there were at least two classes of self-compatible mutants which he equated to mutants at the S and Z loci. The mutants were all pollen-only mutants. The complete analysis of all mutants was not possible because of the difficulties of working with a two-locus system, and with an annual, or at best, a biennial plant. Lundqvist was not able to exclude the possibility that some of the self-compatible mutants were the result of mutations at a third unlinked locus and argued that this was probably the case. This has proved to be a most perceptive comment even though critical evidence was lacking.

Thorogood & Hayward (1991) analysed a self-compatible mutant of *Lolium perenne*. Self-compatibility behaved as a gametophytic character and the pattern of compatibility relationships in the F_1 and an F_2 could not be explained by a mutation at S or Z. It was necessary to postulate a mutation at a third locus. A pollen grain containing a mutation at that locus is compatible irrespective of its S or Z content.

A more extensive study of self-compatible mutations was conducted by Hayman & Richter (1992) in *Phalaris coerulescens*. Since this plant was not grown elsewhere in the area, extensive field plantings were possible in which the only genes segregating were S_1, S_2, Z_1 and Z_2. Heads from plants of the genotype $S_{12}Z_{12}$, interplanted in this population, were harvested and the seeds collected. These seeds could have been produced only where breakdown of the incompatibility system had occurred. Some 508 plants grown from these seeds were examined and 152 were self-compatible. A sample of these was selected for further study. In this sample were plants which had a stigma phenotype $S_{12}Z_{12}$. Since these plants had 50% compatible pollen on selfing they must also have carried a gametophytically acting compatibility factor. Such a mutation could be at the S or Z locus, and, since both S and Z alleles are active in the stigma, the mutation must then affect pollen specificity only. It is also possible that the mutation could have been at a third locus, and that any pollen grain carrying such a mutant was compatible irrespective of its S or Z genotype. Table 3.4 shows the results that would be expected from each of these possibilities when crossing the mutant plant as a pollen parent to self-incompatible tester genotypes.

Two self-compatible $S_{12}Z_{12}$ plants gave results which were like those shown in row (c) of Table 3.4. These results can be explained by the presence of a mutation at a third locus, designated T. It was found that T mutants showed disturbed segregation for a very closely linked leaf peroxidase isozyme and this enabled the identification of further T mutants.

Homozygotes for many of the self-compatible mutations were derived

Table 3.4. *Percentage compatible observed in crosses using pollen from a self-compatible plant* $S_{12}Z_{12}$ *which contains* (a) *a pollen-only* S_1 *mutant,* (b) *a pollen-only* Z_1 *mutant, and* (c) *a mutant at a third unlinked locus* T

Pollen parent	Female testers				
	$S_{12}Z_{12}$	$S_{11}Z_{12}$	$S_{22}Z_{12}$	$S_{12}Z_{11}$	$S_{12}Z_{22}$
(a) $S^*_{12}Z_{12}$	50	100	50	75	75
(b) $S_{12}Z^*_{12}$	50	75	50	100	75
(c) $S_{12}Z_{12}T^*_{11}$	50	75	75	75	75

* Indicates the mutant.

Table 3.5. *The basis of self-compatibility in 38 mutants analysed in* Phalaris coerulescens

S mutants		Z mutants		T mutants	
S_1 pollen only	5	Z_1 pollen only	1	T pollen only	8
S_2 pollen only	5	Z_2 pollen only	5	T complete	0
S complete	1	Z_2 complete	0	T (unresolved)	8
S unresolved	5				

Mutants which it has not been possible to classify as pollen-only or complete are shown as unresolved.
Taken from Hayman & Richter (1992).

and the behaviour of these enabled a classification as to whether a mutant was complete or pollen only. Complete mutants are 100% compatible as both pollen and stigma parents with all tester genotypes. Pollen only mutants are 100% compatible on self-pollination or as pollen parents on the testers, but not as stigma parents. The results of the analysis of mutants obtained in this study are shown in Table 3.5.

These two studies show that, as Lundqvist (1958) suggested, there is at least a third locus operating which influences the compatibility status of pollinations. There is no allelic variability at this locus, as there has never been any evidence of more than two loci controlling incompatibility in any of the grass species studied. It is worth mentioning again that it is, as urged by Lundqvist (1975), necessary to use plants from widely separated populations as parents in incompatibility studies.

While the function of the third locus is not known, there are two genetic possibilities which suggest themselves. The first of these is to suppose that this locus formerly contained allelic variability and contributed to the specification of the pollen and style. The locus then became fixed and is now a 'silent' invariant component. Clearly the locus is still important in that it performs a required function, mutants at the locus resulting in self-compatibility. Such a proposal would require that the S–Z system evolved from a multilocus system – a point which is discussed later. It is also important to realize that formal identity of S and Z in all self-incompatible grass species is lacking. The conserved linkage between S and PGI-2 demonstrated by Leach & Hayman (1987) and described earlier would characterize the S locus but there are no equivalent comparative data for Z. It remains a possibility that different grass species may have arrived at a two-locus system following fixation of different components of a multilocus system.

An alternative hypothesis for the role of the third locus is to suppose that it performs a function which is not connected with specific characterization. If the T locus performs a function which is required to combine the S and Z products, then mutants at T would result in pollen grains which lack the 'S–Z dimer' and so cannot be identified as incompatible. The T locus could also act before S or Z specification, so that in a T mutant pollen grain there is an imperfect S or Z product which fails to characterize the pollen. If there are separate S and Z products prior to individual specification there may well be more than one T type locus at which mutations may occur.

So far the data demonstrate only that pollen-only mutants occur at the T locus in *Phalaris*. It has not been possible to determine the nature of the analogous mutation in *L. perenne*. The absence of complete mutants for T may be a reflection of their relative rarity compared to the pollen-only mutations. At present, however, there is no evidence that the T gene or its mutants has any effect on the incompatibility status of the stigma and its action may indeed be limited to the pollen.

The various types of mutants obtained in studies of single-locus gametophytic systems were explained by suggesting that the S gene consisted of different elements. Some determined the tissue in which the gene was expressed and others determined the allelic specificity. It is likely that a similar means of genetic regulation and specification applies at both the S and Z loci, and perhaps at T, without requiring identity of the recognition mechanism involved.

The demonstration that mutations at the S and Z and a third locus all may confer self-compatibility, raises further very interesting problems. For example, if mutations at any of these three loci have been utilized by a population to adopt self-fertility what happens at the other two loci? These now are no longer required for the function of specifying a pollen grain so that its incompatibility type may be recognized, nor are they needed for the corresponding function in the style. Do these loci degenerate by the accumulation of mutations in the absence of selection? Information on this problem has been obtained by Thorogood & Hayward (1992) in an analysis of hybrids between both the self-incompatible species *L. multiflorum* and *L. perenne* and the self-compatible species *L. temulentum.*

Genetic analyses showed that self-compatibility in *L. temulentum* was determined by a mutation at the *Z* locus or a locus closely linked to it. The mutant is gametophytic in action and homozygotes are viable and 100% compatible on selfing. Thorogood & Hayward (1992) demonstrated that there was a fully active *S* locus present in *L. temulentum* which is capable of interacting in the normal way with a regular *Z* locus introduced from either of the two self-incompatible species. Consequently the problem of the function of the *S* locus and probably that of the *Z* and *T* is highlighted again. Since the *S* and presumably *T* loci persist apparently intact in *L. temulentum*, it is difficult to avoid the conclusion that they have a function which is essential to the plant and that incompatibility specification per se, i.e. the maintenance of allelic diversity, is not an obligatory feature of that function.

Evolution of the S–Z system

The evolution of incompatibility in the angiosperms has yet to be explained. Whitehouse (1950) argued that the evolutionary success of the angiosperms, was, in part, the result of developing superior cross-pollinating devices such as were favoured by the development of the style. Thus the evolution of incompatibility is seen to be unique and coincident with the evolution of the angiosperms. This view has not gone unchallenged (Bateman, 1952; Charlesworth & Charlesworth, 1979), basically because of the diversity of systems with clear differences between them which suggests more than one origin.

Many writers have pointed out that, in order for a single-locus gametophytic system to operate three fully functional alleles are required. It has been difficult to suggest a mechanism whereby such a system could have evolved from a self-compatible species. Charlesworth & Charlesworth (1979) have discussed some of these problems. The difficulties are further compounded in the *S–Z* system, since this will not operate unless there are at least two alleles at one locus and three alleles at the other. A solution of a sort to this problem is to suppose that the *S–Z* system is a derivative from another system.

Early consideration was given to the possibility that the *S–Z* system followed a duplication of the single-locus system. Subsequently one locus differentiated from the other so that the specification became complementary. However, when autotetraploids are made from species with single-locus gametophytic systems, the two *S* genes in the diploid pollen are found to interact. Both dominance and competitive interactions occur and self-fertility may result (Lewis, 1947). Both these features are absent from autotetraploid grasses. Thus, in a duplicative model, this would mean either that there must have been an initial period of instability following duplication, which required the operation of selection to restrict allelic interaction, or that allelic interactions were absent from the single-locus system that led to the evolution of the grass system.

Support for this latter suggestion can be found in the demonstration that in diploid *Tradescantia paludosa* there is a single-locus gametophytic system of incompatibility (Annerstedt & Lundqvist, 1967). The four *S* alleles identified in this study were tested in artificial tetraploids and there was no evidence of allelic interaction in the diploid pollen. As mentioned previously, polyploid *Tradescantia* species and other polyploid species in the Commelinaceae, like autotetraploids in the Poaceae, are self-incompatible. Thus there is a possibility that a different *S* gene is present in *Tradescantia* and this could be a candidate for the position of progenitor of the *S–Z* system. However, the hypothesis of origin by duplication still requires that the initial incompatibility specification consequent upon duplication be modified to become a complementary specification and that there be no epistasis.

The *Tradescantia* results can be accounted for, in formal terms, by supposing that, in addition to the single *S* gene demonstrated by genetic segregation, there is also present a homozygous, i.e. invariant, locus (loci) involved in complementary specification (Lewis, 1979). The effect of polyploidy would then be identical with that observed in the *S–Z* system. It is interesting and perhaps relevant that the genetic results obtained from studying incompatibility in monocotyledons other than the Poaceae suggest that more than one locus is concerned in the incompatibility system. Thus the duplication hypothesis is not the only possibility.

Recently a series of studies by Lundqvist and colleagues has shown that both multilocus control and complementary interaction are involved in specifying incompatibility in the dicotyledonous genera *Ranunculus* and *Beta* (Osterbye, 1975, 1986; Larsen, 1977; Lundqvist, 1990*a*). In both species, four genes, each with a limited number of alleles, are concerned. It is therefore evident that multigene systems of control are not confined to the Poaceae but are distributed among the angiosperms in a way which suggests that such systems may well be of ancient origin and that a duplication hypothesis with all its attendant problems is not required to account for the development of an *S–Z* system.

In theory, if there are more than two loci concerned in specification and if fixation took place at all but two of these loci, an *S–Z* system would operate. Genetic segregation alone would not detect silent homozygous loci. The mutations that occur at such silent loci could well result in self-compatibility. The demonstration described earlier that such self-compatible mutants do occur in *Lolium* and *Phalaris* opens up this possibility. While formally correct, this hypothesis is not helped by the observation that in a study of different populations of *Ranunculus* and *Beta* (Lundqvist, 1990*b*), allelic variability was maintained, although it was possibly of a less extensive nature than that found in the *S–Z* system and in single-iocus systems.

This argument may be extended in a similar manner to suggest that fixation could have occurred at either the *S* or the *Z* loci in the Poaceae and that this would result in an apparent single-locus system. Yet this does not seem to have happened in any of the grasses studied. Is it possible that variability at each of the loci in a multilocus system is required for the system to operate? Are matters more complicated than was first thought? Certainly complexity in the genetic control of specificity in systems which were thought of as single-locus systems with sporophytic determination is much greater than was apparent. (Lewis, Verma & Zuberi, 1988; Zuberi & Lewis, 1988; Lundqvist, 1990*c*). Finally, there remains the hypothesis that incompatibility arose in the Poaceae without direct derivation from a previous system. Speculation about the origin of incompatibility systems has become a 'well worn path where one becomes involved in controversial and unresolvable problems' (Lewis, 1979).

It is to be hoped that solutions to the problems of the *S–Z* system will become apparent when there is more detailed knowledge of the *S* and *Z* and possibly of other gene products and of the structure of the genes

134   *D. L. Hayman*

themselves. Molecular biologists have been studying the single-gene gametophytic and sporophytic systems. In both systems, S-locus-specific glycoproteins have been identified in the style. The subject has been recently reviewed (Haring, Gray, *et al.*, 1990). The amino acid sequences of the S glycoproteins in these two systems are not indicative of homology. There are also considerable differences between the amino acid sequences of the products of alleles within each system. At present no allele-specific pollen product has been identified, so that the actual basis of compatibility *vs* incompatibility is still unknown.

No S-specific or Z-specific glycoproteins have been identified to date. Tan & Jackson (1988) were unable to identify genotype-related glycoprotein differences when comparing style and stigma extracts from plants of *Phalaris coerulescens* with known incompatibility genotypes. It may still be possible that glycoproteins are concerned in the S–Z system, but that they are neither highly expressed nor easily identified in the style and stigma.

Certainly the approach which provides the most hope of answering the problems specific to the S–Z system which have been raised in this review involves the utilization of the techniques of molecular biology. Diverse genetic stocks are already available in a number of species to aid in the dissection of the system. It is also possible that further analyses in the Poaceae may reveal the involvement of additional genetic factors.

References

Adams, M. W. (1953). Cross- and self-incompatibility in relation to seed-setting in *Bromus inermis*. *Botanical Gazette*, **115**, 95–105.
Annerstedt, I. & Lundqvist, A. (1967). Genetics of self-incompatibility in *Tradescantia paludosa* (Commelinaceae). *Hereditas*, **58**, 13–30.
Bateman, A. J. (1952). Self-incompatibility systems in Angiosperms. I. Theory. *Heredity*, **6**, 285–310.
Brandham, P. E. & Owens, S. J. (1978). The genetic control of self-incompatibility in the genus *Gasteria* (Liliaceae). *Heredity*, **40**, 165–169.
Brewbaker, J. L. (1957). Pollen cytology and self-incompatibility systems in plants. *Journal of Heredity*, **48**, 271–277.
Brewbaker, J. L. & Gorrez, D. D. (1967). Genetics of self-incompatibility in the monocot genera *Ananas* (Pineapple) and *Gasteria*. *American Journal of Botany*, **54**, 611–616.
Charlesworth, D. (1979). Some properties of populations with multi-locus homomorphic incompatibility systems. *Heredity*, **43**, 19–25.
Charlesworth, D. & Charlesworth, B. (1979). The evolution and breakdown of S-allele systems. *Heredity*, **43**, 41–56.
Connor, H. E. (1979). Breeding systems in grasses: a survey. *New Zealand Journal of Botany*, **17**, 547–574.

Cornish, M. A. Hayward, M. D. & Lawrence, M. J. (1979). Self-incompatibility in rye grass. I. Genetic control in diploid *L. perenne* L. *Heredity*, **43**, 95–106.

Cornish, M. A., Hayward, M. D. & Lawrence, M. J. (1980). Self-incompatibility in rye grass. III. The joint segregation of *S* and *PGI-2* in *Lolium perenne* L. *Heredity*, **44**, 55–62.

de Nettancourt, D. (1977). *Incompatibility in angiosperms*. Berlin/Heidelberg/New York: Springer Verlag.

Fearon, C. H., Hayward, M. D. & Lawrence, M. J. (1983). Self-incompatibility in rye grass. V. Genetic control, linkage and seed set in diploid *Lolium multiflorum* Lam. *Heredity*, **50**, 35–46.

Fearon, C. H., Hayward, M. D. & Lawrence, M. J. (1984*a*). Self-incompatibility in rye grass. VII. The determination of incompatibility genotypes in autotetraploid families of *Lolium perenne* L. *Heredity*, **53**, 403–413.

Fearon, C. H., Hayward, M. D. & Lawrence, M. J. (1984*b*). Self-incompatibility in rye grass. VIII. The mode of action in the pollen of autotetraploids of *Lolium perenne* L. *Heredity*, **53**, 415–422.

Fearon, C. H., Hayward, M. D. & Lawrence, M. J. (1984*c*). Self-incompatibility in rye grass. IX. Cross-compatibility and seed set in auto-tetraploid *Lolium perenne* L. *Heredity*, **53**, 423–434.

Gertz, A. & Wricke, G. (1989). Linkage between the incompatibility locus *Z* and a β-glucosidase in rye. *Plant Breeding*, **102**, 255–259.

Haring, V., Gray, J. E., McClure, B. A., Anderson, M. A. & Clarke, A. E. (1990). Self-incompatibility: a self-recognition system in plants. *Science*, **250**, 937–941.

Hayman, D. L. (1956). The genetical control of incompatibility in *Phalaris coerulescens*. *Australian Journal of Biological Sciences*, **9**, 321–331.

Hayman, D. L. & Richter, J. (1992). Mutations affecting self-incompatibility in *Phalaris coerulescens* Desf. (Poaceae). *Heredity*, **68**, (in press).

Heslop-Harrison, J. (1975). Incompatibility and the pollen-stigma interaction. *Annual Review of Plant Physiology*, **26**, 403–425.

Larsen, K. (1977). Self-incompatibility in *Beta vulgaris* L. I. Four gametophytic complementary *S*-loci in sugar beet. *Hereditas*, **85**, 227–248.

Leach, C. R. (1988). Detection and estimation of linkage for a co-dominant structural gene locus linked to a gametophytic self-incompatibility locus. *Theoretical and Applied Genetics*, **75**, 882–888.

Leach, C. R. & Hayman, D. L. (1987). The incompatibility loci as indicators of conserved linkage groups in the Poaceae. *Heredity*, **58**, 303–305.

Lewis, D. (1947). Competition and dominance of incompatibility alleles in diploid pollen. *Heredity*, **1**, 85–108.

Lewis, D. (1979). Genetic versatility of incompatibility in plants. *New Zealand Journal of Botany*, **17**, 737–744.

Lewis, D. & Crowe, L. K. (1958). Unilateral interspecific incompatibility in flowering plants. *Heredity*, **12**, 233–256.

Lewis, D., Verma, S. C. & Zuberi, M. I. (1988). Gametophytic-sporophytic incompatibility in the Cruciferae – *Rhaphanus sativus*. *Heredity*, **61**, 355–366.

Lewis, E. J., Humphreys, M. W. & Coton, M. P. (1980). Chromosome location of two isozyme loci in *Lolium perenne* using primary trisomics. *Theoretical and Applied Genetics*, **57**, 237–239.

Lundqvist, A. (1954). Studies on self-sterility in rye, *Secale cereale* L. *Hereditas*, **40**, 278–294.

Lundqvist, A. (1955). Genetics of incompatibility in *Festuca pratensis* Huds. *Hereditas*, **47**, 542–562.

Lundqvist, A. (1957). Self-incompatibility in rye. II. Genetic control in the autotetraploid. *Hereditas*, **43**, 467–511.

Lundqvist, A. (1958). Self-incompatibility in rye. IV. Factors relating to self-seeding. *Hereditas*, **44**, 193–256.

Lundqvist, A. (1961). A rapid method for the analysis of incompatibility in grasses. *Heredity*, **47**, 705–707.

Lundqvist, A. (1962a). The nature of the two-loci incompatibility system in grasses. I. The hypothesis of a duplicative origin. *Hereditas*, **48**, 153–168.

Lundqvist, A. (1962b). Self-incompatibility in diploid *Hordeum bulbosum* L. *Hereditas*, **48**, 138–152.

Lundqvist, A. (1964a). The nature of the two-loci incompatibility system in grasses. III. Frequency of specific incompatibility alleles in a population of *Festuca pratensis* Huds. *Hereditas*, **52**, 189–196.

Lundqvist, A. (1964b). The nature of the two-loci incompatibility system in grasses. IV. Interaction between the loci in relation to pseudo-compatibility in *Festuca pratensis* Huds. *Hereditas*, **52**, 221–234.

Lundqvist, A. (1965). Self-incompatibility in *Dactylis aschersoniana* Graebn. *Hereditas*, **54**, 70–87.

Lundqvist, A. (1968). The mode of origin of self-fertility in grasses. *Hereditas*, **59**, 413–426.

Lundqvist, A. (1969). The identification of the self-incompatibility alleles in a grass population. *Hereditas*, **61**, 345–352.

Lundqvist, A. (1975). Complex self-incompatibility systems in Angiosperms. *Proceedings of the Royal Society of London, B*, **188**, 235–245.

Lundqvist, A. (1990a). The complex *S*-gene system for control of self-incompatibility in the buttercup *Ranunculus*. *Hereditas*, **113**, 29–46.

Lundqvist, A. (1990b). Variability within and among populations in the 4-gene system for control of self-incompatibility in *Ranunculus polyanthemos*. *Hereditas*, **113**, 47–61.

Lundqvist, A. (1990c). One locus sporophytic *S*-gene system with traces of gametophytic control in *Cerastium arvense* spp. *strictum* (Caryophyllaceae). *Hereditas*, **113**, 203–215.

Lundqvist, A. (1991). Four-locus *S*-gene control of self-incompatibility made probable in *Lilium martagon* (Liliaceae). *Hereditas*, **114**, 57–63.

Martin, F. W. (1959). Staining and observing pollen tubes in the style by means of fluorescence. *Stain Technology*, **34**, 125–128.

McGraw, J. M. & Spoor, W. (1983a). Self-incompatibility in *Lolium* species. I. *Lolium rigidum* Gaud and *L. multiflorum* L. *Heredity*, **50**, 21–28.

McGraw, J. M. & Spoor, W. (1983b). Self-incompatibility in *Lolium* species. II. *Lolium perenne*. *Heredity*, **50**, 29–33.

Mogensen, H. L. (1990). Fertilisation and early embryogenesis. In *Reproductive Versatility in the Grasses*, ed. G. P. Chapman, pp. 76–99. Cambridge University Press.

Murray, B. G. (1974). Breeding systems and floral biology in the genus *Briza*. *Heredity*, **33**, 285–292.

Myers, W. M. (1948). Increased meiotic irregularity and decreased fertility accompanying inbreeding in *Dactylis glomerata*. *Journal of the American Society of Agronomy*, **40**, 249–258.

Osterbye, U. (1975). Self-incompatibility in *Ranunculus acris* L. I. Genetic interpretation and evolutionary aspects. *Hereditas*, **80**, 91–112.

Osterbye, U. (1986). Self-incompatibility in *Ranunculus acris* L. III. *S*-loci numbers and allelic identities. *Hereditas*, **104**, 61–73.

Owens, S. J. (1977). Incompatibility studies in the genus *Gibasis* section *Heterobasis* D. R. Hunt (Commelinaceae). *Heredity*, **39**, 365–371.

Schmidt, J. C., Seliger, P. & Schlegel, R. (1984). Isoenzyme als biochemische markerfactoren fur roggenchromosomen. *Biochemie und Physiologie der Pflanzen*, **179**, 197–210.

Spoor, W. & McGraw, J. M. (1984). Self-incompatibility in *Lolium*: A reply. *Heredity*, **53**, 239–240.

Tan, L. W. & Jackson, J. F. (1988). Stigma proteins of the two-loci self-incompatible grass *Phalaris coerulescens*. *Sexual Plant Reproduction*, **1**, 25–27.

Thorogood, D. & Hayward, M. D. (1991). The genetic control of self-compatibility in an inbred line of *Lolium perenne* L. *Heredity*, **67**, 175–182.

Thorogood, D. & Hayward, M. D. (1992). Self-compatibility in *Lolium temulentum* L.: its genetic control and transfer into *L. perenne* L. and *L. multiflorum* Lam. *Heredity*, **68**, 71–78.

Trang, Q. S., Wricke, G. & Weber, W. E. (1982). Number of alleles at the incompatibility loci in *Secale cereale* L. *Theoretical and Applied Genetics*, **63**, 245–248.

Watkins, A. E. (1925). Genetical and cytological studies in wheat. *Journal of Genetics*, **15**, 323–366.

Weber, W. E., Wricke, G. & Trang, Q. S. (1982). Genotypic frequencies at equilibrium for a multi-locus gametophytic incompatibility system. *Heredity*, **48**, 379–383.

Weimarck, A. (1968). Self-incompatibility in the Gramineae. *Hereditas*, **60**, 157–166.

Whitehouse, H. L. K. (1950). Multiple allelomorph incompatibility of pollen and style in the evolution of the Angiosperms. *Annals of Botany N.S.*, **14**, 199–216.

Wricke, G. & Wehring, P. (1985). Linkage between an incompatibility locus and a peroxidase isozyme locus (*Perx-7*) in rye. *Theoretical and Applied Genetics*, **71**, 289–291.

Zuberi, M. I. & Lewis, D. (1988). Gametophytic-sporophytic incompatibility in the Cruciferae – *Brassica campestris*. *Heredity*, **61**, 367–377.

4

Apomixis and evolution
G. P. Chapman

Meiosis and syngamy together fashion the genotypes of new individuals upon which natural selection can operate. The global distribution of grasses combined with their ability to survive in a range of ecological habitats makes them a conspicuously well-adapted family. Since apomixis is thought to be widespread among grasses how does one resolve the paradox of recombination foregone and ecological success? This prompts other questions such as whether apomixis is genuinely common among grasses and if apomixis is what it seems, namely the exclusion of recombination. While there is some understanding of how apomixis is inherited, to what extent can its origin and persistence be explained?

Many of the previous investigations of apomixis in the Poaceae concentrated upon the *Poa pratensis* complex (Nielsen, 1946*a,b*; Akerberg & Bingefors, 1953; Muntzing, 1954; Julen, 1960; Hovin *et al.*, 1976 and Kellogg, 1987, for example) and have perhaps conditioned how we view apomixis more generally. So successful and widely adapted is *P. pratensis* and so versatile its seed reproductive biology that the species appears more to luxuriate in apomixis than to depend crucially upon it as a means of survival. It is probably realistic to argue that for *P. pratensis* we understand its apomixis better cytologically than we do ecologically. If, therefore, we turn attention to a more diverse group of grass apomicts, we may achieve both a wider understanding and, eventually, a new perspective on the remarkable situation that exists in *P. pratensis* and its closer relatives.

Taxonomic distribution of apomixis

How widespread is grass apomixis? Table 4.1 augments the data originally provided by Brown & Emery (1958) and follows the arrangement of Clayton & Renvoize (1986). Generic descriptions are, of

Table 4.1. *Distribution of apoximis among grasses*

Subfamily	Tribe	Subtribe	Genus	References
Bambusoideae			――――― No known examples ―――――	
Pooideae	Nardeae		*Nardus*	Rychlewski (1967)
	Poeae		*Poa*	Nielsen (1946*a,b*)
				Smith *et al.* (1946)
	Aveneae	Phalaridineae	*Hierochloë*	Norstog (1963)
				Weimark (1967)
		Alopecurinae	*Calamagrostis*	Green (1984)
				Tateoka (1984)
	Triticeae		*Agropyron*	Hair (1956)
			Elymus	Crane & Carman (1987)
			Hordeum	Finch (1983)
Centothecoideae			――――― No known examples ―――――	
Arundinoideae	Arundineae		*Cortaderia*	Philipson (1978a)
			Danthonia	Philipson (1986)
			Lamprothyrsus	Connor (1987)
Chloridoideae	Pappophoreae		*Schmidtia*	Brown & Emery (1958)
	Eragrostideae	Uniolinae	*Fingerhuthia*	Brown & Emery (1958)
		Eleusininae	*Eragrostis*	Voight & Bashaw (1976)
				Vorster & Liebenberg (1984)
	Cynodonteae	Sporobolinae	*Muhlenbergia*	Morden & Hatch (1986)

Table 4.1 (*cont.*)

Subfamily	Tribe	Subtribe	Genus	References
		Chloridinae	*Chloris*	Brown & Emery (1958)
		Boutelouinae	*Bouteloua*	Bierzychudek (1985)
			Hilaria	Brown & Emery (1958)
Panicoideae	Paniceae	Setariinae	*Brachiaria*	Brown & Emery (1958)
			Eriochloa	Brown & Emery (1958)
			Panicum	Hanna et al. (1970)
				Usberti & Jain (1978)
				Savidan (1983)
			Paspalum	Bashaw & Holt (1958)
				Quarin (1986)
				Burson (1987)
				Emery (1957)
				Brown & Emery (1958)
			Setaria	Brown & Emery (1958)
			Tricholaena	Pritchard (1970)
			Urochloa	
		Cenchrinae	*Anthephora*	Connor (1981)
			Cenchrus	Bashaw (1962)
			Pennisetum	Dujardin & Hanna (1983)
	Andropogoneae	Saccharinae	*Saccharum*	Bremer (1961)
		Sorghinae	*Bothriochloa*	Harlan et al. (1964)
				Dewald et al. (1985)

Subtribe	Genus	Reference
	(*Capillipedium* can be included with *Bothriochloa*)	Celarier & Harlan (1957)
	Dichanthium	Harlan & de Wet (1963*a*)
		de Wet (1968)
		Knox & Heslop-Harrison (1963)
		Bhanwra & Choda (1981)
	Eremopogon	
	(Now included with *Dichanthium*)	Hanna *et al.* (1970)
	Sorghum	Murty *et al.* (1985)
Ischaeminae	*Apluda*	Brown & Emery (1958)
		Murty (1973)
Andropogoninae	*Shizachyrium*	Carman & Hatch (1982)
Anthistiriinae	*Heteropogon*	Tothill & Knox (1968)
	Hyparrhenia	Brown & Emery (1958)
	Themeda	Brown & Emery (1958)
		Birari (1980)
Tripsacinae	*Tripsacum*	Farquharson (1955)
		de Wet *et al.* (1981)
Coicinae	*Coix*	Brown & Emery (1958)
		Venkateswarlu & Rao (1975)

Amongst 40 tribes, apomixis has been recorded in 10.
Within a tribe, apomixis can be concentrated within subtribes.
Within the tribe Andropogoneae, of 11 subtribes 7 show apomixis.

course, not dependent on reproductive behaviour but conditioned by a range of characters. This can influence our view of apomixis. For example in the tribe Paniceae, subtribe Setariinae, the genus *Urochloa* is entirely apomictic and perhaps therefore a 'terminal' genus. If, however, *Urochloa* were merged with *Brachyaria*, this latter genus having both sexual and apomictic species is made to appear more reproductively diverse.

Present knowledge indicates that apomixis is absent from the Bambusoideae and Centothecoideae. Among other subfamilies, it occurs in some tribes only, being absent from most of the Pooideae and Arundinoideae, about half the Chloridoideae and most of the Panicoideae. There are, however, notable concentrations. Subtribe Setariinae has seven genera wholly or partly apomictic. The tribe Andropogoneae has seven of its subtribes with apomixis or a total of 12 genera.

Although therefore the *Poa pratensis* complex is conspicuous in *Poa*, it is atypical of most of that subfamily and has few, if any, close parallels among other grasses. Apomixis is known among 10 of the 40 tribes and our conclusions about its significance relate, therefore, to a minority of grasses.

Many genera remain unexplored from this viewpoint and some that are exclusively sexual, so far as is known, may yield exceptional individuals eventually. Even so, Table 4.1 probably represents a satisfactory assessment of the extent of apomixis within the Poaceae.

Alternative forms of apomixis

Apospory, diplospory and pseudogamy were described in detail by Bashaw & Hanna (1990) in the companion to this volume. Suffice to mention here that apospory requires the development of an unreduced embryo sac, usually four-nucleate, from a nucellar cell and diplospory the non-reduction of a megaspore mother cell to provide a four-nucleate or an eight-nucleate (asexual) embryo sac. Pseudogamy is the modification of double fertilization whereby only the male gamete (i.e. the 'associate cell' in the terminology of Knox & Singh 1990) is involved in nuclear fusion. This is with the central cell to form the primary endosperm cell. The egg cell is thus stimulated to develop parthenogenetically. Figure 4.1 illustrates apospory in *Pennisetum squamulatum*.

Apospory is the more widespread. Diplospory is reported from (cf. Table 4.1 for references) *Nardus, Poa, Hierochloe, Calamagrostis, Agropyron, Elymus, Eragrostis, Saccharum* and *Tripsacum*, but it does

Fig. 4.1. Apospory in *Pennisetum squamulatum*. The illustration shows three embryo sacs in one ovule, photographed in two planes. The preparation utilized interference microscopy of cleared tissues (Young *et al.*, 1979) and avoided the need for microtome sectioning. In this species, an obligate apomict, the embryo sacs arise from nucellar cells. Solid arrows: embryo sacs; white arrows: egg apparatus/polar nuclei; hollow black arrows: (a) ovule boundary, (b) inner wall of ovary. Photo courtesy of N. Busri.

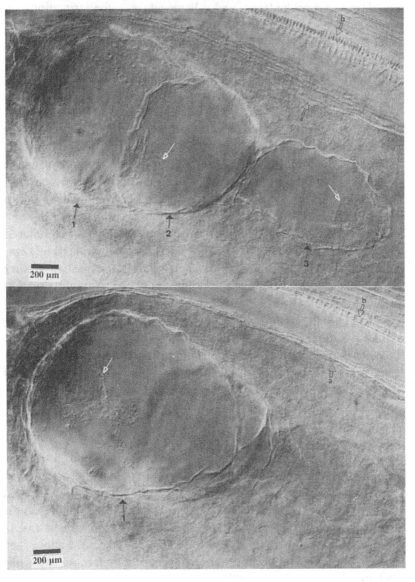

not follow that reports of diplospory automatically preclude the possibility of apospory. *Elymus* can operate two versions of diplospory that of '*Taraxacum*' and of '*Antennaria*', Crane & Carmen (1987).

ARUNDINOIDEAE, ARUNDINEAE, CORTADERIA

In contrast to those just mentioned, drawn from the subfamilies Pooideae, Chloridoideae and Panicoideae, the situation in Arundinoideae is different and thought-provoking. *Cortaderia jubata*, studied by Philipson (1978) is an obligate apomict having aborted stamens that produce no pollen. The archesporial cell disintegrates and somatic embryo sacs arise in the nucellus and two or three that mature intertwine. Each sac contains usually six nuclei, namely two antipodals, two polar nuclei, one synergid and one egg. The polar nuclei fuse and subsequently divide to provide endosperm. Either before or after this the egg cell begins embryogenesis. A similar pattern of apomixis is reported for *C. rudiuscula*. Philipson (1978) noted that embryogenesis began surprisingly early, namely, by the time of panicle emergence.

Cortaderia shows gynodioecism (female and bisexual plants) and in these two species apomixis is independent of pseudogamy but in addition is, curiously, combined with cleistogamy, a situation considered unlikely by Stebbins (1950). Connor (1987) thought apomixis in *Lamprothyrsus* similar to that in *Cortaderia*. *Danthonia spicata*, another arundinoid grass, was examined by Philipson (1986). Here the situation seems more equivocal. Panicles contain both chasmogamous and cleistogamous florets and among the latter pollen viability is appreciably reduced. Several somatic embryo sacs can occur in an ovule and at least occasional apomixis seems likely. The possibility, however, of pseudogamy is not excluded and *Cortaderia* among arundinoid grasses provides the only substantiated case of autogamous, cleistogamous apomixis.

PANICOIDEAE, ANDROPOGONEAE (COICINAE), COIX

Aside from common apospory and the diplospory in *Tripsacum*, *Coix* provides an interesting contrast for this subfamily. Reported by Venkateswarlu & Rao (1975), *Coix aquatica* is both a diploid ($2n=10$) and an autogamous apomict. *Coix* has unusual inflorescences that make it easy to emasculate, has several recognized variants and *C. aquatica* will hybridize with *C. lachryma-jobi* (Job's tears), a minor food grain of SE Asia. Experimentally, it appears therefore an attractive system.

POOIDEAE, POEAE, *POA PRATENSIS*

Based on early papers, Table 4.2 summarizes the cytological understanding of this complex. Later papers add detail but do not appreciably alter the situation. See Muntzing (1954) for *P. alpina*, Akerberg & Bingefors (1953) for *P. pratensis* × *P. alpina* and Julen (1960) and Grazi *et al.* (1961) for studies of irradiation. A description of commercial breeding practice is provided by Bashaw & Funk (1987).

If one adds what is known of the cytology and embryology to that of taxonomic distribution, it emerges for grasses that apospory is more common, diplospory much less so and that *Cortaderia* (together with *Lamprothyrsus*), *Coix* and *Poa* provide, for different reasons, situations exceptional among grasses. If therefore some widely applicable ecological or evolutionary understanding of apomixis were sought it should be, primarily, among aposporous types.

Ecology

Apomixis is regarded, often, as one ingredient in a polyploid complex as with *Calamagrostis hakonensis* (diplosporous) for example studied by Tateoka (1984) at Nynpu in Japan. A similar situation occurs in the *Dichanthium intermedium* complex (aposporous) studied by Harlan & de Wet (1963*a*). In these and other studies, of which there are many, apomixis resolves the problem of seed production for subfertile polyploids and perpetuates the *status quo*. If, however, apomixis is studied at one level of ploidy a different situation can emerge.

Panicoideae, Andropogoneae, (Sorghinae), *Dichanthium*

The following discussion is based on an article by Knox (1967). *Dichanthium aristatum*, native to southern Africa and India, was previously, over many years, introduced into Australia several times. One strain CPI 14366 had been shown by Knox & Heslop-Harrison (1963) to be a versatile apomict. This was planted at six stations in or near Australia over 27% of latitude from Port Moresby (9° 29′) to Bergami (36° 28′). On the basis of a score of more than 1600 embryo sacs, from north to south the proportion of sexual embryo sacs increased. (The situation is not simple since where apomixis is more frequent, there are more embryo sacs per ovule.) Such a situation might be ascribed to differences in temperature, rainfall or photoperiod. By reference to a phytotron experiment, it was subsequently shown that with a photoperiod in excess of 14 hours, the apomictic embryo sacs were 60% of the total, while with less than 14 hours, they were more than 90%.

Table 4.2. *Summary of embryo sac behaviour in* Poa pratensis

1. Normal fertilization occurs to a single embryo sac yielding a single diploid seedling.
 	
2. Occasionally, two megaspore mother cells develop corresponding rows of tetrads (i.e. two meiotic events). In each case three cells abort and two genetically dissimilar embryo sacs result. Separate fertilizations would give diploid twins.
	3. Within a single embryo sac, fertilization of the egg sometimes accompanied by embryogenic development of a synergid resulting in diploid haploid twins.
	
	   If two embryo sacs have developed, fertilization of one egg and haploid development of a synergid in the other embryo sac would also give diploid haploid twins but genetically could not be distinguished from the preceding case.
		4. Parallel development of a gametophytic and a somatic embryo sac observed having haploid and diploid nuclei, respectively. Nielsen surmized the following possibilities.
		
		   The former if fertilized and if both develop would give a diploid diploid twin, one a sexual hybrid and one a maternal copy.
			5. Only somatic embryo sacs completed their development and were subsequently fertilized. Assuming embryogenesis, triploid triploid twins would result.
			
			   Nielsen showed that some diploid progeny was diverse indicating a sexual origin, and some uniform arising apomictically.
			
			   Later Smith *et al.* (1946) found about 84% stable and 16% aberrant types from some thousands of plants. Assuming a diploid can both arise from fertilization and retain the property of apomixis it is possible for a plant to assemble new recombinants and propagate them vegetatively via the seed.

Based on Nielsen (1946*a*, *b*) and Smith, Nielsen & Ahlgren (1946).

Were sampling confined to any one of the six stations, the proportion of sexual embryo sacs need not have had any particular significance. This study showed that apomixis is not a simple alternative to sexuality but an integral part of an elaborate response to environmental change that has parallels in, for example *Bothriochloa decipiens*, which is cleistogamous in short days and chasmogamous in long days (Heslop-Harrison, 1961). Knox (1967) drew a comparison between *D. aristatum* and aphids responding to a seasonal control of the switch from sexuality to apomixis.

Darlington's 'blind alley' reassessed

Hair (1956) examined four populations of *Agropyron scabrum*. Population A was fully sexual, B a facultative apomict, C predominantly apomictic and D an obligate apomict. For populations B, C and D ovule meiosis was disrupted in that crossing over occurred but chromosome number reduction did not, the process being referred to as 'subsexual'. The megaspore mother cell subsequently elaborated a diploid embryo sac for which pseudogamy was required to trigger apomictic seed development.

Darlington (1939) made the remark, often since repeated, that apomixis is 'an escape from sterility but an escape into a blind alley of evolution'. For *Agropyron scabrum* and *Calamagrostis hakonensis* that might be true. For *Dichanthium aristatum* (and similar cases that might emerge) it seems a misjudgement of a subtle and responsive reproductive mechanism, the key point being that, as with aphids, apomixis is not absolute and that even in the shortest day situation this grass does not forego the option of seasonally modulated recombination. Rather, one might suspect that polyploid complexes have, in some instances, driven apomixis into a blind alley.

Evolution and origins

Dichanthium aristatum suggests that apomixis did not arise by some abrupt, once and for all, dislocation of the sexual system. What needs eventually to be explained is how cells near to the archesporium can adopt its potential and become quasi-gametophytic even to the point of receiving a male gamete (pseudogamy), and recreating in its (unfertilized) egg cell the capacity for embryogenesis. The proximity to the archesporial cell suggests some kind of dependence upon it, but the ability of a nearby cell to receive a gamete and differentiate an embryo (or to remain a nucellar cell) implies remarkable developmental plasticity. It is in this milieu that the mutants that might inaugurate apomixis would have to operate. If it is assumed that apospory and diplospory are genuinely absent from bamboos (possibly in floral terms the most primitive subfamily), then the forms of apomixis about which we know might postdate the origin of the grasses and parallel the situation for C_4 photosynthesis where we surmise one or several innovations. Since apospory and diplospory are different each would require a separate innovation. Conversely, neither apospory nor diplospory is radically different from analogues elsewhere in the flowering plants and could conceivably trace directly from their antecedents. This, though, is speculative.

Bashaw & Hanna (1990) invoked a simple two-gene model to explain the inheritance of apomixis but an elaboration would be needed to account for the difference between facultative and obligate apomicts and for the effect of environment. Marshall & Brown (1981) provided a mathematical model, with alternatives, to account for the evolution and establishment of apomixis, but conceded the unlikelihood of the simultaneous mutation at two loci that would be needed to establish even a simple sexual vs apomictic system not incorporating the modifications just mentioned. A curious feature of apomixis is its association with polyploidy and it is possible that 'cryptic' mutants established but not expressed at the diploid level come into a favourable two-gene co-existence that permits expression in (say) an allotetraploid or higher ploid situation. Neither *Triticum* nor *Avena*, for example, are known to contain apomicts but a *Triticum* × *Avena* cross set pseudogamous seeds (Kruse, 1969). A latent capacity for apomixis expressed itself when these (hexaploid) genera interacted.

It is at the diploid level, perhaps, that the origins of a genetic mechanism for apomixis should be sought. Three cases, known to the author, merit consideration. *Nardus stricta* is a monospecific moorland genus distributed in Europe and is diplosporous (Rychlewski, 1967). Since $2n=26$, its diploidy is perhaps secondary, via a $4x=2n=28$ ancestor, as seems likely from its position within the Pooideae. More promising, however, is *Coix aquatica*, a diploid ($2n=10$) autogamous apomict described by Venkateswalu & Rao (1975), which has apparently received no further study. Finally, Finch (1983) studied the *hap* gene in *Hordeum* ($2n=14$) in which apomixis appears not to have been recorded previously. From eight emasculated spikes, 147 florets were hand pollinated with *hap* pollen. Although it was not possible to determine the exact proportion of apomicts arising, in at least 30% of cases there were formed a triploid endosperm (a result of fertilization) and a haploid embryo. Apospory was not reported and Finch considered apomixis to be by facultative pseudogamy and haploid parthenogensis. Some, therefore, but not all of the ingredients for apomixis are in place. Pseudogamy occurred but not the non-reduction of the megaspore mother cell nor its replacement by a diploid nucellar cell. How the situation might change at the tetraploid level is unknown but this model seems the most promising currently available. In nature, its chances of survival and modification would presumably be higher in a perennial, since it would depend less crucially upon seed propagation.

None of the foregoing, regrettably, sheds light on the seasonally responsive system of *Dichanthium aristatum*.

Evolution and survival

Representatives of the Arundinoideae including *Arundo* and *Molinia* are thought to have originated north of the Tethys Sea and to be among the oldest genera in the subfamily (Conert, 1987). *Cortaderia* is among the most recent. Renvoize & Clayton (this volume) view the Arundinoideae as 'relict' and tending to be confined more to the southern hemisphere (although *Aristida* is recognized as more dynamic). The Panicoideae, Chloridoideae and Pooideae, by contrast, are considered to be more recent and successful. Connor (1979) believed that the genus *Lamprothyrsus* consists only of female plants (a situation which would change if it were combined with *Cortaderia*) and reproduces by autogamous somatic apospory. If, long term, the arundinoids genuinely are in decline, their apomixis seems likely only to accelerate the process.

Apomixis among chloridoids and panicoids, for example, seems as readily associated here with success as it was in the previous case with decline. *Chloris*, *Eragrostis* and *Paspalum* are large genera, ecologically aggressive, and known to possess apomicts that may, in part, account for their success.

What seems more significant in understanding the role of apomixis is an association, firstly, with polyploidy and, secondly, with perennialism, the former because among polyploids apomixis finds expression (for whatever reason), and the latter because, among perennials, apomixis has opportunity. Annuals can experiment with recombination but they can hardly experiment so readily with sex.

Sexual variants among apomicts

Bashaw & Hanna (1990) reiterated that *Cenchrus ciliaris* sexual progeny segregating among their crosses tend to be at a disadvantage – either through subsequent inbreeding or because of ecological competition in a pasture. The situation can be read in two ways. Apomicts survive preferentially because of 'hybrid vigour' or sexual individuals are *inherently* weakly and, for practical purposes, transient. In the latter case the sexual variant has almost the status of a drone honeybee – useful in causing genetic turnover but otherwise entirely expendable. What matters for *C. ciliaris* is the existence of robust genotypes that, once assembled, can be perpetuated largely unchanged. The small proportion of sexual variants that occur is, in nature, perhaps a guarantee against elimination through ecological change and, under domestication, an opportunity for the breeder.

On this reckoning apomixis is no mere salvage operation that perpetuates what sexual means cannot. Rather, the emphasis has changed to the point where sexual variants are subordinate to the interests of a (facultative) apomict.

Some provisional conclusions

The literature gives, as shown in Table 4.1, some indication of how apomixis is scattered through the family. The cytological and embryological details provide, in these aspects, an adequate body of knowledge.

Ecologically, the problem is more complex and apomixis is of different consequence for different grasses. In the *Poa pratensis* complex it seems, despite its variety, to be of uncertain significance. One must try not to be dazzled by *Poa* reproductive pyrotechnics so that one may understand what happens in more straightforward cases among other genera. Where apomixis is a feature of complexes such as *Calamagrostis*, it seems to salvage a situation created by sexual mismating leading otherwise to infertility, and thus justifies Darlington's 'blind alley' epigram. Even so, such complexes seem to cause apomixis to surface from the recesses of the genome and so hardly explain how its potential came to be there in the first place.

Apomixis is only one among several features of the genetic system and its influence on a given species might be marginal as in *Sorghum* or more consequential to the point where it dominates as is the case in *Urochloa*. In *Cortaderia* and *Lamprothyrsus*, apomixis in these relict genera is probably a serious liability. For ecologically aggressive

Dichanthium species, apomixis is presumably an asset that mass produces, and thus exploits, the inventiveness of the sexual system (Harlan & de Wet 1963*b*).

Where, as in the case of *Dichanthium aristatum*, the switch from sexuality to apomixis is under an environmental stimulus (but in which neither condition is totally excluded), the conclusion might be that the reproductive system is subtly responsive and highly adapted and akin to the cleistogamy/chasmogamy changes that are similarly conditioned. Changing Darlington's analogy from the roadway to the vehicle, therefore we might say *D. aristatum* has the option of a four-wheel drive when conditions are appropriate.

Apomixis, it will be recalled, can be experimentally induced in grasses such as barley (Finch, 1983) and wheat (Kruse, 1969), where otherwise it does not occur, although the situation is not self perpetuating. Perhaps therefore its potential for expression is much wider than its occurrence suggests. It may be that apomixis is of little use to fertile diploids, is a potentially valuable asset to balanced fertile allotetraploids (where it may take a small or a leading role) and for polyploid complexes is merely one item that serves to prolong the ensuing cytogenetic confusion by ensuring the survival of partially fertile or infertile genotypes.

Particularly when one examines the range of chloridoid and panicoid apomicts, it is apparent that they include grasses which are increasingly recognized now as important in stabilizing fragile environments at low latitudes. Eventually, one assumes, we shall need to manipulate and therefore to understand their genetic systems as part of a wider management strategy.

References

Akerberg, E. & Bingefors, S. (1953). Progeny studies in the hybrid *Poa pratensis* × *Poa alpina*. *Hereditas*, **39**, 125–136.

Bashaw, E. C. (1962). Apomixis and sexuality in buffelgrass. *Crop Science*, **2**, 412–415.

Bashaw, E. C. & Funk, C. R. (1987). Apomictic grasses. In *Principles of cultivar development, vol. 2, Crop Species*, ed. W. R. Fehr, pp. 40–82. New York/London: Macmillan.

Bashaw, E. C. & Hanna, W. W. (1990). Apomictic reproduction. In *Reproductive versatility in the grasses*, ed. G. P. Chapman, pp. 100–130. Cambridge: Cambridge University Press.

Bashaw, E. C. & Holt, E. C. (1958). Megasporogenesis, embryo sac development and embryogenesis in Dallisgrass *Paspalum dilatum* Poir. *Agronony Journal*, **50**, 753–756.

Bhanwra, R. K. & Choda, S. P. (1981). Apomixis in *Eremopogon foveolatus* (Gramineae). *Nordic Journal of Botany*, **1**, 97–101.

Bierzychudek, P. (1985). Patterns in plant parthenogenesis. *Experientia* **41**, 1255–1264.

Birari, S. P. (1980). Apomixis and sexuality in *Themeda* Forssk. at different ploidy levels (Gramineae). *Genetica*, **54**, 133–139.

Bremer, G. (1961). Problems in breeding and cytology of sugar cane. IV. The origin of increase of chromosome number in species hybrids of *Saccharum*. *Euphytica*, **10**, 325–342.

Brown, W. V. & Emery, W. H. P. (1958). Apomixis in the Gramineae: Panicoideae. *American Journal of Botany*, **45**, 253–263.

Burson, B. L. (1987). Pollen germination, pollen tube growth and fertilization following self and interspecific pollination of *Paspalum* species. *Euphytica*, **36**, 641–650.

Carman, J. G. & Hatch, S. L. (1982). Aposporous apomixis in *Schizachyrium* (Poaceae: Andropogoneae). *Crop Science*, **22**, 1252–1255.

Celarier, R. P. & Harlan, J. R. (1957). Apomixis in *Bothriochloa*, *Dichanthium* and *Capillipedium*. *Phytomorphology*, **7**, 93–102.

Clayton, W. D. & Renvoize, S. A. (1986). *Genera graminum: grasses of the world*. Kew Bulletin, Additional Series 13, Royal Botanic Gardens, Kew, London.

Conert, H. J. (1987). Current concepts in the systematics of the Arundinoideae. In *Grass systematics and evolution*, ed. T. R. Soderstrom, K. W. Hilu, C. S. Campbell & H. G. Barkworth), pp. 239–250. Washington DC: Smithsonian Institution Press.

Connor, H. E. (1979). Breeding systems in the grasses: a survey. *New Zealand Journal of Botany*, **17**, 547–574.

Connor, H. E. (1981). Evolution of reproductive systems in the Gramineae. *Annals of the Missouri Botanical Garden*, **68**, 48–74.

Connor, H. W. (1987). Reproductive biology in the grasses. In *Grass systematics and evolution*, ed. T. R. Soderstrom, K. W. Hilu, C. S. Campbell & M. E. Barkworth, pp. 117–132. Washington DC: Smithsonian Institution Press.

Crane, C. F. & Carman, J. G. (1987). Mechanisms of apomixis in *Elymus rectisetus* from Eastern Australia and New Zealand. *American Journal of Botany*, **74**, 477–496.

Darlington, C. D. (1939). *The evolution of genetic systems*. Cambridge: Cambridge University Press.

de Wet, J. M. J. (1968). Diploid–tetraploid haploid cycles and the origin of variability in *Dichanthium* agamospecies. *Evolution*, **22**, 394–397.

de Wet, J. M. J., Timothy, D. H., Hilu, K. W. & Fletcher, G. B. (1981). Systematics of South Americum *Tripsacum* (Gramineae). *American Journal of Botany*, **68**, 269–276.

Dewald, C. L., Sims, P. L., Coyne, P. J. & Berg, W. A. (1985). Registration of W. W. Spar Bluestem. *Crop Science*, **25**, 707.

Dujardin, M. & Hanna, W. V. (1983). Apomictic and sexual pearl millet×*Pennisetum squamulatum* hybrids. *Journal of Heredity*, **74**, 277–279.

Emery, W. H. P. (1957). A study of reproduction in *Setaria macrostachya* and its relatives in the southwestern United States and Northern Mexico. *Bulletin of the Torrey Botanical Club*, **84**, 106–121.

Farquharson, L. I. (1955). Apomixis and polyembryony in *Tripsacum dactyloides. American Journal of Botany*, **42**, 778–784.

Finch, R. A. (1983). The *Hap* gene causes facultative pseudogamy in barley. *Barley Genetics Newsletter*, **13**, 4–6.

Grazi, F., Umaerus, M. & Akerberg, E. (1961). Observations on the mode of reproduction and embryology of *Poa pratensis. Hereditas*, **47**, 489–541.

Green, C. W. (1984). Sexual and apomictic reproduction in *Calamagrostis* (Gramineae) from Eastern North America. *American Journal of Botany*, **71**, 285–293.

Hair, J. B. (1956). Subsexual reproduction in *Agropyron. Heredity*, **10**, 129–160.

Hanna, W. W., Schertz, K. F. & Bashaw, E. C. (1970). Apospory in *Sorghum bicolor* (L.) Moench. *Science*, **170**, 338–339.

Harlan, J. R., Brooks, M. H., Borgaonker, D. S. & de Wet, J. M. J. (1964). Nature and inheritance of apomixis in *Bothriochloa* and *Dicharrthium. Botanical Gazette*, **125**, 41–46.

Harlan, J. R. & de Wet, J. M. J. (1963*a*). Role of apomixis in the evolution of the *Bothriochloa–Dichanthium* complex. *Crop Science*, **3**, 314–316.

Harlan, J. R. & de Wet, J. M. J. (1963*b*). The compilospecies concept. *Evolution*, **17**, 497–501.

Heslop-Harrison, J. (1961). The function of the glume pit and the control of cleistogamy in *Bothriochloa docipiens* (Hack) C. E. Hubbard. *Phytomorphology*, **11**, 378–383.

Hovin, A. W., Berg, C. C., Bashaw, E. C., Buckner, R. C., Dewey, D. R., Dunn, G. M., Hoveland, C. S., Rineker, C. M. & Wood, G. M. (1976). Effects of geographic origin and seed production environments on apomixis in Kentucky bluegrass. *Crop Science*, **16**, 639–638.

Julen, G. (1960). The effect of X-raying on the apomixis in Kentucky bluegrass. *Genetica Agraria*, **13**, 60–65.

Kellogg, E. A. (1987). Apomixis in the *Poa secunda* complex. *American Journal of Botany*, **74**, 1431–1437.

Knox, R. B. (1967). Apomixis, seasonal and population differences in a grass. *Science*, **157**, 325–326.

Knox, R. B. & Heslop-Harrison, J. (1963). Experimental control of aposporous apomixis in a grass of the Andropogoneae. *Botaniska Notiser*, **116**, 127–141.

Knox, R. B. & Singh, M. B. (1990). Reproduction and recognition phenomena in the Poaceae. In *Reproductive versatility in the grasses*, ed. G. P. Chapman, pp. 220–239. Cambridge: Cambridge University Press.

Kruse, A. (1969). Intergeneric hybrids between *Triticum aestivum* L. (v. Kogall 2*n*=42) and *Avena sativa* L. (v. Stal. 2*n*=42) with pseudogamous seed formation. *Preliminary Report in the Royal Vet. & Agric. Coll. Yearbook*, p. 188. Copenhagen.

Marshall, D. R. & Brown, A. H. D. (1981). The evolution of apomixis. *Heredity*, **47**, 1–15.

Morden, C. W. & Hatch, S. L. (1986). Vegetative apomixis in *Muhlenbergia repens* (Poaceae: Eragrostideae). SIDA, *Contrib. Bot.*, **11**, 282–285.

Muntzing, A. (1954). The cytological basis of polymorphism in *Poa alpina*. *Hereditas*, **40**, 459–516.

Murty, U. R. (1973). Polyploidy and apomixis in *Apluda mutica* L. var *aristata* (L.) Pilger. *Cytologia*, **38**, 347–356.

Murty, U. R., Kirit, P. B. & Bharali, S. (1985). The concept of vybrids in *Sorghum*. II. Mechanism of frequency of apomixis under cross pollination. *Zeitschrift für Pflanzenzuchtung*, **95**, 113–117.

Nielsen, E. L. (1964*a*). Breeding behaviour and chromosome numbers in progenies from twin and triplet plants of *Poa pratensis*. *Botanical Gazette*, **108**, 26–40.

Nielsen, E. L. (1946*b*). The origin of multiple macrogametophytes in *Poa pratensis*. *Botanical Gazette*, **108**, 41–50.

Norstog, K. (1963). Apomixis and polyembryony in *Hierochloe odorata*., *American Journal of Botany*, **50**, 815–821.

Philipson, M. N. (1978). Apomixis in *Cortaderia jubata* (Gramineae). *New Zealand Journal of Botany*, **16**, 45–59.

Philipson, M. N. (1986). A reassessment of the form of reproduction in *Danthonia spicata* (L.) Beauv. *New Phytologist*, **102**, 231–243.

Pritchard, A. J. (1970). Meiosis and embryo sac development in *Orochloa mosambicensis* and three *Paspalum* species. *Australian Journal of Agricultural Research*, **21**, 649–652.

Quarin, C. L. (1986). Seasonal changes in the incidence of apomixis of diploid, triploid and tetraploid plants of *Paspalum cromyorrhizon*. *Euphytica* **35**, 515–522.

Rychlewski, J. (1967), Karyological studies on *Nardus stricta* L. *Acta biologica Cracov Séries botanique*, **10**, 55–72.

Savidan, Y. H. (1983). Genetics and utilisation of apomixis for the improvement of guinea grass (*Panicum maximum* Jacq.) In *Proceedings of the 14th International Grassland Congress*, ed. J. A. Smith & V. W. Hayes. Boulder, Colorado: Westview Press.

Smith, D. C., Nielsen, E. L. & Ahlgren, V. (1946). Variation in cytotypes of *Poa pratensis*. *Botanical Gazette*, **108**, 143–166.

Stebbins, G. L. (1950). *Variation and evolution in plants*. New York: Columbia University Press. 643 pp.

Tateoka, T. (1984). *Calamagrostis hakonensis* (Poaceae): distribution and differentation of cytotypes. *Botanical Magazine*, Tokyo, **97**, 247–270.

Tothill, J. C. & Knox, R. B. (1968). Reproduction in *Heteropogon Australian Journal of Agricultural Research*, **19**, 869–878.

Usberti, J. A. & Jain, S. K. (1978). Variation in *Panicum maximum*: a comparison of sexual and asexual populations. *Botanical Gazette*, **129**, 112–116.

Venkateswarlu, J. & Rao, P. N. (1975). Apomixis in *Coix aquatica* Rotb. *Annals of Botany*, **39**, 1131–1136.

Voight, P. W. & Bashaw, E. C. (1976). Facultative apomixis in *Eragrostis curvula*. *Crop Science*, **16**, 803–806.

Vorster, T. B. & Liebenberg, H. (1984). Classification of embryo sacs in the *Eragrostis curvula* complex. *Bothalia*, **15**, 167–174.

Weimark, G. (1967). Apomixis and sexuality in *Hierochloe australis* and in Swedish *H. odorata* on different polyploid levels. *Botaniska Notiser*, **120**, 209–235.

Young, B. A., Sherwood, R. T. & Bashaw, E. C. (1979). Cleared pistil and thick-sectioning techniques for detecting aposporous apomixis in grasses. *Candian Journal of Botany*, **57**, 1668–1672.

Part II

DOMESTICATION

Plate 1

(a) Bamboo scaffolding, *Gigantochloa levis*, shown in Guangzhou (Canton). The structure is flexible and reputed to resist typhoon stress more effectively than steel scaffolding.

(b) Gardener's soil and cuttings basket from southern China woven from *Bambusa textilis*.

(c) Artist's paintbrush holder from Japan carved from a single node and internode of a large-stemmed bamboo. (From Austin, R., Ueda, K. & Levy, D. (1970). *Bamboo*. New York: Weatherhill Publications, with permission.)

(d) *Dichanthium annulatum*, an important pasture grass at low latitudes and one that is an ingredient of the *D. intermedium* compilospecies. See chapter 4.

(e) *Lolium perenne*, a pooid grass, tribe Poeae, in which self-incompatibility has been studied. In temperate countries, among grasses, this species is often the principal cause of hay fever.

(f) Inflorescence of *Panicum turgidum* an important grass of the desert surviving on 30–250 mm of rainfall. It colonizes wind-blown sand dunes, often while they are still unstable.

Plate 2

(a) Fruit of *Sorghum bicolor* (guinea-Kaffir type) showing the swollen caryopses of the sessile spikelet and the persistently green spears of pedicellate (vestigial) spikelets.

(b) Fruit of *Buchloe dactyloides*, buffallograss, a dioecious chloridoid grass important on the Great Plains of the USA. The fruit is a 'burr' formed from compressed female spikelets.

(c) Shelling maize. The cob axis is extremely tough, a feature sometimes attributed to introgression from *Tripsacum*. Binding a group of cobs with a steel packing case band provides a hard surface to dislodge grains from a newly harvested ear and which fall onto a sheet of thick cardboard.

(d) Stressed maize. The distinction between the intercalary (female) and the terminal (male) inflorescences is not absolute and features of the one can be made to appear in the other by both genetic and physiological means. This illustration shows tassels of a drought-stressed maize in which grains were formed.

(e) Microhairs of *Cynodon dactylon* shown here functioning as salt glands. Minute crystals of sodium chloride form on the outside of the terminal cells. See chapter 11 for a discussion of salinization.

(f) Inflorescences of *Coix lachryma-jobi*. The 'utricle', the female spikelet, is in groups of three, of which one is fertile and provides 'adlay' a cereal of South-East Asia. Its relative *C. aquatica* reportedly is unusual in being an autogamous diploid apomict.

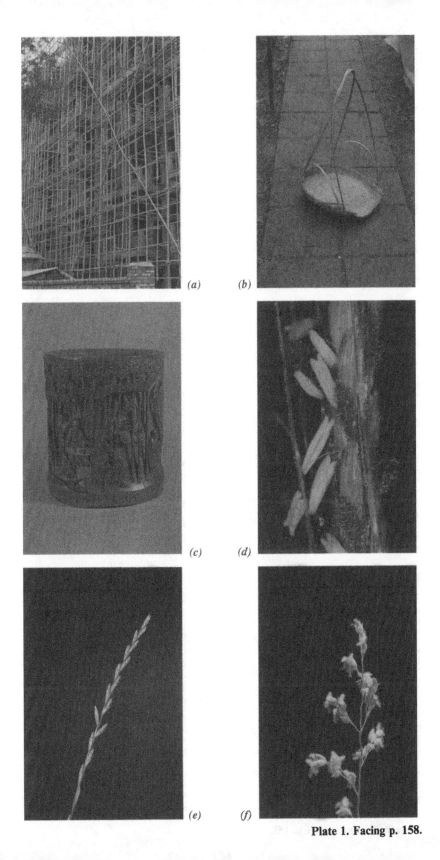

Plate 1. Facing p. 158.

(a)

(b)

(c)

(d)

(e)

(f)

Plate 2. For caption see p. 158.

5

Origins and processes of domestication

J. R. Harlan

The words 'domesticate' and 'domestication' are derived from the Latin *domus* (household). A domestic is a servant who lives in the house of the master. Domesticated plants are those brought into the *domus* which may mean the dooryard, garden, field, orchard, vineyard, pasture or ranch. It may also include yards, parks, cemeteries, golf courses, roadsides, forests and other managed areas. In ecological terms it is the change in habitat that is critical. Genetically, it is the genes conferring fitness to the new habitat that is critical. The genetic architecture of domestication tends in the direction of making the plant populations dependent on human interference and man-made habitats. Since the processes of domestication are evolutionary in nature, all intermediate degrees and conditions may be expected, but a fully domesticated plant is entirely dependent on human intervention for survival.

The cereals
The best known of domesticated grasses are the cereals. Using data from FAO Production Yearbooks and correcting for moisture and wastage, one can show that the cereals, taken together, produce about three times as much edible dry matter as all tubers, pulses, fruits, sugar, meat, milk and eggs taken together. Most of the people on earth live or die according to cereal production. The origins and domestication of these crops should therefore be of vital interest to all.

The cereals, in order of edible dry matter production are: wheat, rice, maize, barley, sorghum, oats, rye and the millets of which there are some 16 or more, only a few of which are important on the world scene. This is not necessarily the order of importance to human nutrition because substantial proportions of maize, barley, sorghum, oats and rye are fed to livestock. Wheat and rice are primarily used as human food. The significance of the conversion of the so-called coarse grains into

159

high quality protein in the form of milk, meat and eggs should not be underestimated, however, in terms of human well-being.

Wild grass seeds have been harvested for food on every inhabited continent, and the practice still continues here and there although not on the scale of even 100 years ago. The harvesting alone apparently has little effect on the plant populations. Grass seeds have been gathered for food for millennia without resulting in domestication. The processes of domestication begin with planting that which has been harvested. But, the seed must be *planted*, not simply sown. A number of Great Basin Indian tribes in North America regularly broadcast grass seeds without ever domesticating any. They practised no tillage. The usual procedure was to burn a patch of vegetation in the fall and broadcast seed in the burned area in the spring. A number of species of native food plants was sown but none domesticated (Steward, 1941).

Planting

As soon as one begins to plant the seed in a seed-bed on a yearly basis and save seed for the following season, selection pressures are automatically set in motion, leading toward domestication (Harlan, de Wet & Price, 1973). The most productive genotypes are likely to contribute more to the next generation than less productive ones. This produces strong selection pressures favouring gene combinations that confer fitness to the environment. The environment involves cultural practices, climate, soil, topography, diseases and pests, human behaviour and probably other features not discerned. In terms of fitness, productivity involves not only number of seeds per genotype but germination and seedling vigour. Competition in the seed-bed can be very intense. The first seeds to sprout and the most vigorous ones, in an average year, have an advantage over dormant seeds and those producing weak or slow-growing seedlings.

Within species, larger seeds are more vigorous and grow faster than smaller seeds (Kneebone & Cremer, 1955). But the cost of larger seeds may be fewer seeds. Some sort of optimum equilibrium will consequently develop between size and number for a given set of environmental conditions. Deep planting will also favour larger seeds, since seedlings of large seeds can emerge from greater soil depths. On the other hand, broadcasting seeds on a cloddy soil or broadcasting followed by harrowing may favour diversity in seed size. Cultural practices, then, are a major part of the environment and diverse forces interact to develop the adapted land races of traditional agriculture.

Shattering

Seed shattering at maturity is an important trait for wild and spontaneous races of a cereal. Whether the inflorescence is a spike, false spike or panicle, the mechanism is the same in all. An abscission layer is formed at a joint of articulation and collapses at maturity permitting the seed unit to fall. In domesticated races, the abscission layer is suppressed or collapse is delayed until harvest. This is the most diagnostic trait distinguishing spontaneous races from domesticated ones and is the one most frequently used by archaeobotanists to classify material recovered from archaeological sites. Because it is diagnostic, it has been perhaps overrated in the domestication process. Rindos (1984, p. 139), for example, claimed that plants were domesticated *before* agriculture through long co-evolution with man and that 'the indehiscent rachis of the small grains is as much the cause as the result of agriculture'. But wild grass seeds can be and have been harvested in commercial quantities and can compete with domesticated cereals in the market-place (Harlan, 1989) and grasses have been co-evolving with Australian Aborigines for some 35 000 years without any being domesticated. It is true, however, that some kind of non-shattering trait shows up in every domesticated cereal and tends to make them dependent on human activity.

Dormancy

Most wild type grass seeds have an appreciable dormancy. Dormancy has fitness value in wild plants but is useful in domesticated cereals only if it breaks down between harvest and seeding time. The dormancy is often conditioned by inhibitors in the glumes, lemma, palea or other appendages. Selection pressures, therefore, serve to reduce appendages in some cereals; in others genotypes with lower concentration of inhibitors will be favoured. In an experiment with a spontaneous finger millet from Africa, Hilu & de Wet (1980) found a very sharp decline in dormancy in only four generations. Simple increase in seed size, common in domesticated grasses, may also reduce the relative amounts of inhibitor.

Hand harvesting

I have harvested wild grass seeds by hand stripping, beating into a basket, cutting with a flint-bladed sickle, steel sickle and scythe and with power equipment such as a binder, stripper, blower and grain combine. They all work, but rarely can one reap more than half the potential seed. Loss from shattering is only part of the reason for this. Wild grasses must be harvested during a short period when the earliest spikelets have shattered and the latest ones are immature. Uneven ripening probably causes greater losses than shattering but, of course, if the spikelets did not shatter, uneven ripening would be less important. In successive reaping and planting, selection pressures favour both uniform ripening and increased seed retention. All domesticated cereals are greatly improved in these respects over their progenitor races.

The suppression of abscission layers is usually under the control of one or two genes and is likely to appear very early in the domestication process. I once obtained substantial improvement in seed retention in *Andropogon hallii* by simply going through my nursery six weeks after most of the seed had shattered and harvesting what remained. I suspect that genes favouring seed retention can be found in almost any large grass population. Non-shattering is probably the simplest and easiest step in the domestication process.

Maturation and tillering

Uniformity of maturation is approached in different ways according to the architecture of the plant. Wild races of sorghum, pearl millet and maize tend to have branched stems with several to many small inflorescences scattered among them, maturing at different times over a considerable period (often a month or more). Evolution toward uniform ripening, then, involves a reduction in the number of inflorescences and a concurrent increase in inflorescence size in the remainder. High-yielding cultivars or hybrids of field maize usually average fewer than two ears per stalk. Sweet and popping types may have more, but not nearly as many as the teosintes. High-yielding sorghums, whether cultivars or hybrids, usually have a single, large terminal inflorescence per stem. Pearl millet has evolved in the same direction under domestication.

Grasses such as wheat, barley, rice, oats and rye normally produce several tillers per plant, each with a single terminal inflorescence. Trends toward even ripening are based mostly on shortening of the tillering period. There may or may not be a reduction in tiller number. Indeed, some cultivars may have more tillers than most wild-type plants, but they are budded off quickly and the inflorescences tend to ripen concurrently. The primary shoot may be somewhat earlier than the lateral ones, but this is likely to be compensated for by it being a somewhat larger inflorescence that is slower to mature, and approximate synchrony is achieved. The millets are too varied a group to generalize very well, but those other than pearl millet tend to follow the pattern just described (Harlan *et al.*, 1973).

Weeds

The evolution of weed races is a nearly universal feature of the processes of domestication. Weeds are adapted to man-made habitats and so are crop plants. They often compete intensely and cost the farmer dearly in control measures. But, the relationships among wild, weed and domesticated races are not as simple as once thought. The notion that crops are derived from weeds was once popular. Sometimes it is true; oats and rye seem to be good examples. But weeds can be derived from crops. Sorghum, common millet and foxtail millet are well-documented examples. Even some forms of rye and oats are derived from cultivars. Plants can be taken into the *domus* and escape again. If the presence or absence of abscission layers is controlled by a single gene, it is not difficult for cultivars to acquire the brittle habit and revert to weed status. Usually there are associated morphological traits that go with the spontaneous habit and reveal the domesticated source, e.g. the *fatua* type of disarticulation in oats, the sucker-mouth articulation in common millet, the secondary dispersal method in sorghum, etc. (Harlan, 1982*a*).

Sorghum provides one of the most interesting examples of domesticated races evolving weed races because the morphological features are unequivocal. In the USA, a race of shattercane evolved which retains the gene suppressing the abscission layer. Grains do not shatter at the normal point of articulation of the spikelet pairs, but panicle branches that support the spikelet pairs become very thin and break. The seed unit, then, has a small piece of panicle branch attached. The shattercane weed carries the domesticated genotype for non-shattering, but seeds are shed anyway by a secondary dispersal system. In addition, in Africa, where wild sorghums are prevalent, I have found shattering types of several domesticated races owing to the transference of brittle alleles.

Recovery of fertility

A rather unexpected feature of cereal domestication is the recovery of fertility of suppressed, even rudimentary florets and spikelets. This is unexpected because evolutionary dogma has it that evolution is irreversible. This is not true of course, and the phenomenon is found in plants other than grasses. Among cereals we find recovery of fertility of sterile spikelets in barley, sorghum and maize and of sterile florets in sorghum, common millet and maize. Grains may be produced in the 'sterile' glumes of wheat in certain genotypes and spikelets may be produced *de novo* from reduced sterile panicle branches in pearl millet (Harlan, 1982*b*). These morphological changes are taxonomically radical but genetically simple; usually one or two genes are involved.

Selection in a barley population

Some insights into the dynamics of crop evolution and the genetic architecture of land race formation have been provided by studies of composite crosses in barley. The most studied is CC II, initially created by my father, H. V. Harlan in 1928. With much deliberation, he chose 28 cultivars out of some 6000 available to represent the barley crop and crossed them in all combinations for a total of 378 crosses. Seeds from the hybrids were bulked as evenly as possible and the mixture grown as a mixture without deliberate selection for successive generations. CC II has now been grown as a bulk population for 60 generations at Davis, California. Large populations were grown to reduce genetic drift. Without intentional selection, the yield increased rapidly from a low level for 15 to 20 generations, then continued to increase at a slow but steady rate. Despite 60 generations of inbreeding, the population is still highly variable, is still evolving; the yields are still improving; the plants are approaching an idiotype suitable for a Davis, California, land race. Spikes have become shorter and denser with more seeds per spike; increase in yield has been primarily in more seeds per plant rather than in larger seeds. The improvement in yield has kept pace with about 95% of the best that plant breeders have been able to do (and at far less cost).

It seems evident that farming, i.e. reaping and planting, *is* plant breeding, and very effective at that. With the advent of chromatography, electrophoresis, DNA restriction fragments and other modern techniques, the changes in genetic architecture can be examined in some detail. It has been established that gene combinations and arrangements are adaptive, and at least some of the increase in fitness depends on gene associations even when the genes are not linked (linkage disequilibrium). A summary of these studies was given by Allard (1990).

Zizania aquatica

A footnote to cereal domestication may be added by current efforts to domesticate the North American wild-rice, *Zizania aquatica*.

The grass grows, or has grown, over large areas of shallow water near the Great Lakes and in fresh water wetlands along the east coast of North America. It was a staple of several Indian tribes who harvested grains by the canoe load. It is considered gourmet food by many people, and the price of wild-rice seed is such that its culture can be very profitable. Efforts at domestication have been undertaken in Minnesota, Canada, California and possibly elsewhere. Non-shattering and semi-shattering genotypes were easily located, but there are more serious problems. The plant is adapted to living in shallow, slowly moving water. Seeds shed in the fall filter down to the mud below and sprout with warming of the water in the spring. If the seed dries out, it dies. This poses serious problems for the management of either the planting stock or the water in the production fields. These studies should be of interest as a model for cereal domestication. There are other experiments in ceral domestication underway in England, France and perhaps elsewhere.

Where and when were cereals domesticated?

In the Near East, emmer and barley were among the first crops to be represented in archaeological contexts, about 8000 BC, or a little earlier. Evidence of cultivated forms was found in Prepottery Neolithic A (PPNA), mostly in the Jordan Valley near and including Jericho. There is another site, Tel-Aswan, in the Damascus Basin at about the same time range. Pea and flax were reported there as well as emmer and barley (Zohary & Hopf, 1988; Bar-Yosef & Kislev, 1989).

WHEAT

The domestication of wheat is a complicated story because of multiple domestications and a long evolution following the initial entry into the *domus*. Both einkorn, a diploid, and emmer, a tetraploid, were taken into the domestic fold very early. Emmer and barley were the main cereals of the Neolithic period that spread from southwest Asia into the Balkans, up the Danube and down the Rhine and around the shores of the Mediterranean. The system, including pulses, flax and other crops spread southward to Ethiopia and eastward to India. Emmer was the primary wheat throughout; einkorn was always less important and did not reach either Ethiopia or India. Emmer is a glume wheat and requires additional processing after threshing to free the grain. Free threshing wheats appeared in the archaeological record quite early. A few grains were reported from Ramad, near Damascus, dated *c.* 7000 BC and more ample remains from sites as far apart as Crete and Baluchistan by 6000 BC. The early finds are presumed to be tetraploid like emmer, but it is difficult or impossible to distinguish tetraploid from hexaploid materials obtained from archaeological sites. Despite the availability of naked wheats, emmer remained dominant for several millennia and was only replaced by bread wheat in Rome and Egypt in historic times (Harlan, 1981).

Bread wheat, a hexaploid, is the most abundant wheat produced today. It is a free threshing derivative of spelt. Botanical evidence suggests that spelt wheat arose through spontaneous hybridization of a domesticated tetraploid wheat and a diploid goatgrass, *Triticum tauschii*, probably somewhere near the southern Caspian sea. Early archaeological finds are scanty, but the evidence we do have indicates that spelt took a route to Europe different from that taken by the other wheats. Remains of spelt are reported from the Caucasus dating to the fourth millennium BC and possibly to the fifth, and other samples from Moldavia are dated between 4000 and 3000 BC. Spelt has not turned up in the Near East or the Mediterranean. Free threshing wheats are found in Europe throughout the Neolithic period, but the difficulty in distinguishing tetraploid from hexaploid has caused serious problems in interpretation. By whatever route, spelt became an important cereal in central Europe. The rise of bread wheat to dominance is a relatively late phenomenon (Harlan, 1981).

RICE

Wild races of Asian rice are widespread and often abundant from Pakistan to northern Australia and northward to southern China. There are both annual and perennial strains in the same species. The perennial races require permanent water and are common in and along the major river systems of southeastern Asia. The annual races are adapted to water holes that fill up during the rainy season and dry out in the dry season. Perennial types are poor seed producers and live in habitats very difficult to exploit without dykes, dams, canals and other water control measures. The annuals set seed abundantly and occupy habitats easily exploited with little or no modification and with the simplest of tools. It seems certain, therefore, that the first rice domesticated was derived from annual races in areas with long dry seasons. These areas are abundant and widespread over the range of wild rice.

Domesticated rice appears archaeologically in both south China and India by 5000 BC, but earlier finds are likely to turn up. Agriculture was well developed by that time. According to Chang (1989), the greatest diversity of cultivated rice today is found in a belt from northeast India to southeast China. Centres of diversity are not necessarily centres of origin, but in this case the area seems reasonable for rice domestication. Whenever a progenitor is widespread, there is always the possibility of multiple domestications, especially of a plant so productive that it is still harvested in the wild, millennia after it was first domesticated.

MILLETS

The common and foxtail millets show up in archaeological contexts first in north central China, *c.* 6500 BC, and become much more abundant a few centuries later (Chang, 1986). There is a sprinkling of sites across Europe with remains of both millets dated through the fourth millennium BC and a few that might be earlier. What this means is not clear at present, but the wild progenitors range across Eurasia, and independent domestications seem likely.

OATS AND RYE

Oats and rye are secondary crops. Their progenitors are in the Near East but the scenario indicates they spread as weeds in emmer and barley fields into central and northern Europe where they were later domesticated. The earliest evidence of cultivated forms is about 1000 BC (Zohary & Hopf, 1988).

SORGHUM

Sorghum is at present an anomaly. It is clearly an African domesticate but archaeology has not yet turned up early remains in Africa. It seems to have reached India by 1000 BC. African finds, so far, are after AD 500. More research should eventually resolve the problem. While sorghum was undoubtedly domesticated in Africa, there is a possibility that it was domesticated independently in the Far East. There is a wild diploid race in southeast Asia (race *propinquum*) cytogenetically compatible with domesticated sorghum, and the cultivated sorghums of Indonesia and China are both quite different from African races. The problem has been little studied. An early find was once reported from China, but the identification has been questioned. Again, more research may unravel the puzzle.

PEARL MILLET

Pearl millet has about the same history as African sorghum, reaching India at roughly the same time. Archaeological evidence suggests that it became established as a crop in Mauritania about 1000 BC, replacing wild grass seed harvests (Munson, 1976). The distribution of wild races in the Sahara suggests that it may have been domesticated there during a pluvial when the Sahara was more widely inhabited than in recent millennia.

AFRICAN RICE

The progenitor of African rice is widespread in the African savannas, but the distribution of rice agriculture indicates that it was domesticated somewhere in West Africa. The bend of the Niger has been suggested, but no really early sites have been excavated yielding rice remains. Accelerator mass spectroscopy dating of the Phillipson find is *c*. 820±200 bp (Phillipson, 1977; Gowlet *et al.*, 1987). African archaeology has not been of much help, so far, in establishing times and places of domestication of indigenous crops. What evidence there is has been reviewed by Harlan, de Wet & Stemler (1976) and Clark & Brandt (1984).

MAIZE

In the Americas, plant cultivation seems to have begun at least as early as in the Near East, that is before 8000 BC, but the first crops were cucurbits, beans and chilli peppers. The earliest evidence of maize, so far, is from Tehuacan Valley, Puebla, Mexico, at about 5500 BC (Byers, 1967). Earlier finds are likely to turn up, since a maize-based agriculture can be demonstrated for Panama by 5000 BC (Piperno, 1989). In addition to maize, there were some minor millets, some of which were abandoned after more productive cereals became available.

The domestication of maize has raised a great deal of controversy and the whole story is not yet in. Despite Mangelsdorf's repeated and eloquent pleas for a wild maize progenitor that later became extinct, there is little or no doubt that some sort of annual teosinte was the ancestor of maize. Teosinte behaves exactly like wild wheat, wild barley, wild rice, wild sorghum or any other cereal progenitor. There are truly wild races, weed races, and domesticated races as with other crops. The male inflorescence of teosinte is too much like that of maize to raise any questions. The female inflorescence of teosinte is, however, radically different in appearance from the ear of maize, causing endless speculation and argument. A review of the lengthy discussion is not warranted here, but genetic studies by Beadle (1980), Galinat (1983), Iltis (1983), the Crop Evolution Laboratory, University of Illinois, and others have shown that the genetic differences are much less than the morphological differences would imply. A single gene could produce most of the transformation observed. Iltis suggested possibilities along this line in 1983, and recently, such a gene has been located in *Tripsacum dactyloides*, a wild relative which can be hybridized with maize (Dewald *et al.*, 1987). The ground-plan of the female flowers is the same in teosinte, maize and *Tripsacum*, and a change in gene regulation could convert one morphology into the other. The *Tripsacum* mutant can survive in the wild but a similar one in teosinte could well render the plant dependent on human intervention for survival. A case could be made for instant domestication.

Forage and turf grasses

FORAGE GRASS

Actual domestication of forage grasses is a recent development. Meadows and forage crops were important in the agriculture of classical times. Pliny, Columella and Varro each mention sowing 'hayseed' along with vetch, bitter vetch, lupine, lucerne, fenugreek and other soil builders and fodder crops. The ryegrasses, cock's foot, timothy and other grasses were well known and appreciated in Europe before history records them. If sown, however, the seed was taken from local meadows, often as hay in seed. Seed from different regions performed differently, and traditional farmers were well aware of it. The different sources were, in fact, land races that had evolved over long periods of time and were adapted to local environments and management systems. The exploitation of elite land race populations continues to this day.

Serious breeding to improve on land races is very much a product of this century. A popular method has been to develop polycross populations among elite, tested clonal selections. This is most useful in outcrossing species. Composite crosses can be generated in self-pollinating species, but large scale crossing is difficult and expensive because of the small flowers of most forage grasses. Hot water and chemical emasculation are sometimes appropriate. In our work with Bothriochoininae, we found hand emasculation was to be quite feasible despite the very small flowers. This was possible, however, because of a determined, dedicated and skilful technician. It can be done but it is not for everyone.

Cultivars of forage grasses are domesticated in the sense that they cannot reproduce themselves as defined cultivars without human management. For sexually reproducing species, a limitation of generations is usually imposed. Systems vary from country to country, depending on the laws and regulations established, and supervision is usually by some state or federal agency. Typically, breeder's seed is supplied to a seed multiplication system by the plant breeders who generated the cultivar. The first increase may produce registered seed, the second increase certified seed which the farmer may purchase to seed pastures or forage crop fields. In most cases, seed from certified seed is not certifiable, since the population may deviate too much from the originally constituted population. This is especially true if the seed production region is not the same as the region of usage. There are variations on this theme but the objective of generation restriction is to maintain a cultivar as closely as practicable to the described material developed by the plant breeders.

While these procedures are followed by most countries with advanced agriculture, it should be noted that population improvement may take place without the intervention of plant breeders. Much valuable material has been selected from old pastures. Genotypes surviving over long periods of time may be much better fitted as pasture plants than many in the original population. Old pastures, closely cropped by sheep for many years have provided useful genotypes for lawns and putting surfaces.

TURF GRASS

Turf-grass domestication is even more recent than that of forage grasses. As an industry, lawn and turf-grass culture has mushroomed in recent decades among the more affluent nations. Home owners usually have lawns, and there are many home owners. The use of turf for recreation has expanded enormously even in less developed countries where the value of grassed playing surfaces is more and more appreciated. A well kept soccer, cricket or baseball field is an important asset for even the poorest of villages. Special clones have often been selected for the purpose.

REPRODUCTION

Forage grasses that can be propagated vegetatively have a great advantage in terms of maintenance of hybrid vigour and maintenance of purity of the elite cultivars. Much has been done with *Cynodon dactylon* for that reason. It can be propagated by both stolons and rhizomes. 'Coastal' Bermudagrass was a chance nursery hybrid selected by Burton (1973) and has been established on some millions of hectares. This is a case of instant domestication. Other clones superior in various ways have followed.

'Coastal' is only one generation removed from the parents, taken from spontaneous or naturalized sources, but it is fully domesticated in the sense that it cannot reproduce itself without human intervention. It does produce some seed but these do not breed true and do not reproduce 'Coastal'. 'Coastal' has been used as a parent to produce other improved cultivars but these are not 'Coastal' either and are given other names. There is by now a substantial inventory of named clones of *Cynodon dactylon* ranging from very coarse, bulky forage and hay types to very fine, dense turf types used for putting greens in climates where bentgrass greens are impracticable.

Facultative apomixis can fix desirable genotypes and exploit hybrid vigour in grasses that reproduce by seed rather than stolons or rhizomes. Cultivars of *Cenchrus, Paspalum, Panicum* and of the *Bothriochloa–Dichanthium* complex have been developed through this mechanism. Apomixis is usually dominant to sexuality, so that selected apomictic F_1 plants can be propagated from seed and breed true. Again, there is instant domestication in a single generation, although some of these apomictic lines have become naturalized and spread on their own.

Several schemes have been devised to exploit heterosis in sexually reproducing forage grasses. In one system, Burton (1973) found that growing selected inbreds of pearl millet together resulted in a high proportion of hybrid seed, and that the relatively small percentage of inbreds in the mixture did not reduce yield over controlled hybrids. Later, he developed a cytoplasmic sterile system with some advantage in hybrid seed production, but with the risks that cytoplasmic systems have in conferring susceptibility to disease. Hybrids have been exploited in other species, and such products can be considered domesticated because they require human management for survival.

Forage grass domestication and improvement has been led and dominated by Europe and Europeanized parts of the world, and the first domesticates were of cool-season grasses such as *Lolium, Festuca, Dactylis, Phleum, Poa, Agrostis*, etc. In recent decades great interest has developed in improving tropical pastures, and cultivars of warm-season grasses are being developed. In general, the most useful grasses for tropical pastures come from Africa and the most useful legumes from tropical America. *Pennisetum, Panicum, Paspalum, Brachiaria, Cynodon* and others are being exploited and domesticated cultivars developed.

One curious exception is Guatemalagrass (*Tripsacum andersonii*). It is sterile and cannot reproduce itself. The chromosome number is 64; a triploid of $x=18$ would have 54 chromosomes. The extra 10 chromosomes appear to be from some kind of *Zea*. Where and when this remarkable hybrid originated, we do not know, but it was propagated by the American Indians, at least to some extent. It has been suggested that it was grown to feed guinea pigs. The only other herbivores domesticated by the American Indians were the llama and alpaca, but they were not raised in the area of Guatemalagrass culture (de Wet *et al.*, 1983).

Grasses for special uses

Here, I shall avoid comment on the bamboos which have enormous value for almost innumerable uses and are a whole topic in their own right. A number of grasses have been domesticated for their chemical content. In terms of value and socio-political impact, sugarcane is, by far, the most important. To the best of our knowledge, the first clones domesticated were mutants of *Saccharum robustum* in New Guinea. They have very stout canes and $2n=80$ chromosomes. Wild *S. robustum* (also $2n=80$) has stout canes that are very hard and woody and are not sweet. They have been used by natives as living fences to keep pigs in their pens or to keep pigs out of the gardens. The mutation(s) blocked the pathway from sugar to starch and resulted in canes with high sugar content. These 'noble canes' were rather widely distributed by Polynesians in their travels and were the first to be known and used by Europeans. They were the basis of slavery plantations in the New World. Later, they fell victim to a series of diseases and wild relatives were called to the rescue. Modern clones are complex hybrids, usually with more than 100 chromosomes and germplasm from several species. Since sugarcane is propagated vegetatively, domestication of sweet mutants was probably accomplished in one generation. Improvement over the orginal clones has been substantial.

Several species of grass have been grown and selected for the aromatic oils they contain. Lemon grass (*Cymbopogon citratus*), *C. martini*, that produces palmerosa oil and ginger oil, citronella grass (*C. nardus*) and vetiver (*Vetiveria zizanloides*) are the most important commercially.

Infusions of lemon grass are popular among Africans and people of African descent in various parts of the New World. Citronella extract has been used as an insect repellent with some success. Vetiver oil, extracted from the roots is used in perfumery; indeed, most of the essential grass oils are used in some fashion in perfumery.

Vetiver has recently been vigorously promoted by the World Bank for control of soil erosion and moisture conservation in developing countries around the world. It is a tall (to 2 m) coarse tufted perennial and can be grown under a very wide range of conditions. It is propagated vegetatively, although it does produce some seed. It is recommended that hedges of it be established at intervals along the contour to slow the runoff of water during rains and to trap soil that would be washed away with the runoff. In time, the hedges build small terraces behind them, and since the roots mostly grow straight down, the hedges do not compete severely with interplanted crops. The technique has been in use for a long time in India where some special clones have been selected.

Grasses have long been used for land stabilization and erosion control. Traditional farmers in both the Old and New Worlds used them on terraces on mountainsides. In the USA, many millions of hectares were established as artificial stands of grass in the aftermath of the dust-bowl days of the 1930s. Selected (i.e. domesticated) strains were used in some cases and populations from wild harvests in others. Road construction in recent decades has posed enormous problems for roadside stabilization and erosion control around the world and the role of grasses for the purpose has come into sharp focus. Roadside verge stabilization has become a new, important, and exacting form of urban agriculture. Construction may be completed at any time of year, often out of the growing season, leaving soil and (worse) subsoil exposed to the elements. Prompt and effective stabilization measures can save enormous sums of money in damage repair. Emergency measures may be required, but the long term goal is a permanent, easily maintained stabilization system and perennial grasses are usually the primary components of such systems. Airports, canals, levée embankments and other structures that disturb large land areas pose the same problems.

While legumes such as *Lespedeza, Pueraria, Vicia, Coronilla, Trifolium, Medicago, Centrosema* and others have been used to some extent in stabilization work, perennial grasses have generally been the mainstay. The species and genera vary according to climatic and edaphic factors, but some of the most exploited are species of *Agropyron, Bromus, Festuca, Eragrostis, Panicum, Paspalum, Cynodon, Pennisetum* and *Cenchrus*. Cereal rye and oats are often used as temporary or emergency control plants.

Conclusions

Bringing grasses into the *domus* is not difficult. We are doing it all the time with grasses for forage, erosion control, recreation and garden purposes. Our major cereals, however, have long, venerable and sometimes complex histories, have provided the foundation for all our civilizations, and supply more food for humanity, by far, than all other sources combined. The minor millets are becoming even more minor and wheat and rice ever more dominant. It is unlikely that any of the major cereals will ever be replaced, but there is a place for development of gourmet and speciality grains.

References

Allard, R. W. (1990). Future directions in plant population genetics, evolution and breeding. In *Plant population genetics, breeding and genetic resources*, ed. A. D. H. Brown, M. T. Clegg, A. L. Kahler & B. S. Weir, pp. 1–19. Sunderland, MA: Sinauer Associates Inc.

Bar-Yosef, O. & Kislev, M. E. (1989). Early farming communities in the Jordan Valley. In *Foraging and farming: the evolution of plant exploitation*, ed. D. R. Harris & G. C. Hillman, pp. 632–642. London: Unwin Hyman.

Beadle, G. W. (1980). The ancestry of corn. *Scientific American*, **242**, 112–119.

Burton, G. W. (1973). Bermuda grass. In *Forages*, ed. M. E. Heath, D. S. Metcalfe, L. R. E. Barnes, p. 755. Ames, Iowa.

Byers, D. S. (ed.) (1967). *The prehistory of the Tehuacán Valley*, 2 vols. Austin: University of Texas Press.

Chang, K. C. (1986). *The archaeology of ancient China*, 4th ed. New Haven: Yale Univ. Press.

Chang, T. T. (1989). Domestication and the spread of the cultivated rices. In *Foraging and farming: the evolution of plant exploitation*, ed. D. R. Harris & G. C. Hillman, pp. 408–417. London: Unwin Hyman.

Clark, J. D. & Brandt, S. A. (eds) (1984). *From hunters to farmers: the causes and consequences of food production in Africa*. Berkeley, CA: University of California Press.

Dewald, C. L., Burson, B. L., de Wet, J. M. J. & Harlan, J. R. (1987). Morphology, inheritance and evolutionary significance of sex reversal in *Tripsacum dactyloides* (Poaceae). *American Journal of Botany*, **74**, 1055–1059.

de Wet, J. M. J., Fletcher, G. B., Hilu, K. W. & Harlan, J. R. (1983). Origin of *tripsacum andersonii* (Gramineae). *American Journal of Botany*, **70**, 706–711.

Galinat, W. C. (1983). The origin of maize as shown by key morphological traits of its ancestor, teosinte. *Maydica*, **28**, 121–138.

Gowlett, J. A. J., Hedges, R. E. M., Law, I. A. & Perry, C. (1987). Radiocarbon dates from the Oxford AMS System: datelist 5. *Archaeometry*, **29**, 125–155.

Harlan, J. R. (1981). The early history of wheat: earliest traces to the sack of

Rome. In *Wheat science – today and tomorrow*, ed. L. T. Evans & W. T. Peacock, pp. 1–19. Cambridge: Cambridge University Press.

Harlan, J. R. (1982*a*). Relationships between weeds and crops. In *Biology and ecology of weeds*, ed. W. Holzner & M. Numata, pp. 91–96. The Hague: W. Junk.

Harlan, J. R. (1982*b*). Human interference with grass systematics. In *Grasses and grasslands: systematics and ecology*, ed. J. R. Estes, R. J. Tyre & J. N. Brunken, pp. 37–50. Norman: University of Oklahoma Press.

Harlan, J. R. (1989). Wild grass-seed harvesting in the Sahara and Sub-Sahara of Africa. In *Foraging and farming: the evolution of plant exploitation*, ed. D. R. Harris & G. C. Hillman, pp. 79–98. London: Unwin Hyman.

Harlan, J. R., de Wet, J. M. J. & Price, E. G. (1973). Comparative evolution of cereals. *Evolution*, **27**, 311–325.

Harlan, J. R., de Wet, J. M. J. & Stremler, A. B. L. (eds.) (1976). *Origins of African plant domestication*. The Hague: Mouton.

Hilu, K. W. & de Wet, J. M. J. (1980). The effect of artificial selection on grain dormancy in *Eleusine* (Gramineae). *Systematic Botany*, **5**, 54–60.

Iltis, H. (1983). From teosinte to maize: the catastrophic sexual transmutation. *Science*, **222**, 886–894.

Kneebone, W. R. & Cremer, C. L. (1955). The relationship of seed size to seedling vigor in some native grass species. *Agronomy Journal*, **47**, 472–477.

Munson, P. J. (1976). The origins of cultivation in the south western Sahara. In *Origins of African plant domestication*, ed. J. R. Harlan, J. M. J. de Wet & A. B. L. Stemler, pp. 187–209. The Hague: Mouton.

Phillipson, D. W. (1977). The excavation of Gobedra rock shelter, Axum: an early occurrence of cultivated finger millet in northern Ethiopia. *Azania*, **12**, 53–82.

Piperno, D. R. (1989). Non-affluent foragers: resource availability, seasonal shortages and the emergence of agriculture in Panamanian tropical forests. In *Foraging and farming: the evolution of plant exploitation*, ed. D. R. Harris & G. C. Hillman, pp. 538–554. London: Unwin Hyman.

Rindos, D. (1984). *The origins of agriculture: an evolutionary perspective*. New York: Academic Press.

Steward, J. H. (1941). Culture element distribution: xiii. Nevada Shoshoni. *California Anthropological Record*, **4**, 209–359.

Zohary, D. & Hopf, M. (1988). *Domestication of plants in the old world*. Oxford: Clarendon Press.

6

The three phases of cereal domestication

J. M. J. de Wet

The caryopses of grasses have been harvested as human food since long before the advent of agriculture. Numerous species are still regularly harvested by nomadic tribes in Africa, and by farmers in Africa and Asia during times of scarcity. Among the many hundreds of species harvested as wild cereals, 33 species (Table 6.1) belonging to 20 genera were domesticated (de Wet, 1979). They lost the ability to disperse seeds naturally, and became adapted to cultivated fields. Domesticated cereals totally depend on agriculture for survival. These 33 species were not all domesticated at the same time, in the same geographical region, or by the same people (Table 6.1).

There were three major phases of cereal domestication. The first phase represents domestication during the beginnings of agriculture by people who were experimenting with plant husbandry. The second phase was initiated as knowledge of agriculture spread beyond these early centres of domestication and indigenous species were consciously brought into cultivation. The third phase started with crop improvement through modern plant breeding.

This chapter discusses the origins and evolution of those cereals not treated in chapters 8, 9 and 10.

The first phase of cereal domestication

Cereals are among the oldest of plant domesticates. They appear in the archaeological records of the earliest known farming settlements. These records suggest that farming started independently in at least three widely separated regions more than 7000 years ago. The three most important cereals grown today were independently domesticated in these three centres of early agriculture. They are wheat that was domesticated in South West Asia (Harlan, 1981), maize that was domesticated in the tropical highlands of Mexico (Mangelsdorf,

Table 6.1. *Domesticated cereals, their area of origin, and their scientific and common names*

Region of origin	Species	Common name
North America	Zizania aquatica L.	American rice
South America	Bromus mango Desv.	Mango
Mexico	Zea mays L.	Maize
	Panicum sonorum Beal	Sauwi
Africa	Brachiaria deflexa (Schumach.)	
	C.E. Hubb. ex Robyns	Animal fonio
	Digitaria exilis (Kippist) Stapf	Fonio
	D. iburua Stapf	Black fonio
	Eleusine coracana (L.) Gaertn.	Finger millet
	Eragrostis tef (Zucc.) Trotter	Teff (T'ef)
	Oryza glaberrima Steud.	African rice
	Pennisetum glaucum L. Rich	Pearl millet
	Sorghum bicolor (L.) Moench	Sorghum
Europe	Avena sativa L.	Oats
	A. strigosa Schrebn.	Sand oats
	Digitaria sanguinalis (L.) Scopali	Manna
	Phalaris canariense L.	Canary millet
	Triticum timopheevi (Zuck.) Zuck.	Timopheev wheat
South West Asia	Hordeum vulgare L.	Barley
	Secale cereale L.	Rye
	Triticum aestivum L.	Bread wheat
	T. turgidum L.	Macaroni wheat
	T. monococcum L.	Einkorn wheat
China	Oryza sativa L.	Rice
	Panicum miliaceum L.	Proso millet
	Setaria italica (L.) P. Beauv.	Foxtail millet
	Echinochloa crus-galli (L.) P. Beauv.	Japanese millet
India	Coix lacryma-jobi L.	Adlay
	Digitaria cruciata (Nees) A. Camus	Raishan
	Brachiaria ramosa (L.) Stapf	
	Echinochloa colona (L.) Link	Sawa
	Panicum sumatrense Roth.	
	ex Roem. et Schult.	Sama
	Paspalum scrobiculatum L.	Khodo
	Setaria pumila (Poir.)	
	Roem. et Schult.	Korali

MacNeish & Galinat, 1967), and rice that was domesticated in China (Ho, 1975).

Domestication is a slow process. This is not surprising. Domestication is a natural evolutionary process resulting from selection pressures associated with harvesting and sowing (Harlan, de Wet & Price, 1973). The earliest farmers probably started sowing to increase population density and facilitate harvesting of their favourite wild cereals. They must have started cultivation to reduce competition between sown cereals and other species.

Cereals were cultivated for generations before they became dependent on harvesting and sowing for survival. The archaeological record indicates that a species of *Setaria* was grown by early farmers in Mexico for perhaps 1500 years without losing its ability to disperse seeds naturally (Callen, 1965, 1967). It is only when harvesting and sowing follow one another in successive generations that selection pressures favour domestication.

CEREALS OF SOUTH WEST ASIA AND EUROPE

Wheat and barley are the oldest, and probably the only cereals domesticated in South West Asia during the early phase of agricultural origins (see later chapters). Rye and oats are also natives of South West Asia, but were probably domesticated after the cultivation of wheat and barley spread from the fertile river valleys of Jordan north into the plateau regions of northwestern Iran and eastern Turkey. Evans (1976) suggests that the progenitor of rye (*Secale cereale* L.) entered wheat and barley fields as a weed, and eventually became domesticated. Rye reached Europe as a cultivated cereal around 2500 BC (Khush, 1963). Oats (*Avena sativa* L.) first occurred as a weed in barley fields in Jordan dating back to about 6500 BC, and this weed appeared in Europe only during the late Neolithic period (Villaret von Rochow, 1971). The first cultivated oats appeared at about 2000 BC in Central Europe (Helbaek, 1959), and became a widely grown cereal in Europe during the first century AD (Holden, 1976). The diploid sand oats (*Avena strigosa* Schrebn.) is a native of Spain and was domesticated after the introduction of cereal cultivation from South West Asia into Europe. Sand oats was later introduced into Britain, but is no longer cultivated outside Spain.

The widely distributed temperate weed *Digitaria sanguinalis* (L.) Scopali (manna, blut hirse) became an important cereal in Europe during Roman times (Matthiolus, 1565). It was still extensively harvested as a semi-domesticated weed in southern Europe during the first quarter of the nineteenth century (Ascherson & Graebner, 1890). It is now grown as a cereal only in the Caucasus and in Kashmir (Henrard, 1950).

CEREALS OF EURASIA

Rice was not the only cereal domesticated in China (see elsewhere in this volume). Proso millet (*Panicum miliaceum* L.) and foxtail millet (*Setaria italica* (L.) P. Beauv.) are the oldest known cultivated cereals in China, dating back to the earliest agricultural settlements of the Yang-shao culture phase, about 6000 years ago (Cheng, 1973). Wild proso millet is native to northeastern China (Kitagawa, 1937), and wild foxtail millet occurs across temperate Eurasia (de Wet, Oestry-Stidd & Cubero, 1979). Ho (1975) suggested that these two cereals were cultivated in southern Shansi shortly after 5000 BC. Both cereals reached Europe about 3000 years ago (Neuweiler, 1946) and became widely distributed during the Bronze Age (Werth, 1937). Proso and foxtail millets remained important cereals in eastern Europe until the beginning of the present century and are still widely grown in Temperate Asia. Foxtail millet is also an important cereal in parts of tropical south Asia.

Foxtail millets were classified into a European complex (race moharia) and an Asian complex (race maxima) by Dekaprelevich & Kasparian (1928). The European complex includes plants with relatively small and erect inflorescences, while the Asian complex is characterized by large and pendulous inflorescences. Gritzenko (1960) indicated that within race maxima, cultivars from northwestern China and Mongolia have compact, and often erect, inflorescences, while the inflorescences of those from eastern China and Japan are typically large and pendulous. Cultivars from India are distinct from those of Eurasia (Prasada Rao *et al.*, 1987). These cultivars are robust, with loose inflorescences that are erect or slightly bent. They were probably derived from Asian cultivars that entered India during the expansion of rice cultivation into south Asia.

Cultivated proso millets are extensively variable. Lyssov (1975) recognized five races on the basis of inflorescence morphology. These races have little geographic unity. Race miliaceum has large inflorescences with spreading branches, and is grown across Eurasia. It resembles *P. miliaceum* var. *ruderale* Kitagawa from Manchuria that probably represents the progenitor of proso millet. Race patentissimum with open inflorescences and slender panicle branches also occurs across temperate Eurasia. The highly evolved races contractum, compactum and ovatum resemble one another in having compact inflorescences, and differ primarily in inflorescence shape. They are grown in China and Japan.

CEREALS OF THE AMERICAS

In the Americas the oldest known cultivated cereal was a species of *Setaria* (Callen, 1965, 1967). Archaeological records from the tropical highlands of Mexico indicate that this millet was an important source of food. It occurs in the oldest occupation layers in the Valley of Mexico dating back almost nine millennia. This foxtail millet was eventually replaced by maize as a cereal in 5000 BC but later enjoyed a temporary resurgence in importance. At this time it was probably harvested from weed populations that invaded maize fields. It was also used as a cereal in northwestern Mexico, where it was sown by the people who occupied the Ocampo caves some 6000 years ago. Callen (1967) demonstrated a steady increase in size of caryopses of this millet over a period of 1500 years, but it never lost its ability for natural seed dispersal.

Two cereals besides maize were fully domesticated in the Americas, mango (*Bromus mango* Desv.) in Central Chile (Parodi & Hernandez, 1964) and sauwi (*Panicum sonorum* Beal) in northwestern Mexico (Nabhan & de Wet, 1984).

The botanist Claudio Gay was probably the last European explorer to see mango grown extensively. Gay (1865) recorded that this cereal was grown at Chiloe in the department of Castro, Central Chile. He described mango as resembling barley, with caryopses similar to but smaller than those of rye. Earlier, Laet (1633) described this cereal as having leaves similar to those of barley, stems like oats and grains a little smaller than those of rye.

Mango was a remarkable cereal. It is the only known cereal that was grown as a biennial. Gay (1865) noted that farmers allowed livestock to graze on mango fields during the first year, and harvested it as a cereal at the end of the next summer. It is surprising that people should have domesticated a biennial grass, particularly since these farmers did not have livestock before the New World was colonized by people from Europe. It may have been the only grass in the region that lent itself to domestication. Ball (1884) indicated that the people of North Patagonia and adjacent territories harvested a *Bromus* as a wild cereal. Mango was replaced as a cereal by wheat that was introduced by European settlers to Chile in the eighteenth century.

Sauwi, the other native American domesticated millet, was once widely grown by the Papago and Juma people of Arizona, and by the Warihio tribe of northwestern Mexico and adjacent California (Nabhan & de Wet, 1984). The species is native to flood plains of southern

Arizona, and to the western slopes of the Sierras from Arizona to El Salvador.

Sauwi was domesticated by farmers who also grew other crops. It is commonly found in an archaeological context associated with beans and cucurbits (Kaemlein, 1936). Hernando Alarcon recorded in 1540 that the Colorado river Yuma tribes had maize, cucurbits and millet (Elsasser, 1979). Sauwi was still extensively grown in the late nineteenth century (Palmer, 1871). The species was grown along the flood plains of the Rio Grande. Kelly (1977) records that an informer, Sam Spa, described an area of 500 metres wide and eight kilometres long being sown to sauwi along the lower flood plains of the Colorado river. The cereal was sown as soon as the water receded. Grains were carried in a gourd tied around the neck, and sowing was done by blowing the grain from the mouth. Today it is grown only by the Warihios of the southeastern Sonora and adjacent Chihuahua, where it survives in competition with maize.

The archaeological record suggests that at least one other cereal was cultivated in the Americas before the introduction of maize. In the southeastern United States, maygrass (*Phalaris caroliana* Walt.) is a common food component of early agricultural settlements (Chomko & Crawford, 1978). It is assumed that maygrass was cultivated, since these early agricultural settlements are outside the natural range of the species (Cowan, 1978).

The second phase of cereal domestication

The archaeological record suggests that from the three major centres of early plant domestication the concept of agriculture was transferred to the rest of the world. Rice agriculture reached southern Asia from China, wheat and barley cultivation were introduced into Europe, temperate Asia and North Africa from the Near East, and maize spread across the Americas from Central Mexico.

There was little need to domesticate other cereals in places where domesticated wheat, maize and rice could be grown successfully. Maize is an amazingly adaptable cereal. It is native to the tropical highlands of Central America, but soon after domestication was grown also in the lowland tropics of South and Central America, north into what is now southeastern Canada and as far south as Argentina. Wheat and barley rapidly replaced the wild cereals that were harvested in Europe, and rice became a major cereal across tropical Asia.

Wheat and rice, however, have specialized habitat requirements.

Wheat is adapted to a Mediterranean climate, and rice is grown mostly where irrigation is available. Where these cereals could not be grown successfully, native cereals were consciously domesticated by farmers with this newly acquired skill.

Racial evolution of domesticated cereals speeded up dramatically during this second phase of domestication. After generations of growing crops, farmers became conscious of the effectiveness of selection. They now selected phenotypes to suit their fancies and needs, and these selections were grown and isolated. This makes cultivated cereals the most variable of species in the plant kingdom (de Wet, 1981). As an example, Snowden (1936) recognized 28 taxonomic species of cultivated sorghum, all bred by traditional farmers.

CEREALS THAT WERE DOMESTICATED IN AFRICA

Agriculture became established in North Africa during the late fifth millennium BC (Clark, 1962). The North African region was at that time wetter than it is at present. Pollen analyses from beds of former lakes indicate that the Sahara entered the present dry phase about 5000 years ago (Clark, 1976). Wheat and barley were grown well inland in Africa from the Mediterranean coast. Clark (1964) proposes that farmers who grew wheat and barley were pushed south by the expanding desert into the tropics, where they were forced to abandon the cultivation of these cereals. It seems logical to assume that these farmers would adapt indigenous grasses for cultivation. Numerous species were harvested as wild cereals by nomadic tribes of the region when these farmers arrived, and several species were domesticated.

The major African cereals, pearl millet (*Pennisetum glaucum* L. Rich) and sorghum (*Sorghum bicolor* (L.) *Moench*), were domesticated along the southern fringes of the present Sahara. Wild pearl millet is adapted to the Sahelo–Sudanian climatic zone with an annual rainfall of between 350 and 600 mm (Clayton, 1972). Wild sorghum is widely distributed across the Sudanian zone with 600–800 mm of annual rainfall (de Wet, 1978). It also extends south across the savanna (Snowden, 1955). Differences in adaptation allowed for the independent domestication of pearl millet and sorghum. They are still cultivated in Africa primarily in the climatic zones where they were domesticated, and in similar agroecological zones on the Indian subcontinent.

Wild pearl millet is an aggressive colonizer of disturbed habitats. Its colonizing ability allows for large populations that are still harvested for food in times of scarcity. Another grass that is commonly harvested as a wild cereal in this arid zone is *Cenchrus biflorus* Roxb. (kram-kram). This species is a poor candidate for domestication because of the burr-like involucres around the spikelets.

Archaeological evidence presented by Davies (1968) and by Munson (1976) suggests that pearl millet was domesticated at least 3000 years ago. Remains of pearl millet have also been found in farming settlements from Gujarat and the Deccan of India that are dated at about 1000 BC (Rao *et al.*, 1963; Rao, 1967). Since wild pearl millet is not native to India, it must have been introduced into India after it was domesticated in Africa.

Pearl millet is the dominant cereal in the Sahelo–Sudanian zone from the coasts of Senegal and Mauritania to the Sudan (Brunken, de Wet & Harlan, 1977). It is also extensively grown in arid regions of Zambia, Zimbabwe, Namibia, Angola, India and Pakistan. Pearl millet is morphologically variable, but races are not clearly defined, probably because of its narrow adaptation to areas with less than 800 mm of annual rainfall. Brunken *et al.* (1977) pointed out that in Africa pearl millets from the western Sahel and the eastern Sahel can be distinguished on the basis of caryopsis morphology, but that inflorescence morphology does not allow for distinguishing races. Little racial differentiation took place in pearl millet after its introduction into India and southern Africa.

The fonios are two other cultivated cereals of the Sahelo–Sudanian climatic zone of West Africa (Chevalier, 1950; Porteres, 1976). True fonio (*Digitaria exilis* (Kippist) Stapf) is grown from Senegal to Lake Chad, and black fonio (*D. iburua* Stapf) is grown in the Togo highlands and in Nigeria (de Wet, 1977). These two fonios are often grown together in the same field. Stapf (1915) suggested close genetic affinities between *D. ternata* (A. Rich) Stapf and black fonio, and between *D. longifolia* (Retz) Pers. and true fonio.

The fonios probably were extensively harvested as wild cereals before pearl millet was domesticated. They are still frequently grown as encouraged weeds in fields of other crops, and provide a harvest long before pearl millet or sorghum matures. Porteres (1976) recorded that fonio is harvested from some 721 000 acres annually, providing food to over three million people during the most difficult months of the year. Fonios were important food crops in the fourteenth century when the traveller Ibn Batuta recorded that these cereals were extensively available in the markets between Outala in Mauritania and Bamako in Mali (Lewicki, 1974).

Another weed, animal fonio (*Brachiaria deflexa* (Schumach.) C. E. Hubb. *ex* Robyns), is commonly harvested as a wild cereal across the savanna of Africa. Porteres (1951) indicated that it is grown as a cereal in the West African Futa Jalon highlands. It is semi-domesticated in many parts of the African savanna. Farmers often encourage its invasion into sorghum or maize fields where it matures and is harvested about two months after the onset of rains, allowing the main cereal to mature without competition.

Grass weeds differ from domesticated cereals primarily in the ability to dispose seeds naturally. They are spontaneous in cultivated fields and do not require harvesting and sowing for survival. In Ethiopia, *Avena abyssinica* Hochst. is on obligate weed (Ladizinsky, 1975). It is not consciously sown, but has lost the ability to disperse seeds naturally. The species is harvested, threshed, used and sown with the wheat or barley it accompanies as a weed.

Sorghum was domesticated in the Sudano–Guinean climatic zone south of the zone where pearl millet is native, probably at about the same time as pearl millet (Harlan, 1971; Harlan & Stemler, 1976). It reached India as a cereal during the first millennium BC (Vishnu-Mittre, 1968). From India, sorghum spread across tropical Asia and into China, where it became an important crop after the Mogul conquest (Hagerty, 1940). Since the sixteenth century sorghum has also been widely introduced as a cereal into the Americas.

Little is known about the antiquity of sorghum in Africa, except that it was domesticated before 1000 BC when it is known also to have been grown in India. Doggett (1976) suggests that sorghum could have been domesticated 5000 years ago in the Sudan–Ethiopia region. This seems a logical date and region for sorghum domestication. Impressions of sorghum spikelets occur on potsherds from Sudan that date back to 3000 BC (Klichowska, 1984). It is not possible to determine whether these sorghums were harvested from wild or cultivated populations.

Racial evolution in sorghum is associated with geographic isolation (de Wet, 1978). Five basic races are recognized by Harlan & de Wet (1972). Bicolor sorghums resemble their wild progenitor in phenotype but have lost the ability for natural seed dispersal. This race is widely distributed in Africa and India, and represents relics of the first cultivated sorghums. Race kafir is the common sorghum of eastern and southern Africa. Sorghums grown in Chad, Sudan, Uganda and northern Nigeria commonly belong to the race caudatum (Stemler, Harlan & de Wet, 1975). This is an old race. Carbonized grains from Daima, dated to the ninth century (Connah, 1967), belong typically with race caudatum. Race durra is grown across arid West Africa, the Yemens and in India. In West Africa it is primarily grown in the Sudanian zone. It may have evolved in India from where it was introduced into Africa during the expansion of Islam. A sheath of sorghum from a building at Qasr Ibrim and dated to Meroitic times (Plumley, 1970) represents race durra. Sorghum is absent from early farming settlements in South West Asia. The often cited carved reliefs from the palace of Senacherib (Piedallu, 1923) represent the common reed *Phragmites*, and not sorghum (de Wet & Harlan, 1971). Guinea sorghums are grown in the Guinean zone of West Africa with between 900 and 1200 mm of annual rainfall, and in the high rainfall regions of East Africa. Where sorghum races overlap, hybridization occurs and several hybrid races became established. Most significant are guinea–caudatum in West Africa, kafir–caudatum in East Africa and guinea–kafir in parts of southern Africa.

Finger millet (*Eleusine coracana* (L.) Gaertn.) is another African cereal that was introduced into India during the first millennium BC (Vishnu-Mittre, 1968). This millet was domesticated in eastern Africa, probably in Uganda or Ethiopia. It became widely distributed in the African savanna south of the equator during the Iron Age, but its cultivation never extended west beyond Uganda, Rwanda and Burundi.

Finger millet has been identified as impressions on potsherds from the Neolithic settlement of Kadero in Central Sudan dating back about 5000 years (Klichowska, 1984). If this date, and the date for finger millet at Axum (3000 BC) are correct (Hilu, de Wet & Harlan, 1979), it may be the oldest indigenous domesticated cereal in Africa. The archaeological remains from Axum resemble the highly evolved race plana that is still grown in Ethiopia, suggesting a much earlier date than 3000 BC for initial domestication. This is not impossible. Agriculture could have been introduced into East Africa from Egypt or directly from South West Asia before the onset of the present dry phase of North Africa. Vishnu-Mittre (1968) presented archaeological evidence to show that finger millet has been grown in India for at least 3000 years. Finger millets grown in India today resemble those from eastern Africa, suggesting that the cereal was introduced into India after racial evolution was well established.

Inflorescence morphology of finger millet is not closely correlated with geographical distribution (Hilu & de Wet, 1976). Inflorescence morphology nevertheless allows for the recognition of five races (de Wet *et al.*, 1984*a*). Race coracana is the most widely distributed finger millet. It is particularly well adapted to agriculture in the East African highlands and the Ghats of India. Race elongata is grown in Malawi and adjacent Zambia, and in the eastern Ghats of India. Race plana is primarily grown in Ethiopia and Uganda. Its presence in the eastern and western Ghats of India may represent recent introductions. Races compacta and vulgaris are widely grown in Africa and India.

Two other native African cereals, teff and African rice, should be mentioned. Teff (*Eragrostis tef* (Zucc.) Trotter) is endemic to the highlands of Ethiopia, where it competes successfully with barley and wheat (Costanza, de Wet & Harlan, 1979). The cultivation of teff may represent an independent origin of agriculture from that of South West Asia or North Africa. It seems unlikely that people would have domesticated teff if they already had wheat and barley (Clark, 1984).

African rice (*Oryza glaberrima* Steud.) is indigenous to the swamps of West Africa, where wild and cultivated kinds are frequently harvested together (Harlan, 1969). This species is further discussed in chapter 10 on the origins of rice.

CEREALS THAT WERE DOMESTICATED IN INDIA

A number of cereals were domesticated in India. This is not surprising. Wheat can be grown successfully in India only along the southwestern fringes of the Himalayas, and rice cultivation is restricted to areas with irrigation. What is surprising is that these indigenous cereals continue to compete successfully with introduced pearl millet, sorghum and finger millet. This would suggest that the Indian millets were domesticated before the introduction of these African cereals.

Raishan (*Digitaria cruciata* (Nees) A. Camus) and adlay (*Coix lacryma-jobi* L.) were domesticated in the wet tropics of northeastern India. Raishan is grown by the Khasi people of Assam (Singh & Arora, 1972). It is also grown by hill tribes in Vietnam (Veldkamp, 1973). Raishan is commonly grown as a second crop in fields that were previously planted with vegetables. It probably survives as a cereal in Assam because the hay provides excellent feed for livestock.

Adlay is grown under shifting cultivation by hill tribes from Assam to the Philippines (Arora, 1977). It was probably introduced into South

East Asia during the expansion of Buddhism from India. It is also possible that adlay was independently domesticated as a cereal in India and in the Philippines. The greatest diversity of cultivated adlay occurs in the Philippines (Wester, 1920).

The wild adlay (Job's tears) was probably used as beads long before the species became a cereal. Fertile female spikelets of wild *Coix* species are individually enclosed by an involucre that is indurated, glossy and from white to black in colour. These are commonly used as beads to make necklaces. The involucres of cultivated adlay are papery, allowing for easy removal of the caryopses from the fruit cases.

Five minor cereals, *Panicum sumatrense* (Roth.) *ex* Roem. *et* Schult., *Echinochloa colona* (L.) Link, *Paspalum scrobiculatum* L., *Setaria pumila* (Poir.) Roem. *et* Schult. and *Brachiaria ramosa* (L.) Stapf, are grown and were probably domesticated in semi-arid India.

Sama (*Panicum sumatrense*) is widely grown in India, Nepal, Sikkim and western Mayanmar (de Wet *et al.*, 1984*b*). It is an important cereal in the eastern Ghats of Andhra Pradesh and adjacent Orissa. It is drought tolerant and will produce a crop even on the poorest agricultural soil. Sama is commonly sown as a mixture with foxtail millet in sorghum or pearl millet fields. Sama usually matures first, followed by foxtail millet, and sorghum or pearl millet is harvested at the end of the rainy season. This provides a supply of freshly harvested cereals starting about two months after planting. Primitive cultivars resemble the widely distributed *Panicum psilopodium* Trin., except for their persistent spikelets. The highly evolved race robusta is often planted as a single crop.

Sawa (*Echinocloa colona*) is grown in India, Nepal and Sikkim (de Wet *et al.*, 1983*a*). It is closely allied to Japanese millet (*Echinochloa cruss-galli* (L.) P. Beauv.), but tropical rather than temperate in distribution. These two *Echinochloa* species are genetically isolated (Yabuno, 1966). Sawa was domesticated in India and Japanese millet in northwestern China. Primitive races of sawa grown in India differ from the wild *E. colona* only in the tardy disarticulation of their spikelets at maturity. These weedy sawas are often grown as mixtures with the cultivated races laxum and robustum. Sawa has so far not been recorded in the archaeological remains from early farming settlements in India. It may have been domesticated from a weed in cultivated fields of other cereals.

Khodo (*Paspalum scrobiculatum*) is widely grown in India. The species occurs wild across the Old World tropics (Clayton & Renvoize,

1982). It is an aggressive colonizer and often invades agricultural fields as a weed. The species lends itself to domestication, and occurs in the agricultural record of India starting 3000 years ago (Kajale, 1977; Vishnu-Mittre, 1977). In southern India small-seeded and large-seeded cultivars are recognized by farmers. It is an important cereal in Kerala and is also grown from Tamil Nadu north to Rajasthan, Uttar Pradesh, Bihar and West Bengal (de Wet *et al.*, 1983*b*).

Farmers believe that khodo grains are poisonous when harvested after a rain. Bhide & Aimen (1959) proposed that the spikelets produce a poisonous alkaloid. It is more likely that poisoning occurs because of infection by the ergot fungus. In dry weather infection does not occur.

Korali (*Setaria pumila*) and *Brachiaria ramosa* are domesticated weeds. These species are widely distributed in tropical Africa and Asia. They are commonly harvested as wild cereals, but grown as domesticated cereals only by hill tribes of the eastern and western Ghats of India. Weedy kinds are commonly found growing with these cereals. Farmers tend to keep the domesticated kinds pure through selection.

The third phase of cereal domestication

Domestication is evolution under a selection regime of harvesting and sowing. This process continues with each generation a crop is grown. The third phase of cereal evolution was introduced through conscious breeding for increase in yield potential.

Racial boundaries within cereal species are rapidly disappearing as plant breeders recombine the genetic variation of once isolated land races, and traditional cultivars are being replaced by improved cultivars. This trend is not likely to be reversed. Cereal production is not keeping pace with demands from an ever-expanding human population. New cultivars are needed with high yield potential not under only conditions of high agricultural inputs, but also when grown in marginal agricultural lands. Wheat, rice and maize today feed over half the human population. Other important cereals are barley, sorghum, pearl millet and finger millet.

It is sometimes proposed that plant breeders should domesticate new cereals for cultivation in marginal agricultural lands. There are numerous wild grass species that are aggressive colonizers and potential domesticated cereals. It seems a more promising strategy, however, to improve the existing minor cereals already adapted to marginal agricultural lands, than to domesticate new ones.

Only one grass species *Zizania aquatica* L. (American wild rice) has been domesticated in historical times. It was domesticated because production from wild populations did not meet demand.

American wild rice has been harvested as a cereal from rivers and lakes in the United States and adjacent Canada since long before recorded history (Coville & Coves, 1894). Charred remains of caryopses from threshing pits are known to date from well before contact with Europeans (Ford & Brose, 1975).

Wild rice was domesticated in the 1970s (de Wet & Oelke, 1979). It has traditionally been abundantly available from wild populations. There was no need to practise any plant husbandry, except to expand the distribution of the species through sowing.

Harvesters of wild rice belonged traditionally to people of the Algonquian and Siouan linguistic groups. Numerous different tribes harvested this cereal, and the Chippewa and the Manomini relied on it as a staple food. Indian legends refer to wild rice as a gift from Manabush. He created the bear out of earth, from the bear he made Skekatchekenan, and to him Manabush gave the Manomine river with all the kinds of fish in it, the sugar maples that grow along its shores, manomin the rice that grows in its shallows, and an assurance that the Manomini tribe will always have these things (Jenks, 1900). Today the Manomini are lumbermen, but the Chippewa remain commercial harvesters of this wild cereal.

Early European explorers in North America were impressed with the extensive use of this wild rice (Charlevoix, 1763). Carver (1778) reported that wild rice is the most valuable of all the native wild food plants of the country. The first known attempt to introduce *Zizania* into Europe was by the Swedish naturalist Peter Kalm who travelled in the country during 1748, and brought home to Sweden seed of 126 different plant species. He recorded that in North America where the species grows wild, the natives collect it as food, and that if it could be made to grow in Europe, the wettest place could be made productive (Larsen, 1939). This attempt at introducing the species into Europe failed. Caryopses rapidly lose viability when not kept wet. Lambert (1804), however, reported that seeds he sent to the Royal Botanic Gardens at Kew in a bottle of water germinated well, and that the species was successfully introduced into ponds on several English country estates.

The species does not lend itself to domestication. Its habitat requirements are precise, as was discovered by modern plant breeders. A specialized cropping system had to be developed before this wild rice could be grown commercially.

Seed viability and dormancy were major problems to be solved in domesticating the species. Others were habitat adaptation and natural seed dispersal. Northern wild rice, the race commonly harvested, grows naturally along the margins of rivers and marshes with muddy bottoms that are flooded with 40–100 cm of water during spring and maintain a depth of at least 10 cm until the plants mature.

Domestication involved a combination of selection for spikelets that were persistent at maturity, and the development of a cropping system that took advantage of the natural adaptation of the species. The species is now successfully grown commercially. Paddies are constructed so that a minimum water depth of 15 cm can be maintained. Paddies are flooded and seeded in early spring. Germination is rapid, and water level is maintained until August when the fields are drained and the crop is mechanically harvested. The cultivated kind retains some degree of natural seed dispersal, and sufficient spikelets fall to the ground to make future sowing unnecessary. The straw and self-sown spikelets are incorporated into the wet soil during the fall, where the grains lie dormant until the next spring when the paddies are again flooded. Selection continues for genotypes that have lost the ability to disperse seeds naturally, and with caryopses that will remain viable without being immersed in water (de Wet & Oelke, 1979).

The future of sorghum and millets

The demand for cereals is increasing rapidly. In developed countries this demand is as animal feed and raw materials in industry. In other countries demand is as human food. The major rainfed cereals will continue to be maize, sorghum, pearl millet and finger millet.

Sorghum is grown on about 18 million hectares in Africa, where maize frequently fails to produce a harvest because rainfall is unpredictable and variable. Pearl millet is grown in Africa on about 16 million hectares, mostly on sandy soils in areas with less than 600 mm of annual rainfall, where no other cereal can consistently produce a harvest because of frequent droughts, and high air and soil temperatures. In Asia sorghum is grown on about 20 million hectares and pearl millet on about 13 million hectares. Finger millet competes successfully with maize in parts of Africa and India on marginal agricultural land. Sorghum is also grown on about 9 million hectares in the Americas, and on about one million hectares in Oceania, mostly in Australia. As in Africa and Asia, sorghum production in the Americas is confined to areas with less than 1000 mm of annual rainfall, or to marginal agricultural land in higher rainfall regions. These include the poorly drained, seasonally flooded areas of northeastern South America. The minor millets of Africa and Asia are essential components of rainfed agriculture on marginal farming areas, and their production is expected to increase as population increases.

Sorghum and pearl millet compare favourably with major cereals in yield potential under high-input agriculture, and are superior to the major cereals under low-input agriculture. Their grain and nutritional quality also compare favourably or even exceed those of major cereals.

Major genetic constraints on increasing cereal production in marginal agriculture areas are poor seedling establishment, poor seed set under terminal heat and drought stress, and susceptibility to diseases, insect pests, and the parasitic weed, *Striga*. Rapid progress is being made to eliminate these constraints from breeding populations with high yield potential.

Genetic resistance to regionally important fungal diseases has been identified. These resistances are heritable and readily introduced into breeding populations. Transfer of resistance to downy mildew (*Sclerospora graminicola*) from African populations to Indian cultivars of pearl millet saved this cereal when a particularly virulent race of the pathogen threatened production in 1982. Progress is also being made to introduce resistance to *Striga asiatica* and *S. hermonthica*, and resistance to midge (*Contarinia sorghicola*) into sorghum breeding populations. Resistance to stemborers in sorghum and pearl millet, shootfly in sorghum, the earhead caterpillar in pearl millet and grain moulds in sorghum are difficult to manipulate genetically. So also is breeding for effective seed set under heat and drought stress. In pearl millet also, genetic resistance to *Striga* appears to be lacking.

Research at the International Crops Research Institute for the Semi-Arid Tropics (ICRISAT) predicts that modern plant breeding techniques will overcome these constraints to sorghum and pearl millet production. It is possible to breed cultivars that will mature within the period of adequate moisture and still produce an acceptable harvest. Cultivars of pearl millet were bred that mature within the 65 day assured growing season of Rajasthan in India with less than 500 mm of annual rainfall.

Techniques of embryo rescue through *in vitro* culture and the manipulation of chromosome pairing are being used to transfer resistance to shootfly from wild to cultivated sorghums. Molecular markers, particularly restriction fragment length polymorphisms

(RFLPs), are being identified for use in tracing quantitative and recessive genetic traits in segregating populations. This will facilitate genetic transfer of traits such as seed set under stress and insect resistance from land races to cultivars with high yield potential.

Developmental biological studies indicate that susceptibility to *Striga* in sorghum is controlled by root-produced substances that independently control germination of *Striga* seeds, orientation of the *Striga* root towards the sorghum root, and haustorium formation. Identification of these substances is a first step in identifying and cloning genes controlling resistance, and transferring resistance from sorghum to pearl millet using genetic transformation.

Research to improve disease and pest resistance and regional adaptation of minor millets holds tremendous promise. Selection for resistance of finger millet to blast caused by the fungus *Pyricularia* has been successful. Hybrids among Eurasian cultivars of foxtail and proso millets have shown vigour that increases yield several fold. The fonios are among the most drought tolerant of cereals. They are increasing in importance as crops to help to stabilize the southward expansion of the Sahel. Research directed towards increasing their yield potential under extreme heat and increasing drought stress is underway.

Sorghum, pearl millet and other millets are important components of traditional agriculture in Africa and Asia. Their cultivation is increasing in importance as crop production moves into marginal agricultural areas because of population pressure. Sorghum is also becoming a major cereal in the Americas, where it is in great demand as animal feed. Research into increasing production under low-input farming, and into fitting these cereals into modern agriculture is proving successful. They will continue to feed the rapidly expanding populations of semi-arid sub-Saharan Africa, South Asia and semi-arid China.

Summary

This survey allows for the formulation of a number of concepts related to cereal domestication.

1. There were three phases of cereal domestication. The first phase started with the origins of agriculture. The second phase was initiated when the concept of agriculture spread from these primary centres of plant domestication into areas where the available domesticated cereals were not adapted. The third phase started with the development of modern plant breeding.

Most cereals were domesticated during the second phase of domestication.

2. The progenitors of dometicated cereals were aggressive colonizers of disturbed habitats.
3. These species were harvested as wild cereals for millennia before they became domesticated.
4. Loss of natural seed dispersal is a basic characteristic of all domesticated cereals.
5. Domestication depends on selection pressures associated with a combination of harvesting and sowing.
6. Racial evolution in cereals is primarily associated with conscious selection by farmers for specific uses and for phenotypes to suit their fancies.
7. Yield potential of cereal cultivars has been greatly increased by modern plant breeding at the expense of racial evolution.

References

Arora, R. K. (1977). Adlay (*Coix*) crop in Maghalaya. *Journal of the Indian Botanical Society*, **52**, 95–98.

Ascherson, P. & Graebner, P. (1890). *Synopsis der Miteleuropaischen Flora*, 2. Berlin.

Ball, J. (1884). Contributions to the flora of North Patagonia and the adjacent territory. *Journal of the Society of Botany, London*, **21**, 203–240.

Bhide, N. V. & Aimen, R. R. (1959). Pharmacology of a tranquilizing principle in *Paspalum scrobiculatum* grains. *Nature*, **153**, 1735–1736.

Brunken, J. N., de Wet, J. M. J. & Harlan, J. R. (1977). The morphology and domestication of pearl millet. *Economic Botany*, **31**, 163–174.

Callen, E. O. (1965). Food habits of some pre-Columbian Mexican Indians. *Economic Botany*, **19**, 335–343.

Callen, E. O. (1967). The First New World cereal. *American Antiquity*, **32**, 535–538.

Carver, J. (1778). *Travels through interior parts of North America in the years 1766, 1767 and 1768*. London: printed for the author.

Charlevoix, P. de (1763). *Letters to the Duchess of Lesdiguives, giving an account of a voyage to Canada and travels through that vast country and Louisiana to the Gulf of Mexico*. London: R. Goodby.

Cheng, K. (1973). Radio carbon dates from China: some initial interpretations. *Current Anthropology*, **14**, 525–528.

Chevalier, A. (1950). Sur l'origine des *Digitaria* cultives. *Revue International Botanique Appliqué d'Agriculture Tropical*, **12**, 669–919.

Chomko, S. A. & Crawford, G. W. (1978). Plant husbandry in pre-historic eastern North America: new evidence for its development. *American Antiquity*, **43**, 405–408.

Clark, J. D. (1962). The spread of food production in sub-Saharan Africa. *Journal of African History*, 3, 211–228.

Clark, J. D. (1964). The prehistoric origins of African culture. *Journal of African History*, 5, 161–183.

Clark, J. D. (1976). Prehistoric populations and pressures favoring plant domestication in Africa. In *Origins of African plant domestication*, ed. J. R. Harlan, J. M. J. de Wet & A. B. L. Stemler, pp. 67–105. The Hague: Mouton Publishers.

Clark, J. D. (1984). The domestication process in northeastern Africa: ecological change and adaptive strategies. In *Origin and early development of food-producing cultures in north-eastern Africa*, ed. L. Krzyzaniak & M. Kobusiewicz, pp. 27–41. Poznan: Polish Academy of Sciences.

Clayton, W. D. (1972). Gramineae. In *Flora of tropical Africa*, vol. 3, part 2, ed. F. N. Hepper, pp. 459–469. London: Crown Agents.

Clayton, W. D. & Renvoize, S. A. (1982). Gramineae. In *Flora of tropical East Africa*, part 3, ed. R. M. Polhill, pp. 607–612. Rotterdam: A. A. Balkema.

Connah, G. (1967). Progress report on archaeological work in Bornu 1964–1966 with particular reference to the excavations at Daima mound. *Northern History Research Scheme, 2nd Interim Report*. Zaria.

Costanza, S. H., de Wet, J. M. J. & Harlan, J. R. (1979). Literature review and numerical taxonomy of *Eragrostis tef* (t'ef). *Economic Botany*, 33, 413–414.

Coville, F. V. & Coves, E. (1894). The wild rice of Minnesota. *Botanical Gazette*, 19, 504–506.

Cowan, C. W. (1978). The prehistoric use and distribution of maygrass in eastern North America: culture and phytogeographical implications. In *Nature and Status of Ethnobotany*, University of Michigan Anthropological papers No. 67, ed. R. I. Ford, pp. 263–288. Ann Arbor: University of Michigan.

Davies, O. (1968). The origins of agriculture in West Africa. *Current Anthropology*, 9, 479–482.

Dekaprelerich, L. L. & Kasparian, A. S. (1928). A contribution to the study of Foxtail millet (*Setaria italica* P.B. *maxima* Alf.) cultivated in Georgia (Western Transcaucasia). *Bulletin of Applied Botany and Plant Breeding*, 19, 533–572.

de Wet, J. M. J. (1977). Domestication of African cereals. *African Economic History*, 3, 15–32.

de Wet, J. M. J. (1978). Systematics and evolution of *Sorghum* sect. *Sorghum* (Gramineae). *American Journal of Botany*, 65, 477–484.

de Wet, J. M. J. (1979). Principles of evolution and cereal domestication. In *Broadening the genetic base of crops*, ed. A. C. Zeven & A. M. Harlen, pp. 269–282. Wageningen: PUDOC.

de Wet, J. M. J. (1981). Species concepts and systematics of domesticated cereals. *Külturpflanzen*, 29, 177–198.

de Wet, J. M. J. & Harlan, J. R. (1971). Origin and domestication of *Sorghum bicolor*. *Economic Botany*, 25, 128–135.

de Wet, J. M. J. & Oelke, E. A. (1979). Domestication of American wild rice

(*Zizania aquatica* L. Gramineae). *Journal d'Agriculture Traditional et Botanique Appliqué*, **30**, 159–168.

de Wet, J. M. J., Oestry-Stidd, L. L. & Cubero, J. I. (1979). Origins and evolution of foxtail millets (*Setaria italica*). *Journal d'Agriculture Traditional et Botanique Applique*, **30**, 53–64.

de Wet, J. M. J., Prasada Rao, K. E. & Brink, D. E. (1984*b*). Systematics and domestication of *Panicum sumatrense* (Gramineae). *Journal d'agriculture traditional et botanique Appliqué*, **30**, 159–168.

de Wet, J. M. J., Prasada Rao, K. E., Brink, D. E. & Mengesha, M. H. (1984*a*). Systematics and evolution of *Eleusine coracana* (Gramineae). *American Journal of Botany*, **71**, 550–557.

de Wet, J. M. J., Prasada Rao, K. E., Mengesha, M. H. & Brink, D. E. (1983*a*). Domestication of sawa millet (*Echinochloa colona*). *Economic Botany*, **37**, 283–291.

de Wet, J. M. J., Prasada Rao, K. E., Mengesha, M. H. & Brink, D. E. (1983*b*). Diversity in Kodo millet, *Paspalum scrobiculatum*. *Economic Botany*, **37**, 159–163.

Doggett, H. (1976). Sorghum, *Sorghum bicolor* (Gramineae-Andropogoneae). In *Evolution of crop plants*, ed. N. W. Simmonds, pp. 112–117. London: Longman.

Elsasser, A. B. (1979). Explorations of Hernando Alarcon in the lower Colorado river, 1540. *Journal of California Great Basin Anthropology*, **1**, 8–39.

Evans, G. M. (1976). Rye, *Secale cereale* (Gramineae–Triticinae). In *Evolution of crop plants*, ed. N. W. Simmonds, pp. 108–111. London: Longman.

Ford, R. I. & Brose, D. S. (1975). Prehistoric wild rice from Dunn farm site, Leelanau Country, Michigan. *Wisconsin Archaeologist*, **56**, 9–15.

Gay, C. (1865). *Historia Fisica y Politica de Chile, Agricultura Chilena*, 2 vols. Paris.

Gritzenko, R. J. (1960). Chumiza (Italian millet), taxonomy (*Setaria italica* (L.) P.B. subsp. *maxima* Alef.). *Bulletin of Applied Botany, Genetics and Plant Breeding*, **32**, 145–182.

Hagerty, M. (1940). Comments on writing concerning Chinese sorghums. *Harvard Journal of Asiatic Studies*, **5**, 234–260.

Harlan, J. R. (1969). Evolutionary dynamics of plant domestication. *Japanese Journal of Genetics*, **44**, Suppl. 1, 377–343.

Harlan, J. R. (1971). Agricultural origins: centers and non-centers. *Science*, **174**, 463–474.

Harlan, J. R. (1981). The early history of wheat: earliest traces to the sack of Rome. In *Wheat science today and tomorrow*, ed. L. T. Evans & W. J. Peacock, pp. 1–19. Cambridge: Cambridge University Press.

Harlan, J. R. & de Wet, J. M. J. (1972). A simplified classification of cultivated sorghum. *Crop Science*, **12**, 172–176.

Harlan, J. R., de Wet, J. M. J. & Price, E. G. (1973). Comparative evolution of cereals. *Evolution*, **27**, 311–325.

Harlan, J. R. & Stemler, A. B. L. (1976). The races of sorghum in Africa. In *Origins of African plant domestication*, ed. J. R. Harlan, J. M. J. de Wet & A. B. L. Stemler, pp. 465–478. The Hague: Mouton Publishers.

Helbaek, H. (1959). Domestication of food plants in the Old World. *Science*, **130**, 365–372.

Henrard, J. Th. (1950). *Monograph on the genus* Digitaria. Leiden: University of Leiden Press.

Hilu, K. W. & de Wet, J. M. J. (1976). Domestication of *Eleusine coracana*. *Economic Botany*, **30**, 199–208.

Hilu, K. W., de Wet, J. M. J. & Harlan, J. R. (1979). Archaeobotanical studies of *Eleusine coracana* subsp. *coracana* (finger millet). *American Journal of Botany*, **66**, 330–333.

Ho, Ping-Ti (1975). *The cradle of the East.* Chicago: Chicago University Press.

Holden, J. H. W. (1976). Oats, *Avena* spp. (Gramineae–Aveneae). In *Evolution of crop plants*, ed. N. W. Simmonds, pp. 86–90. London: Longman.

Jenks, A. E. (1900). The wild rice gatherers of the upper lakes. *Annual Report of the American Bureau of Ethnology* 1989 part 2, **19**, 1013–1137.

Kaemlein, W. (1936). A prehistoric twined-woven bag from the Trigo mountains, Arizona. *Kiva*, **28**, 1–13.

Kajale, M. P. (1977). Ancient grains from excavations at Nevassa, Maharashtra. *Geophytologia*, **7**, 98–106.

Kelly, W. H. (1977). Cocopa ethnography. *Anthropological Papers of the University of Arizona*, **29**, 1–150.

Khush, G. S. (1963). Cytogenetic and evolutionary studies on *Secale*. III. Cytogenetics of weedy ryes and origin of cultivated rye. *Economic Botany*, **17**, 60–71.

Kitagawa, M. (1937). Contributio ad Cognitionem Florae Manshuricae. *Botanical Magazine, Tokyo*, **51**, 150–157.

Klichowska, M. (1984). Plants of the Neolithic Kadero (Central Sudan): a palaeobotanical study of the plant impressions on pottery. In *Origins and early development of food-producing cultures in north-eastern Africa*, ed. L. Krzyzaniak & M. Kobusiewicz, pp. 321–326. Poznan: Polish Academy of Sciences.

Ladizinsky, G. (1975). Oats in Ethiopia. *Economic Botany*, **29**, 238–241.

Laet, J. de (1633). *Novus orbis sen descriptionis Indiae occidentalis.* Batavia: Elzevirios El Libro XII.

Lambert, A. B. (1804). Observations on the *Zizania aquatica. Transactions of the Linnean Society of London*, **7**, 264–265.

Larsen, E. L. (1939). Peter Kalm's short account of the natural use and care of some plants, of which the seeds were recently brought home from North America to the service of those who take pleasure in experimenting with the cultivation of the same in our climate. *Agricultural History*, **13**, 34, 43–44.

Lewicki, T. (1974). *West African food in the Middle Ages.* London: Cambridge University Press.

Lyssov, B. H. (1975). Proso (*Panicum* L.). In *Flora of cultivated plants. III. Croat Crops*, ed. A. S. Krotov. Leningrad: Kolos.

Mangelsdorf, P. C., MacNeish, R. S. & Galinat, W. C. (1967). Prehistoric wild and cultivated maize. In *The prehistory of the Tehuacan Valley*, ed. D. S. Byers, pp. 178–200. Austin, Tx: University of Texas Press.

Matthiolus, P. A. (1565). *Commentarii in Libros ex Pedacii Dioscoridis.* Venice: Anazarbei de Medica Materia.

Munson, P. J. (1976). Archaeological data on the origins of cultivation in the southwestern Sahara and their implications for West Africa. In *Origins of African plant domestication*, ed. J. R. Harlan, J. M. J. de Wet & A. B. L. Stemler, pp. 187–209. The Hague: Mouton Publishers.

Nabhan, G. & de Wet, J. M. J. (1984). *Panicum sonorum* in the Sonoran desert agriculture. *Economic Botany*, **38**, 65–82.

Neuweiler, E. (1946). Nachtrage Urgeschichtlicher Pflanzen. *Vierteljahrsschriften der Natureforschung Gesellschaft, Zurich*, **91**, 122–236.

Palmer, E. (1871). Food products of the North American Indians. *United States Commerce and Agriculture reports* 1870, pp. 404–428.

Parodi, L. R. & Hernandez, J. C. (1964). El Mango, cereal extinguado en cultivo, sobre vive en estado salvage. *Ciencia et Investigado*, **20**, 543–549.

Piedallu, A. (1923). *Le Sorgo, son historie, ses applications.* Paris: Challamel Press.

Plumley, J. M. (1970). Qusr Ibrim 1969. *Journal of Egyptian Archaeology*, **56**, 12–18.

Porteres, R. (1951). Une cereale mineure cultivée dans l'Ouest-Africain (*Brachiaria deflexa* C. E. Hubbard var. *sativa nov. var.*). *L'Agronomique Tropicale*, **6**, 39–42.

Porteres, R. (1976). African cereals: *Eleusine*, fonio, black fonio, teff, *Brachiaria, Paspalum, Pennisetum* and African rice. In *Origins of African plant domestication*, ed. J. R. Harlan, J. M. J. de Wet & A. B. L. Stemler, pp. 409–452. The Hague: Mouton Publishers.

Prasada Rao, K. E., de Wet, J. M. J., Brink, D. E. & Mengesha, M. H. (1987). Infraspecific variation and systematics of cultivated *Setaria italica*, foxtail millet (Poaceae). *Economic Botany*, **41**, 108–116.

Rao, S. R. (1967). *Lothal and Indus Valley civilization.* Bombay: Asia Publishing House.

Rao, S. R., Lal, B. B., Nath, B., Ghoshe, S. S. & Lal, K. (1963). Excavations at Rangpur and other explorations in Gujarat. *Bulletin of the Archaeological Survey of India*, **18/19**, 5–207.

Singh, H. B. & Arora, R. K. (1972). Raishan (*Digitaria* sp.), a minor millet of the Khasi Hills, India. *Economic Botany*, **26**, 376–380.

Snowden, J. D. (1936). *The cultivated races of sorghum.* London: Adlard & Son.

Snowden, J. D. (1955). The wild fodder sorghums of the section *Eu-Sorghum*. *Journal of the Linnean Society, London*, **55**, 191–260.

Stapf, O. (1915). Iburu and fundi, two cereals of Upper Guinea. *Kew Bulletin*, 1915, 381–386.

Stemler, A. B. L., Harlan, J. R. & de Wet, J. M. J. (1975). Caudatum sorghums and speakers of Chari-Nile languages in Africa. *Journal of African History*, **16**, 161–183.

Veldkamp, J. K. (1973). A revision of *Digitaria* Haller (Gramineae) in Malesia. *Blumea*, **21**, 1–80.

Villaret von Rochow, M. (1971). *Avena ludoviciana* Dur. in the late Neolithic of

Switzerland, a contribution to the origin of the oats (*Avena sativa* L.). *Berichte der Deutsche botanische Gessellschaft,* **84**, 243–248.

Vishnu-Mittre. (1968). Prehistoric records of agriculture in India. *Transactions of the Bose Research Institute,* **31**, 87–106.

Vishnu-Mittre. (1977). Changing economy in ancient India. In *Origins of agriculture,* ed. C. A. Reed, pp. 569–588. The Hague: Mouton Publishers.

Werth, E. (1937). Zur Geographie und Geschichte der Hirzen. *Angewandte Botanik,* **19**, 42–88.

Wester, P. S. (1920). Notes on adlay. *Philippine Agricultural Review,* **13**, 217–222.

Yabuno, T. (1966). Biosystematic studies of the genus *Echinochloa* (Gramineae). *Japanese Journal of Botany,* **19**, 277–323.

Note added in proof: Vernacular names of cereals can cause considerable uncertainty, the term 'millet' by itself being obviously insufficient except in a very restricted context. To uncertainty can be added confusion by similar names being applied to different cereals as the following brief selection will indicate which includes those given in chapter 6.

Echinochloa colona   – jungle rice, Kudiravali, Shama millet, sawa.
Panicum sumatrense   – little millet, Mutki, samai, sama
Echinochloa crus-galli   – sawa millet, sawan
Echinochloa frumentacea – Sanwa millet, sawan

(D. V. Field, pers. comm.). The first of these has also, for example, five Arabic names in the Sudan all rather similar Anon (1984) *Forage and browse plants for arid and semi-arid Africa.* IBPGR/Royal Botanic Gardens, Kew. [G.P.C.]

7

Domestication of cereals
M. S. Davies and Gordon C. Hillman

Humans have long exploited the Gramineae as a food source. Ancient systems of harvesting the grains of wild populations of grasses, which may have originated in Palaeolithic times, are still utilized today by groups in Africa and other continents (Harlan, 1975; Bohrer 1991). Contrary to the widespread misconception that wild-grain harvesting is a last resort to avert starvation, natural stands of grasses can give comparatively high yields of grain (Harlan, 1989; Zohary, 1969) and they have provided a central pillar of subsistence for a broad spectrum of hunter–gatherer societies, even in recent times (for a useful compendium of examples, see Maurizio, 1927).

Harvesting of wild grass seed by beating with paddles into a container or by using the 'swinging basket' technique can result in a particularly high collection rate per unit time (see below) and is particularly efficient when the harvestable yield is considered in terms of the energy expended in collecting it. Harlan (1989) cited many quoted examples of the harvesting of wild grass seeds in Africa in the nineteenth and early part of the twentieth century, particularly in desert, savanna and swamp-land regions. However, the spread of agriculture and crop cultivation has led to a decline in the harvesting of grass seeds as a staple food source, although grass seeds still constitute a valuable supplemental food supply in many areas.

Wild-grass harvesting is not likely to have favoured the genetic changes necessary to have brought about the evolution of domesticated cereals from their wild ancestors (see below) and whereas such harvesting may be a preliminary stage, the process of domestication itself does not begin until the seed which is harvested is planted as the seed corn for next year's crop. The shift from hunting and gathering to cultivation and pastoralism represents one of the most dramatic changes that human society has experienced. However, the reasons for the shift continue to be the subject of active debate and we will not attempt an overview here. One of the practical consequences of this change was the domestication of crop plants and it is this event which forms the subject of our present outline.

In this chapter we adhere to a narrow classical definition of domestication. Here 'domestication' refers to that process which (a) occurs under cultivation in populations of early wild-type crops (sown originally from seed gathered from wild stands), (b) selectively advantages rare mutant plants lacking features (particularly reproductive features) necessary for their survival in the wild and (c) continues until these mutant phenotypes dominate the crop population and the original wild-type phenotypes are eliminated or maintained at very low frequency. Thus the domestication process, because of the loss of wild-type adaptive features, renders the domestic type dependent upon human intervention for its survival.

Domestication was recognized as an example of accelerated evolution by both Darwin (1859, 1868) and de Candolle (1886), but it was Vavilov (1917, 1926) who first postulated specific evolutionary pathways for the domestication of cereals such as the wild wheats and barleys and for the secondary domestication of cereals such as rye and oats. Definition of these evolutionary pathways formed an integral part of Vavilov's analysis and identification of centres of origin of cultivated plants (Harris, 1990; Hawkes, 1990). Since then many other studies have extended our understanding of the possible processes involved (e.g. Darlington, 1969, 1973; de Wet & Harlan, 1975; Hawkes, 1969; Heiser, 1988; Pickersgill, 1989; Zohary, 1969, 1984, 1989*a*).

Wheats and barley

The seminal paper by Wilke *et al.* (1972) suggested a mechanism for domestication in the wild wheats and barleys. However, their hypothesis overlooked some key factors necessary for domestication to have occurred and they made no estimate of the time taken for domestication to occur under their proposed system of productive cultivation. The few published estimates of the time-scale of domestication range from 1 to 1000 years, but many crop geneticists such as Harlan (1975) and Zohary (1969, 1984, 1989*b*) have recognized that domestication of these crops could have occurred extremely rapidly. The first cereal crops must have been sown from seed gathered from wild stands, and these must have been essentially of the wild type. Domestication thus occurred during the course of cultivation. In the case of einkorn wheat, for example, the domestic form (*Triticum monococcum* L. subsp. *monococcum*) emerged from crops of its immediate anestor – wild einkorn (*T. monococcum* L. subsp. *boeoticum* (Boiss.) A. & D. Löve), which still grows wild in the Near East – mainly in the ecotone between oak forest and steppe. (Some authors have suggested that domestication might have occurred in the wild, prior to any cultivation (e.g. Ladizinsky, 1987; Blumler & Byrne, 1991). However, their proposed models seem highly implausible and we do not consider them in this chapter.)

In this chapter we consider the likely advantages and disadvantages of domestic mutants of wild-type diploid wheat and barley under various systems of primitive cultivation and consider the outcome of a simple mathematical model to simulate the increase in frequency, with time, of the domestic phenotype, starting in a wild-type stand, at various levels of advantage to the domestic phenotype.

DIFFERENCES BETWEEN WILD AND DOMESTICATED FORMS OF WHEAT AND BARLEY

In cereals such as wheat, even the most primitive domesticated forms today differ from their wild progenitors in a number of polygenically determined grade characters such as awn robustness, glume rigidity, grain size, number of fertile florets, tillering tendency, uniformity of grain ripening, photosynthetic rate, and the abundance of barbs and hairs on the rachis and glumes (see, for example, Percival, 1921; Schwanitz, 1937; Schiemann, 1948; Darlington, 1969, 1973; Zohary, 1969, 1984, 1989*a*,*b*; Harlan, de Wet & Price, 1973; Harlan, 1975; Evans, 1976; de Wet, 1977; Sharma & Waines, 1980; Hammer, 1984; Ladizinsky, 1985, 1987, 1989; Miller, 1986, 1992; Heiser, 1988). However, all these authors noted that the most critical adaptive differences involve loss of wild-type seed dormancy and rachis fragility and of these, only rachis fragility is readily apparent morphologically.

DIFFERENCES IN RACHIS FRAGILITY

In the wild wheats and barley, the mature rachis disarticulates between each of the fertile spikelets, thereby allowing them to be shed spontaneously. Disarticulation occurs from the top of the ear downwards (Fig. 7.1). The arrow-like morphology of the spikelets with their smooth points, springy awns, long straight glumes and backward-pointing barbs and hairs thereafter ensures that they quickly penetrate any

Fig. 7.1. Diagrams showing the features affecting seed dispersal and spikelet implantation which distinguish the wild and domestic forms of einkorn wheat. (*a*) Wild einkorn (*Triticum boeoticum*) showing the brittle-rachis ear and arrow-shaped spikelets adapted for penetrating

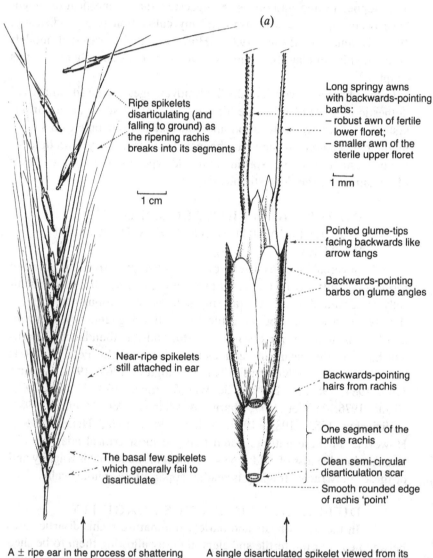

(*a*)

Ripe spikelets disarticulating (and falling to ground) as the ripening rachis breaks into its segments

1 cm

Long springy awns with backwards-pointing barbs:
– robust awn of fertile lower floret;
– smaller awn of the sterile upper floret

1 mm

Pointed glume-tips facing backwards like arrow tangs

Backwards-pointing barbs on glume angles

Near-ripe spikelets still attached in ear

Backwards-pointing hairs from rachis

One segment of the brittle rachis

The basal few spikelets which generally fail to disarticulate

Clean semi-circular disarticulation scar

Smooth rounded edge of rachis 'point'

A ± ripe ear in the process of shattering and thereby shedding its spikelets. The ear ripens from the top downwards

A single disarticulated spikelet viewed from its inside (adaxial) face. In this case, the spikelets contained a single grain

surface litter and cracks in the ground. (*b*) Domestic einkorn (*T. monococcum*) showing the semi-tough-rachis ear and plumper spikelets which have lost some of the key features necessary for self-implantation. (Drawings: G.C.H.).

(*b*)

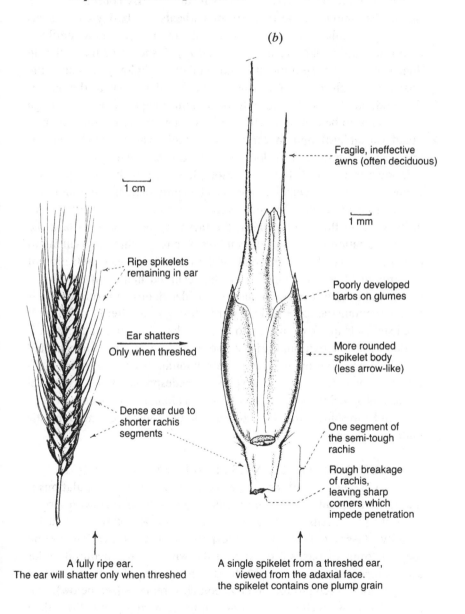

Fragile, ineffective awns (often deciduous)

1 cm

1 mm

Ripe spikelets remaining in ear

Poorly developed barbs on glumes

Ear shatters
Only when threshed

More rounded spikelet body (less arrow-like)

Dense ear due to shorter rachis segments

One segment of the semi-tough rachis

Rough breakage of rachis, leaving sharp corners which impede penetration

A fully ripe ear.
The ear will shatter only when threshed

A single spikelet from a threshed ear, viewed from the adaxial face. the spikelet contains one plump grain

surface litter and wedge themselves in cracks in the ground where at least a proportion of them remain relatively safe from birds, rodents and seed-eating ants. (In the Near East where wild wheats and barley are native, pressure from these predators is intense.) By contrast, in even the most primitive of the domesticated wheats and barleys, the rachis fails to disarticulate spontaneously, and the ear remains intact until the crop is harvested and threshed (Fig. 7.1(*b*)). It should be noted that in domestic emmer (tetraploid) and einkorn (diploid) wheats, the relatively tough rachis nevertheless disarticulates when the ear is threshed, and it is therefore termed 'semi-tough'. This semi-tough rachis is not to be confused with the fully tough rachis of, for example, bread (hexaploid) and macaroni (tetraploid) wheats which remains intact when threshed, as it does in all the domestic barleys.

If sown in the wild, these domestic plants are unable to reproduce themselves, as their spikelets are not efficiently disseminated and protected from predation. Indeed, even if their ears were eventually to disintegrate in the autumn rains, the fact that their spikelets lack the self-implantation features of the wild types, ensures that they will fail to bury themselves: the rough break in the rachis impedes penetration of the ground litter, the glumes lack the recurved hairs and prominent barbs, and the awns are weak and readily deciduous and so fail in their task of pointing the spikelet downwards through leaf litter (Fig. 7.1(*b*)). Such spikelets quickly fall prey to birds, rodents and ants, a fact that is evident from the extreme rarity in the Near East of feral domesticated cereals in anything other than modern habitats (such as the first half-metre of motorway verges), despite the widespread spillage of cereal grains and spikelets which regularly occurs along the waysides between field and threshing yard. Outside cultivation, therefore, the domestic mutant is doomed.

CHOICE OF DOMESTICATION CRITERION

One measure of the 'degree of domestication' in populations is the ratio of plants with semi-tough rachises to plants with brittle rachises. The merits of using this criterion are as follows: (a) rachis fragility plays a crucial role in the process of domestication, as the tougher forms of rachis are lethal in the wild but are favoured under certain forms of cultivation; (b) the different states of rachis toughness are potentially recognizable in archaeological remains (see below); and (c) the rachis is much easier to study in modern crop populations than characters such as seed dormancy (see above). However, it should be pointed out that domestication involves changes in many different features, of which rachis fragility is only one.

ARCHAEOLOGICAL EVIDENCE OF DOMESTICATION: RACHIS REMAINS

Rachis remains of cereals such as wheat, barley, rye and oats are quite commonly preserved in late archaeological sites by virtue of having been charred by fire immediately prior to their deposition. When preservation is good, these rachis remains can often provide clear evidence of whether the cereal was (a) of the wild type with a fully brittle rachis (as in wild einkorn and emmer), (b) of the domestic type with a semi-tough rachis (as in domestic einkorn and emmer), or (c) of the domestic type with a fully tough rachis (as in bread or macaroni wheats). Distinguishing states (a) and (b) can prove difficult, but generally disarticulation in wild cereals leaves a clean, semi-circular or reniform scar, while in domestic derivatives with a semi-tough rachis, the scar is linear, jagged and irregular, with no abscission surface (Fig. 7.1; see also Willcox, 1992). In addition, differences in the histology of the abscission zone can often be recognized microscopically, even in the charred remains of immature ears (Kislev, 1989). However, all forms of rachis remains are remarkably rare at early archaeological sites, dating from the beginnings of agriculture (i.e. from the earliest phases of the Aceramic Neolithic period). In consequence, archaeobotanists generally attempt to distinguish wild wheats and barleys from their domestic derivatives using secondary features such as grain shape. In the wheats, these grain features are often unsatisfactory, although in the barleys they can perform a useful role in distinguishing six-rowed and naked domesticates from the wild type.

On the basis of these often problematic grain-based characteristics, the earliest appearance of seemingly fully domesticated cereals in western Eurasia is currently dated to *c.* 8800 BC (in calendar years) at Neolithic Aswad in southwestern Syria, and fractionally later in Jericho, Gilgal and Netiv Hagdud in Palestine, the Neolithic occupation at Abu Hureyra in northern Syria, and slightly later again at Çayónú in southeastern Turkey. In each case, the cereals identified were emmer wheat and barley, except at Neolithic Abu Hureyra and Çayónú where einkorn was also found (see Hillman & Davies, 1992). Of these sites, only Netiv Hagdud and Neolithic Abu Hureyra produced useful quantities of rachis remains. However, such finds do not necessarily date the beginnings of cultivation; they merely date the completion of the ensuing process of domestication. If we are to date the beginnings of cultivation, we must take the earliest date for the emergence of ostensibly domesticated forms (currently c. 8800 BC – calibrated) and add to it that block of time required to achieve full domestication once the crop was under cultivation ('pre-domestication cultivation'). However, not all forms of primitive husbandry advantage tougher-rachised forms (see below), and it is therefore possible that many of the earliest farmers would have applied practices of this 'ineffective' type for an indefinite period of 'non-domestication cultivation' before eventually adopting harvesting techniques which favoured the domestic phenotype and eventually led to full domestication.

To date the beginnings of even 'pre-domestication cultivation', we therefore need to know (a) how quickly semi-tough rachis phenotypes could have appeared in early wild-type crops, (b) precisely which combination of husbandry methods would have effected domestication in a wild-type crop (including the state of maturity at which it had to be harvested), (c) whether other forms of husbandry would have been completely ineffective in this role, (d) whether the initial husbandry methods used by the first farmers were most likely to have been the 'ineffective' ones, and, if so, how quickly they would have converted to the 'effective' methods; and (e) how long the process would have taken once the effective methods were applied. All five questions are addressed below.

A THEORETICAL APPROACH TO DOMESTICATION

The domestic forms of einkorn, emmer and barley originated from recessive mutants exhibiting the semi-tough rachis which were produced in (and still are being produced in) populations of brittle-rachised wild forms.

In einkorn, Sharma & Waines (1980) showed that two independent recessive loci control rachis fragility. However, they suggest that simultaneous mutations for a 'domestic allele' at both loci was an improbable event and that, initially, mutation at one locus might have resulted in phenotypes that had a degree of rachis toughness. Mutation at the second locus might have been an independent event that occurred later in an already domesticated line or indeed may have occurred in another *boeoticum* line and the two mutant *boeoticum* lines might have hybridized and resulted in the present-day *monococcum*, homozygous at both loci.

Our analysis, therefore, is based upon mutation occurring spontaneously at a single locus controlling rachis toughness. We have assumed a low net forward mutation rate of fragile to semi-tough rachis alleles of 10^{-6} per generation. Since the semi-tough allele is recessive to the wild type it is manifested only in the homozygous state. In wild stands there would be rapid elimination of the domestic mutant phenotypes which have zero fitness in the wild. Since wild wheats and barley are largely self pollinating (D. Zohary personal communication), half the mutant alleles would be represented in homozygotes in the first generation and would thus be eliminated, and there would be few alleles maintained in the heterozygous state, unlike the situation in an outbreeding population. Thus wild-type populations would be almost entirely homozygous for the brittle-rachised allele with nominally only one or two heterozygotes per two million brittle-rachised plants. However, it must be stressed that this scenario is a convenient simplification. Not only was more than one locus eventually involved but modifier genes could also have altered patterns of dominance and expression at these loci during the past 11 000 years. Consequently our mathematical model has incorporated a very broad range of values for both the fitness of the domestic phenotype under cultivation and the breeding system of the early crop populations.

In the early stage of cultivation domestication would have been delayed owing to the absence of mutant phenotypes in the first crops. However, it is likely that the size of the crop stands required to meet the calorific needs of the early farmers would have been sufficient to ensure the generation of mutant phenotypes within a very few years (Hillman & Davies, 1990*a*,*b*, 1992).

All the available evidence suggests that in the early stages of cultivation any selection for the semi-tough rachis phenotype would have been entirely unconscious since their frequency in the crop population would have been so low that they would have been unnoticed by the farmer. Only when the frequency of semi-tough rachis phenotypes had risen to a level at which they were obvious in the crop stand (say 1–5%) would conscious selection have been likely to have been applied.

In trying to model the time required for the semi-tough rachis phenotypes to increase to a level at which conscious selection would have occurred, it is necessary to consider (a) the harvesting methods which might have been used by the early cultivator and (b) the selective advantage/disadvantage which each harvesting method may have conferred on the domestic phenotype.

HARVESTING METHODS

There are five main harvesting methods with which the earliest cereal cultivators were familiar from their previous experience as foragers. These are (a) *beating* ripe spikelets into baskets, applied to partially or near ripe stands, (b) *reaping* with sickles or other cutting implements, (c) *uprooting*, (d) *plucking* or *hand-stripping*, and (e) *harvesting by burning*. A fuller description of each of these methods and the evidence of their viability as methods used by the early hunter–gatherer before cultivation is given by Hillman & Davies (1990*b*, 1992); although for the most comprehensive review, see Bohrer (1991).

Hillman & Davies (1990*a*) reported the results of preliminary field trials in which some of these harvesting methods, namely, beating, sickle harvesting and uprooting, were applied to stands of near-ripe wild einkorn in Turkey. A similar experiment was carried out on a population of a cultivated glume wheat emmer, to determine the effect of these harvesting methods on phenotypes with a semi-tough ear, since semi-tough rachis phenotypes are too rare in wild stands to be measurable. In each harvesting method, counts were made of the number of spikelets harvested as a proportion of the spikelets available. In this way it was possible to obtain a measure of the relative fitness of the two phenotypes in the various harvesting regimes.

Harvesting by beating, whether by single passes or repeated passes clearly favoured the fragile-rachis phenotype with a recovery rate of 44% and 84%, respectively, compared with *c*. 5% for the tough-rachis type. In complete contrast, sickle-reaping and uprooting produced a recovery rate of 100% for the tough-eared type compared with *c*. 40% for the fragile-rachis type. Clearly the domestic phenotype would only be advantaged if the earliest cultivators had used either of these latter methods. It must be stressed, however, that these measures of relative recovery of spikelets under the three harvesting regimes are not definitive. Attempts to gain more reliable estimates in mixed populations of the domestic and wild type einkorn in properly designed and replicated experiments in Cardiff have failed owing to the failures of the brittle-rachis ear to disarticulate because of wet weather during the trials. Therefore, in our computer simulation of the domestication process, we have incorporated a broad range of much more conservative values for the fitness of the wild type relative to the domestic type, ranging from 0.4 to 0.95, i.e. the lowest limit of the disadvantage of the wild type is that estimated from the Turkish trials.

The computer simulation model also incorporated varying degress of inbreeding from complete inbreeding to complete outcrossing. However, Zohary (personal communication) informed us that the rate of outcrossing in wild wheats is probably less than 1%. This opinion is based upon the floral biology (anther dehiscence occurs within florets prior to lodicule inflation and anther emergence), the absence of intermediates when different forms are grown together and more recent electrophoretic evidence which reveals the predominance of homozygosity for protein markers. He considered it safe to conclude that *T. dicoccoides, T. boeoticum* and *Hordeum spontaneum* are predominantly self pollinating. However, Willcox (1992) cited what seems to be an aberrant exception to this pattern observed by Boyeldieu (10–15% outcrossing under hot conditions in North Africa). Our computer model has therefore incorporated a broad range of outcrossing frequencies (0–100%) to allow for such eventuality and for the unlikely possibility that there has been a major shift in breeding behaviour of these cereals during the past 11 millennia.

The simulation necessarily assumed that the husbandry methods used were those capable of selecting for phenotypes with a semi-tough rachis in a brittle-rachis wild-type crop, i.e. harvesting when partially ripe by sickle-reaping or uprooting (combined with sowing on virgin plots every year with seed corn taken from last year's plots). The simulation also assumed that the initial frequency of the domestic allele in the wild-type population was 10^{-6} and that the allele was present only in heterozygotes, since homozygotes for this tough-rachis allele would be lethal in wild populations (see above).

RESULTS

The results of the computer simulation are presented in Fig. 7.2 which shows the pattern of increase in the semi-tough rachis (domestic) phenotype under a range of selective intensities, but with inbreeding at a constant 100%. With a selection coefficient of 0.6 against the wild type (fitness 0.4), equivalent to that measured in the Turkish trials, domestication occurs within 20 generations (i.e. within 20 years if the crop is sown annually). Even with a selection coefficient as low as 0.1 against the brittle-rachis phenotype, (fitness 0.9) domestication is still complete within 200 generations. Table 7.1 gives the number of generations required for the domestic phenotype to reach a level of 99% in the crop population with a range of selection cofficients against the wild type and also with a range of values for the degree of inbreeding ranging from complete outcrossing to complete inbreeding. It could be argued that

Fig. 7.2. The frequency of the semi-tough-rachis domestic phenotypes in populations of brittle-rachis wild-type einkorn under a range of selection coefficients against the wild-type phenotype. The rate of selfing was assumed to be 100% and the initial frequency of the allele for semi-tough rachis was assumed to be 10^{-6}, with the allele present only in heterozygotes in generation 0.

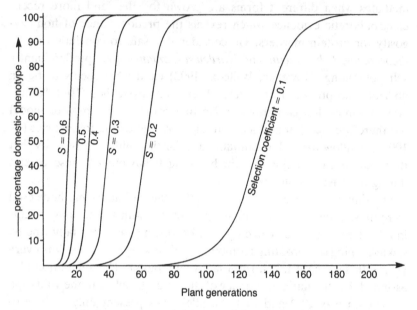

Plant generations

the more realistic estimates of the rate of domestication are those in the bottom right section of the table, i.e. the higher levels of selection against the wild type 0.3–0.6 (0.6 measured in the Turkish trials) and a higher level of inbreeding (50–100%; Zohary's estimate of probable inbreeding frequency is more than 99%). This gives the number of generations requir. d to achieve domestication as being from 100 to 21. However, it is probably wise to err on the side of even lower selection pressures against the wild type, say 0.1; these would correspond to harvesting the crop when it is much less ripe. G. H. Willcox (personal communication, 1990) found that the bulk harvesting of wild-type einkorn was much more efficient when the crop was in this state.

Thus the model indicates that domestication (as defined by the fixation of the semi-tough rachis in an initially brittle-rachis crop population) could be achieved within a few decades provided that the crop was harvested when near-ripe by means of sickle reaping or uprooting and also provided that some of the harvested seed was sown on virgin land each year and taken from the previous year's plots. Even if the advan-

Table 7.1. *The number of generations required for the semi-tough rachis phenotype to reach a frequency of 99% at various levels of selection against the wild-type phenotype and at various levels of inbreeding*[1]

% inbreeding	Selection coefficient against fragile-rachis phenotype						
	0.05	0.1	0.2	0.3	0.4	0.5	0.6
0	*	*	*	*	*	*	*
10	4090	1940	860	510	330	220	150
20	2130	1010	450	270	170	120	77
30	1440	680	310	180	120	78	53
40	1070	510	230	140	87	60	42
50	840	380	180	100	71	49	34
60	690	330	150	89	59	42	30
70	580	280	130	76	51	37	27
80	490	240	110	67	45	33	24
90	420	210	95	59	41	30	23
100	360	180	84	53	37	28	21

The initial frequency of the semi-tough rachis allele was taken as 10^{-6}, and the allele was assumed to be present only in heterozygotes in generation 0. In cases marked *, no homozygotes with semi-tough rachis were produced, even after 8000 generations.

tage of the semi-tough phenotype was less, as would be the case if crops were harvested when less ripe, the process could still be complete within two centuries.

The results broadly concur with the rapid rates of domestication proposed by Zohary (1969, 1984) and Ladizinsky (1987). Even if the longer time-scale of two to three centuries were the case, domestication would have been such a rapid event that it is unlikely that it would be evidenced as a clinal process in samples of plant remains recovered from archaeological sites.

LIMITATIONS OF THE COMPUTER MODEL

It should be stressed that the computer model is deterministic and takes no account of random processes. It was designed merely to give a general idea of the possible time-scales of the domestication process. It also made a number of simplifications. For example, the model only considered a single locus controlling the rachis character, whereas in fact two independent loci are involved in modern einkorns (Sharma & Waines, 1980). However, as discussed above, Sharma & Waines suggested that initially only one locus was involved in rachis toughness. The model takes no account of introgression of wild-type alleles from nearby wild-type stands, which would slow down the domestication rate. However, given the high degree of selfing in wheats and barleys (see above), the effect of introgression is likely to have been modest during the short period involved in domestication. We have also disregarded the possibility, albeit a remote one (Hillman & Davies, 1992), that when sowing virgin land the farmers might have incorporated in their 'seed corn' some wild-type spikelets gathered from the ground of their old plots where brittle-rachis ears had disarticulated during harvesting. This would clearly have retarded the domestication rate.

We have not measured the effects of harvesting the crop at different stages of average ripeness. However, the large range of fitness values of the domestic type in the model should allow for these effects. (For further discussion, see Hillman & Davies, 1990*a*,*b*, 1992.) It is also possible that modifier genes might influence the expression of the gene for rachis toughness. The time required for full fixation of such modifier genes might have considerably extended the period required for full domestication and might have resulted in a fluctuation in the frequency at the top end of the sigmoid curve of the increase in frequency of the domestic type (G. Wilkes, personal communication, 1986 – see Hillman & Davies, 1992). Such fluctuations could conceivably account for the remarkable mixture of 'primitive' and 'advanced' forms which continued to occur in essentially domesticated crops for a millennium or more, well into the Aceramic Neolithic period (see Hillman & Davies, 1992).

However, other factors might have acted dramatically to speed up the domestication process. The major factor might have been conscious selection by farmers, taking the semi-tough reduced ears, sowing the spikelets in separate plots and multiplying the seed stock to a level at which there was sufficient to sow all their cereal plots (perhaps 3–4 years – Hillman & Davies, 1991). This process would only have begun as the frequency of the semi-tough rachis type increased to a level at which it became noticeable (perhaps around the 1–5% level), but, once started, it would have resulted in virtually instantaneous domestication.

Our model assumed a frequency of the domestic allele in the initial population of 10^{-6}. This is probably an underestimate, since mutation would have occurred over many generations in wild-type stands and, although there would be loss of alleles as lethal homozygotes every generation, at least some heterozygotes would have survived. It might have been more realistic to allow a diploid heterozygote frequency of perhaps 10^{-4} to 10^{-5}, thus allowing domestication to occur more rapidly.

Essential to the domestication process is the assumption that the early farmers would have harvested their crops by sickle-reaping or uprooting rather than by beating. The shift to these alternative methods is explained by the fact that they would have allowed a greater yield per unit area of land than beating, which gives the greatest yield per unit time. This deserves explanation. Hunter–gatherers using gregarious wild cereals such as einkorn, emmer or barley as dietary staples would generally have had access to very extensive wild stands (well in excess of their needs), and their utilization would have involved expending energy only in travel, harvesting and processing. It was therefore in their interest to use the least energy-expensive method of harvesting, namely beating, even if it meant leaving behind a large proportion of ears that were too unripe to be harvested by this method. By contrast, the ensuing farming settlements got much or all of their grain from cultivated plots and they invested heavily in clearing and tilling these plots, as well as in harvesting and processing. It therefore made sense for the farmers to use harvesting methods which maximized the harvested yields per unit area of these expensively cultivated plots and this meant harvesting the crops in the partially ripe state with sickles or uprooting. The shift in harvesting methods was therefore dictated by the need to continue to maximize the calorific return per unit of energy expended on this resource.

Sickle-reaping/uprooting also provided the farmer with straw which was likley to have been a valuable commodity. Indeed, archaeological evidence reveals that this secondary product was quickly used by the first farmers as tinder, as temper for making mud bricks (which we first see being used in buildings dating from the Neolithic, i.e. from the beginnings of agriculture), and, within a relatively short time, as fodder for newly domesticated sheep and goats. (A fuller discussion of factors affecting changes in harvesting methods is given by Hillman & Davies (1990b, 1992).)

Our model has also assumed that seed corn for each year's crop was taken from the previous year's crop and sown on virgin land. If crops were regularly sown on old plots it would seem likely that the domestic phenotypes in the seed corn would be 'swamped' by self-sown wild-type spikelets shed spontaneously at the preceding harvest and that domestication would not proceed. However, the relative proportion of wild-type phenotypes in the new crop would depend on the relative number of the previous year's shed spikelets which had survived predation: if the proportion of plants derived from these in the new crop population were less than that derived from the sown spikelets, domestication would still proceed, albeit at a slower rate (for a fuller discussion, see Hillman & Davies, 1990*a,b*). However, there are also reasons to suppose that the early farmers may have extended cultivation annually onto virgin land. This might have been done because of reductions in yield on old plots owing to depletion of soil nutrients or because of the build up of weedy contaminants. Annual extensions of cultivation could also have occurred without the abandonment of old plots, because of pressures from expanding populations (see Hillman & Davies, 1990*a,b*, 1992).

We have so far considered domestication merely in terms of the fixation of the semi-tough rachis phenotypes in crop populations, using einkorn wheat as our prime example. However, domestication in wheat went a step further. Tetraploid and hexaploid wheats carry mutant forms ('*Q*') in a supergene complex of the speltoid '*q*' alleles, producing a fully tough rachis that fails to disarticulate even when thoroughly threshed. They also have thin deciduous glumes which release the grain during threshing (Miller, 1986; Muramatsu, 1986). These 'free thresh-ing' and 'naked-grain' types include both bread wheat (*Triticum aestivum*) and macaroni wheat (*T. turgidum* var. *durum*) and, because they emerged from already domesticated crops (such as emmer or spelt), they are termed 'secondary crops'. Rye and oats are also tradi-tionally thought to be secondary crops, evolving, supposedly by unconscious selection, from weeds infesting already domesticated crops, (Vavilov, 1917; Hillman, 1978; Sencer & Hawkes, 1980). For the free-threshing wheats, however, there are no convincing explanations of their possible mode of emergence and conscious selection may well have played a central role. Either way, present archaeological evidence ten-tatively suggests that the free-threshing wheats and oats emerged initially as 'domesticated weeds' of primary founder crops complete with fully tough rachises (or in oats, with tough rachillas) before becom-ing established as crops in their own right.

Rye

It was noted above that rye is traditionally regarded as a secondary crop (Vavilov, 1917; Sencer & Hawkes, 1980; Hillman, 1978). However, recent archaeological evidence suggests that rye was one of the primary Near Eastern founder crops, and that it was domesti-cated at approximately the same time as einkorn, emmer and barley

(G. C. Hillman & F. McLaren, unpublished results; Moore *et al.*, 1992). This new evidence requires explanation, as present-day rye is an out-breeder, and our model above suggests that fixation of the tough-rachis phenotype in wild-type crops of an outbreeder would have required millennia. On the basis of our model, therefore, the synchronous domestication of an outbreeder such as rye and inbreeders such as wheat and barley is difficult to explain. In reality, however, the first ryes taken into cultivation were probably *inbreeders* after all, and it was in these inbreeding crops that domestication occurred.

The immediate wild (brittle-rachis) ancestor of domestic rye is thought by workers such as Miller (1987) and D. Zohary (personal communication, 1990) to be *Secale vavilovii*, which is an inbreeder like wheat and barley (Hammer, 1990; Hammer, Skolinowska & Knupffer, 1987). Thus, if crops of brittle-rachis *S. vavilovii* were exposed to the pattern of unconscious selection proposed in our model above, it would probably have been domesticated just as rapidly as wheat and barley, and, on this basis, our model can be extended to all four cereals: einkorn, emmer, wheat, barley and rye. If, however, the immediate ancestor of *S. cereale* was the outbreeder *S. montanum* (syn. *S. strictum*), as proposed by Hammer (1990) on the basis of anther morphology, then the rapid appearance of domestic rye (*S. cereale*) in the archaeological record is inexplicable, unless we assume carefully controlled (and highly improbable) systems of conscious selection and genetic isolation. Present archaeological evidence, therefore, favours *S. vavilovii* as the immediate ancestor of domestic rye.

A derivation of *S. cereale* from *S. vavilovii* is also supported by the fact that *S. vavilovii* is an annual while *S. montanum* is a perennial. Grain yield per unit of vegetative growth is correspondingly greater in *S. vavilovii*, and this species would have been more amenable to cultivation of the sort applied to the other wild annual cereals: einkorn, emmer and barley. *S. vavilovii* was therefore more likely than *S. montanum* to have been exposed to the selective pressures of cultivation which favoured the fixation of the tough rachis which today characterizes *S. cereale*. On this basis also, it could thus be argued that *S. vavilovii* is more likely to have been the ancestor.

In earlier accounts of our model, we did not include rye with einkorn, emmer and barley because of contradictions in the literature regarding the breeding behaviour of *S. vavilovii*. It now transpires that these contradictions arise from the fact that several laboratories accessed seed stocks of what they thought was *S. vavilovii* (as originally collected in

Transcaucasia and described by Grossheim), but which was in fact a brittle-rachis, outbreeding form of rye, which is sometimes called *S. iranicum* and was collected from a roadside in northern Iran by Kuckuck but has never been observed since (D. Zohary personal communication). Publications based on studies of these accessions therefore refer wrongly to *S. vavilovii* as an outbreeder (F. McLaren, in preparation).

   On the basis that the genuine *S. vavilovii* was, indeed, the immediate ancestor of *S. cereale* as suggested by Miller & Zohary, we can assume that the earliest crops of rye were *S. vavilovii* (= *S. cereale* subsp. *vavilovii* in Zohary & Hopf, 1988), and that fixation of tough-rachis phenotypes occurred in some of these crops when they were exposed to harvesting in the partially ripe state by reaping or uprooting (as described in our model above). The earliest domestic rye would therefore have been a tough-rachis form of *S. vavilovii*, and the shift to outbreeding would have occurred after this. It is suggested by Miller (1987, personal communication 1990) and D. Zohary (personal communication, 1990) that the outbreeding habit is the product of introgression from the outbreeder *S. montanum* (syn. *S. strictum*), which occurred when the cultivation of domesticated *S. vavilovii* was extended into areas in which *S. montanum* was abundant. It is to the resultant outbreeding (and already domesticated) ryes that we can first apply the name *S. cereale* in the strict sense (i.e. *S. cereale* subsp. *cereale* as used by Zohary & Hopf (1988)). However, we cannot as yet be certain whether the charred remains of early domestic rye from Neolithic Abu Hureyra in Syria (Hillman, 1975; Moore *et al.* 1992) and Neolithic Can Hasan III in Turkey (Hillman, 1978) represent the initial domesticated inbreeding *S. cereale* subsp. *vavilovii* or the ensuing outbreeding *S. cereale* subsp. *cereale*.

   (Note that rye and barley differ from the wheats in that there is no clear intermediate semi-tough rachis state, merely (a) brittle-rachis wild types, (b) fully tough rachis domesticates (including some tough-rachis weed forms) and (c) occasional weed forms in which the base of the ear is tough and the apex brittle (Zohary & Hopf, 1988).)

Maize

   This history of domestication of maize (*Zea mays*) contrasts in many respects with that of wheat and barley. Maize is thought to have originated in the highlands of Mexico and appears to have become domesticated within the area of the wild species early in American agricultural history (*c.* 8000–10 000 years BP). It has since been spread by humans from one continent to another.

Maize is unique among the grasses in carrying its male and female flowers on separate parts of the plant; the terminal tassel bears the male flowers and the female ear is positioned laterally on the plant. In modern maize the individual female inflorescence bears seeds which, again unusually for a grass, are not generally covered by floral bracts (glumes or lemmas) but are exposed on the cob, which itself is covered by large modified leaf sheaths.

Despite nearly a century of research, there is still considerable disagreement about the origin and ancestry of maize. There are two principal schools of thought about its domestication. One theory proposes that maize was domesticated from the Mexican annual teosinte (*Zea mexicana* ($2n=20$), which has six recognized races, although there is some disagreement about the taxonomic classification (Beadle, 1972, 1980; Harlan & de Wet, 1972; Galinat, 1978). There are also considerable arguments within this school about the cultural and genetic pathways involved in the domestication process and the source populations involved (Galinat, 1984). The other theory proposes that modern maize originated as a result of hybridization between an early domesticated maize and one of the teosinte species (e.g. Randolph, 1976; Bird, 1979; Wilkes, 1989). There is again argument as to which teosinte species was involved and which modern maize races are most similar to the early domesticates (Bird, 1991).

The best available source of evidence for determining maize phylogeny seems to be archaeobotanical evidence, but morphological, cytological and, more recently, molecular biological techniques have been used to investigate present-day relationships within the genus *Zea*, although it is not possible to cover these aspects here.

The oldest and most complete archaeological sequence of maize remains occurs at Tehuacán in south-central Mexico and spans a sequence from what is considered by many to be a wild maize in the oldest remains to modern races which are similar to those currently cultivated in the region, though Beadle (1980) is of the view that the earliest cobs can be considered to be domesticates. These earliest cobs, dating to *c.* 5000 BC, are of a pod corn, with each kernel individually enclosed in long glumes. The ears are uniform in size and bisexual, with the male staminate spikes located in the tassel tip at the top of each ear. The seed was probably dispersed by disarticulation of the thin and fragile rachilla and not the rachis. These are characteristics of a wild plant and are indeed those possesed by teosinte (the closest modern-day relative of maize), a wild plant often occurring in maize fields in Mexico (Wilkes, 1989). However, in teosinte the rachis, and not the rachilla, is brittle. Later maize cobs in the Tehuacán sequence are larger and more varied.

However, comparison between modern and archaeological maize and teosinte reveals that, apart from their small size the small number of rows on the cob, the archaeological cobs differ from teosinte in many respects (Bird, 1991). Most of the main taxonomic traits distinguishing maize from teosinte are multigenically controlled and only three characters: abscission *vs* non-abscission, paired *vs* solitary spikelets and two *vs* four rows are controlled by three or fewer loci (Manglesdorf, 1974; Galinat, 1978), with complete or partial dominance in the direction of the maize traits. Given the outbreeding nature of maize, the probability of fixation of these dominant traits, arising as spontaneous mutants, in the homozygous state in wild stands of teosinte is extremely small (Bird, 1991). Furthermore, 30–40 of the distinguishing loci are linked genetically in four to five linkage groups (Manglesdorf, 1974; Galinat, 1978), so the potential for recombination to produce plants possessing all the modern traits would have been much reduced unless the linkage arose after domestication had occurred. Thus the evidence seems to point against teosinte being the direct ancestor of modern maize. Manglesdorf (1974) suggested that teosinte is derived from wild maize as is cultivated maize. Following the discovery of *Zea diploperennis*, the diploid perennial form of teosinte (Iltis *et al.*, 1979), Manglesdorf (1986) modified his view and hypothesized that the perennial teosintes were the more ancient wild plants and that annual teosinte may have originated through the hybridization of maize and perennial teosinte. Magoja *et al.* (1985) have examined biometrical characters in the genus *Zea* and have generated further support for the view that cultivated maize is descended directly from a wild maize, now extinct.

Wilkes (1989) argued that early maize and teosinte populations would have been spatially separated by altitude and that, only when the domestication of maize had begun and cultivators began extending its cultivation into lower altitudes, would the distribution of the two species have substantially overlapped. The effects of hybridization and introgression would have resulted in a broadening of the gene pool of both species to allow maize to become adapted to lower elevations and teosinte to the higher altitudes, further increasing the degree of overlap and the possibility of introgression. Introgression between teosinte and modern maize populations continues to occur today where they are sympatric in South America. The largest teosinte populations, however, are restricted to sparsely populated areas and this, together with the fact that teosinte tends to flower later than maize, serves to produce a degree of isolation. Since teosinte and maize are fully interfertile, any theory which proposes the origin of maize from teosinte (see above) should also consider how genetic isolation of the two taxa might have occurred to prevent the genetic swamping of early maize by the teosinte populations in which it originated. The simplest way in which this might have been achieved would have been by geographical isolation, or by domestication of maize outside the natural range of teosinte. Unfortunately, there is no archaeological evidence of any transitional forms between teosinte and maize pre-dating the oldest maize finds. Galinat (1983) put forward a rapid-selection theory which envisaged that the transition to domesticated maize occurred quickly (within a 100 years), thus accounting for the lack of intermediates in the archaeological record (cf. our model for wheat and barley above). However, the problem still remains of how, in an outbreeder such as maize, such rapid fixation of domestic traits would have occurred in the absence of isolation from the wild-type ancestor.

Maize is characterized by a high degree of genetic heterozygosity built up by persistent outcrossing. As the cultivation of early domesticated maize spread to different regions, there was a rapid evolution of many races adapted to a wide variety of growing conditions. This was reinforced by the general tolerance of maize cultivators, such as those in present-day Mexico, of a wide variation of forms within their crop, in the belief that such genetic diversity in the crop improves the degree of buffering against unpredictable environmental variation. Added to this is the continued outcrossing to teosinte which would have occurred throughout the evolutionary history of maize, thus further widening the gene pool. Thus today there is a bewildering complex of natural geographical races and local cultivars of maize which fall under the general subdivisions: pop, flint, floury, dent and sweet, based mainly on the nature of the endosperm. Inbreeding experiments, which began in the early part of this century, have produced a massive array of inbred lines from the original eighteenth century open-pollinated dent corn.

There is controversy over whether the domestication of maize occurred solely in Middle America or whether there was a secondary domestication in the Andean region of South America (for the opposing arguments, see Bonavia & Grobman, 1989; Pickersgill, 1989). However, data from morphology, karyotyping and mitochondrial DNA analyses indicate that, even if maize was introduced into South America, it has undergone considerable evolution in this region, independently from the Middle American populations.

The evolution of cultivated maize from its ancestors has resulted from a sequence of genetic changes over time rather than from the equivalent of the fixation of a single trait such as the semi-tough rachis in diploid wheat. The majority of the characteristics that have changed are polygenically controlled (see above). The species now consists of a multitude of diverse forms reflecting the diversity of selection pressures in its wide range of habitats and continued hybridization with closely related species. The wild form is now extinct and maize has so altered in morphology and physiology that it can survive only under cultivation, thus conforming to our definition of domestication (see above).

Conclusions

The precise pathway and time-scale for the evolution and domestication of the major cereal crops is still largely unresolved. Many of the evolutionary pathways that have been proposed are based upon archaeological evidence which is often incomplete and open to varying interpretation. Most of the theories are based largely on speculation.

Our attempt to model the increase in the domestic phenotype of wild wheat and barley under primitive cultivation is open to criticism. The model is simplistic and is based upon many assumptions, not all of which may have operated. However, we hope that this attempt at modelling the process will lead to the development by others of more rigorous models that will include a stochastic element and that will further define the optimum conditions under which domestication could occur and the effects that departure from these conditions may have on the process.

However, the predictive value of mathematical models is only as good as the assumptions on which they are based. Their predictive accuracy can perhaps be tested in long-term experiments where the changes in frequency of wild and domestic phenotypes are followed under various mangement régimes. Our predictions of the increase in frequency of the tough-rachis phenotype of einkorn wheat is currently being tested in the Cultures Préhistoriques Expérimentales programme at Jalès, Ardèche, in the south of France, under the supervision of Patricia Anderson-Gerfaud and George Willcox (Anderson-Gerfaud, de Aprahamiyan & Willcox, 1990; Willcox, 1992). However, such a trial can demonstrate merely that it was *feasible* for unconscious domestication of wheat and barley to have occurred in the way we have proposed. It cannot prove that it actually happened that way 11 millennia ago.

References

Anderson-Gerfaud, P. C., de Aprahamiyan, A. P. & Willcox, G. H. (1990). Cultures de céréales sauvages et primitives au Proche-Orient Neolithique: Résultats préliminaires d'experiens à Jalès (Ardèche), France. *Cahiers de l'Euphrate*, **5**.

Beadle, G. W. (1972). The mystery of maize. *Field Museum of Natural History Bulletin*, **43**, 2–11.

Beadle, G. W. (1980). The ancestry of corn. *Scientific American*, **242**, 112–119.

Bird, R. M. C. K. (1979). The evolution of maize: a new model for the early stages. *Maize Genetics Cooperation Newsletter*, **55**, 53–54.

Bird, R. M. C. K. (1991). Comment on: "The ecological genetics of domestication and the origins of agriculture" by Blumler, M. A. & Byrne, R. *Current Anthropology*, **32**, 36–37.

Blumler, M. A. & Byrne, R. (1991). The ecological genetics of domestication and the origins of agriculture. *Current Anthropology*, **32**, 23–54.

Bohrer, V. L. (1991). The relation of grain and its method of harvest to plants in prehistory. *Reviews in Anthropology*, **16**, 149–156.

Bonavia, D. & Grobman, A. (1989). Andean maize: its origins and domestication. In *Foraging and farming – the evolution of plant exploitation*, ed. D. R. Harris & G. C. Hillman, pp. 456–470. London: Unwin & Hyman.

Darlington, C. D. (1969). *The evolution of man and society*, 1st edn. London: Allen & Unwin.

Darlington, C. D. (1973). *Chromosome botany and the origin of cultivated plants*, 2nd edn. London: Allen & Unwin.

Darwin, C. (1859). *On the origin of species by means of natural selection*. London: John Murray.

Darwin, C. (1868 & 1975). *The variation of animals and plants under domestication*, 1st and 2nd edn, vol. 1. London: John Murray.

de Candolle, A. (1886). *Origin of cultivated plants* (English translation of the 2nd edn, New York & London: Hafner, 1967).

de Wet, J. M. J. (1977). Increasing cereals yields: evolution under domestication. In *Crop resources*, ed. D. S. Siegler, pp. 111–118. New York, San Francisco & London: Academic Press.

de Wet, J. M. J. & Harlan, J. R. (1975). Weeds and domesticates: evolution in the man-made habitat. *Economic Botany*, **29**, 99–107.

Evans, L. T. (1976). Physiological adaptation to performance as crop plants. *Philosophical Transactions of the Royal Society London, B*, **275**, 71–83.

Galinat, W. C. (1978). The inheritance of some traits essential to maize and teosinte. In *Maize breeding and genetics*, ed. D. B. Walden, pp. 99–111. New York: John Wiley.

Galinat, W. C. (1983). The origin of maize as shown by key morphological traits of its ancestor, teosinte. *Maydica*, **28**, 121–138.

Galinat, W. C. (1984). The origin of maize. *Science*, **225**, 1093–1094.

Hammer, K. (1984). Das Domestikationssyndrom. *Die Kulturpflanze*, **32**, 11–34.

Hammer, K. (1990). Breeding system and phylogentic relationships in *Secale* L. *Biologische Zentralblatt*, **109**, 45–50.

Hammer, K., Skolimowska, E. & Knupffer, H. (1987). Vorarbeiten zur monografischen Darstellung von Wildpflanzensortimenten: *Secale* L. *Die Kulturpflanze*, **35**, 135–177.

Harlan, J. R. (1975). *Crops and Man*. Madison: American Society of Agronomy.

Harlan, J. R. (1989). Wild grass-seed harvesting in the Sahara and Sub-Sahara of Africa. In *Foraging and farming: the evolution of plant exploitation*, ed. D. R. Harris & G. C. Hillman, pp. 79–98. London: Unwin & Hyman.

Harlan, J. R. & de Wet, J. M. J. (1972). Origins of maize – the tripartite hypothesis. *Euphytica*, **21**, 271–279.

Harlan, J. R., de Wet, J. M. J. & Price, E. G. (1973). Comparative evolution in cereals. *Evolution*, **27**, 311–325.

Harris, D. R. (1990). Vavilov's concept of centres of origin of cultivated plants: its genesis and its influence on the study of agricultural origins. *Biological Journal of the Linnean Society*, **39**, 7–16.

Hawkes, J. G. (1969). The ecological background to plant domestication. In *Domestication and Exploitation of Plants and Animals*, ed. P. J. Ucko & G. W. Dimbleby, pp. 17–29. London: Duckworth.

Hawkes, J. G. (1989). The domestication of roots and tubers in the American tropics. In *Foraging and farming: the evolution of plant exploitation*, ed. D. R. Harris & G. C. Hillman, pp. 481–503. London: Unwin & Hyman.

Hawkes, J. G. (1990). N. I. Vavilov – the man and his work. *Biological Journal of the Linnean Society*, **39**, 3–6.

Heiser, C. B. (1988). Aspects of unconscious selection and the evolution of domesticated plants. *Euphytica*, **37**, 77–81.

Hillman, G. C. (1975). The plant remains from Tell Abu Hureyra. In *The Excavation of Tell Abu Hureyra in Syria; A Preliminary Report*, ed. A. M. T. Moore. *Proceedings of the Prehistoric Society*, **41**, 70–73.

Hillman, G. C. (1978). On the origins of domestic rye – *Secale cereale* L.: the finds from Aceramic Can Hasan III in Turkey. *Anatolian Studies*, **28**, 157–174.

Hillman, G. C. & Davies, M. S. (1990a). Domestication rate in wild-type wheats and barley under primitive cultivation. *Biological Journal of the Linnean Society*, **39**, 39–78.

Hillman, G. C. & Davies, M. S. (1990b). Measured domestication rates in wild wheats and barley under primitive cultivation and their archaeological implications. *Journal of World Prehistory*, **4**, 157–222.

Hillman, G. C. & Davies, M. S. (1992). Domestication rates in wild wheat and barley under primitive cultivation: preliminary results using field measurements of selection coefficients and their archaeological implications. In *Préhistoire de l'agriculture: nouvelle approches expérimentales et ethnographiques*, ed. P. C. Anderson-Gerfaud. Valbonne: Monographies du Centre de Recherches Archéologiques, **6**, pp. 1–46.

Iltis, H. H., Doebly, J. F., Guzmán, R. M. & Pazy, B. (1979). *Zea diploperennis* (Gramineae): a new teosinte from Mexico. *Science*, 203, 186–188.
Kislev, M. E. (1989). Pre-domesticated cereals in the pre-pottery Neolithic A period. In *Man and culture in change*, ed. I. Herschkovitz, pp. 147–151. British Archaeological Reports (International Series) Oxford. 508 pp.
Ladizinsky, G. (1985). Founder effects in crop evolution. *Economic Botany*, 39, 191–199.
Ladizinsky, G. (1987). Pulse domestication before cultivation. *Economic Botany*, 41, 60–65.
Ladizinsky, G. (1989). Origin and domestication of SW Asian grain legumes. In *Foraging and farming: the evolution of plant exploitation*, ed. D. R. Harris & G. C. Hillman, pp. 374–389. London: Unwin & Hyman.
Magoja, J. L., Palacios, I. C., Bertoia, L. M. & Streitenbger, M. E. (1985). Evolution of *Zea*. *Maize Genetics Cooperation Newsletter*, 50, 61–67.
Manglesdorf, P. C. (1974). *Corn: its origin, evolution and improvement*. Cambridge, MA: Harvard University Press.
Manglesdorf, P. C. (1986). The origin of corn. *Scientific American*, 255, 72–78.
Maurizio, A. (1927). *Die Geschichte Unserer Pflanzennahrung*. Berlin: Parey.
Miller, T. E. (1986). Systematics and evolution. In *Wheat breeding: its scientific basis*, ed. F. G. H. Lupton, pp. 1–30. London: Chapman & Hall.
Miller, T. E. (1992). A cautionary note on the use of morphological characters for recognising forms of wheat (genus *Triticum*). In *Préhistoire de l'agriculture: nouvelle approches expérimentales et ethnographiques*, ed. P. C. Anderson-Gerfaud. Valbonne: Monographies du Centre de Recherches Archéologiques.
Moore, A. M. T., Hillman, G. C. & Legge, A. J. (1992). *Abu Hureyra and the advent of agriculture*. New Haven, CN: Yale University Press (in press).
Muramatsu, M. (1986). The *vulgare* supergene, *Q*: its universality in *durum* wheat and its phenotypic effects in tetraploid and hexaploid wheats. *Canadian Journal of Genetics and Cytology*, 28, 30–41.
Percival, J. (1921). *The wheat plant*. London: Duckworth. (1974 reprint).
Pickersgill, B. (1989). Cytological and genetical evidence for the domestication and diffusion of crops within the Americas. In *Foraging and farming: the evolution of plant exploitation*, ed. D. R. Harris & G. C. Hillman, pp. 426–439. London: Unwin & Hyman.
Randolph. L. F. (1976). Contribution of wild relatives of maize to the evolutionary history of domesticated maize. A synthesis of divergent hypotheses. I. *Economic Botany*, 30, 321–345.
Schiemann, E. (1948). *Weizen, Roggen, Gerste: Systematik, Geschichte und Verwendung*. Jena: Gustav Fischer.
Schwanitz, F. (1937). *The origin of cultivated plants*. Cambridge, MA: Harvard University Press. (1966 translation from the German original).
Sencer, H. A. & Hawkes, J. G. (1980). On the origin of cultivated rye. *Biological Journal of the Linnean Society*, 13, 299–313.
Sharma H. C. & Waines, J. G. (1980). Inheritance of tough rachis in crosses of *Triticum monococcum* and *T. boeoticum*. *Journal of Heredity*, 71, 214–216.

Vavilov, N. I. (1917). On the origin of cultivated rye. *Bulletin of Applied Botany and Plant Breeding*, **10**, 561–590. (In Russian with an English summary).

Vavilov, N. I. (1926). *Studies on the origin of cultivated plants*. Leningrad: Institute of Applied Botany and Plant Breeding.

Wilke, P. J., Bettinger, R., King, T. F. & O'Connell, J. F. (1972). Harvest selection and domestication in seed plants. *Antiquity*, **46**, 203–209.

Wilkes, G. (1989). Maize: domestication, racial evolution and spread. In *Foraging and farming: the evolution of plant exploitation*, ed. D. R. Harris & G. C. Hillman, pp. 440–455. London: Unwin & Hyman.

Willcox, G. H. (1992). Archaeobotanical significance of growing Near Eastern progenitors of domestic plants at Jalès. In *Préhistoire de l'agriculture: nouvelle approches expérimentales et ethnographiques*. Valbonne: Monographies du Centre de Recherches Archéologiques.

Zohary, D. (1969). The progenitors of wheat and barley in relation to domestication and agricultural dispersal in the Old World. In *The domestication and exploitation of plants and animals*, ed. P. J. Ucko & G. W. Dimbleby, pp. 47–66. London: Duckworth.

Zohary, D. (1984). Modes of evolution of plants under domestication. In *Plant biosystematics*, ed. W. F. Grant, pp. 579–586. Montreal: Academic Press.

Zohary, D. (1989a). Domestication of the Southwest Asian crop assemblage of cereals, pulses and flax: the evidence from the living plants. In *Foraging and farming: the evolution of plant exploitation*, ed. D. R. Harris & G. C. Hillman, pp. 359–373. London: Unwin & Hyman.

Zohary, D. (1989b). Pulse domestication and cereal domestication: how different are they? *Economic Botany*, **43**, 31–34.

Zohary, D. & Hopf, M. (1988). *Domestication of plants in the Old World*. Oxford: Oxford University Press.

8

Wheat as a model system

E. S. Lagudah and R. Appels

Introduction

Wheat is grown throughout temperate, subtropical and tropical highland regions. It is the most extensively cultivated crop among the cereals in terms of land area and production. The fact that wheat is grown under a wide range of ecological conditions is due partly to its inherent wide adaptability and partly to the intense plant breeding efforts which have been targeted at either broad or specific environments. Wheat has been a major component of farming systems spanning primitive to modern agriculture. It has provided a major source of energy, protein and dietary fibre in human nutrition. The progressive domestication of wheat has made it a major commodity in international trade, providing a source of income to rural industries. The wheat trade represents a significant component of the balance of trade of national economies. The economic significance of wheat has led to extensive cytogenetic and, more recently, molecular studies on this plant to the extent that wheat can now be considered a model system for studying chromosome manipulation and the evolution of polyploids.

The polyploid nature of wheat is an advantage in chromosome manipulation studies

Bread wheat (*Triticum aestivum* L.) is a polyploid species with 42 chromosomes derived from three genomes designated A, B and D, respectively; they each contribute seven pairs of chromosomes to the total genome. Bread wheat originated from the natural synthesis of a polyploid formed between tetraploid wheat, *T. turgidum* (AABB), and the diploid species *T. tauschii* (DD) (Kihara, 1944; McFadden & Sears, 1946; see Fig. 8.1). Throughout this chapter, species names of the Triticineae subtribe used by Morris & Sears (1967) have been adopted. The chromosomes of the D genome of bread wheat are considered relatively

Fig. 8.1. Evolution of hexaploid wheat derived from the natural synthesis of tetraploid wheat (AABB) and *Triticum tauschii* (DD). It is postulated that the tetraploid wheat (*T. Turgidum*) originated as an amphidiploid between *T. urartu* (AA) and a diploid species related to the Sitopsis sectional taxa (for example *T. speltoides*).

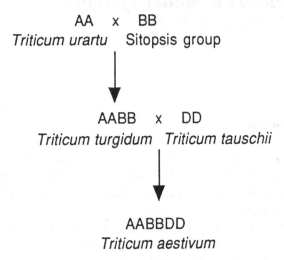

recent additions and they have diverged less from their putative donor than have the chromosomes of the A and B genomes.

The large genome size of wheat and the frequent occurrence of triplicate gene loci associated with its three constituent genomes make wheat less favourable than other crops as a system for gene manipulation. However, the ability of wheat to tolerate a wide range of aneuploidy (Sears, 1954) has made it one of the best studied crops for cytogenetic analysis and chromosome manipulation. The 42 chromosomes of bread wheat are organized as 21 homologues that are made up of seven pairs from each of the A, B and D genomes, respectively. Although the genomes are distinct, a varying degree of genetic similarity occurs between them. The relationships between genetically equivalent chromosomes in each genome set have therefore been described as homoeologous (Table 8.1) rather than strictly homologous. In a given homoeologous set, such as the group 1 chromosomes, the complete absence (nullisomy) of chromosome 1A is compensated for by an additional pair (tetrasome) of either chromosome 1B or 1D (Table 8.2). In this way chromosome homoeology was demonstrated in the complete set of 42 nulli–tetrasomic combinations in the wheat variety Chinese

Table 8.1. *The genome constitution and concept of chromosome homoeology in hexaploid wheat*

Homoeologous group	Genome		
	A	B	D
1	1A	1B	1D
2	2A	2B	2D
3	3A	3B	3D
4	4A	4B	4D
5	5A	5B	5D
6	6A	6B	6D
7	7A	7B	7D

Wheat has a euploid chromosome number of $2n = 42$. These are divided into three groups, A, B and D, with each group being $2n = 14$.

Table 8.2. *Aneuploids in wheat involving wheat group 1 chromosomes[a]*

	Chromosome constitution		
Euploid state	1A1A	1B1B	1D1D
Monosomic 1D	1A1A	1B1B	1D
Nulli-1A tetra-1B	—	1B1B1B1B	1D1D
Ditelo-1D short arm	1A1A	1B1B	1DS. 1DS

[a]The euploid states of the chromosomes in groups 2–7 remain unchanged in this example.

Spring by Sears (1966) (Table 8.2). These observations were an indication of the similarities in genetic content of homoeologous sets of chromosomes. The variation in the degree of genetic relatedness was revealed by the fact that while a complete compensation for the group 1 and 3 homoeologues could be obtained, the tetrasomic homoeologues could not adequately compensate for the nullisomics 2B, 4A (redesignated 4B, see Dvorak, 1983; Chen & Gill, 1984; Naranjo *et al.*, 1988; Dvorak *et al.*, 1989) and 6B. The genetic similarity of homoeologous groups is also shown by assaying specific genes such as isozymes and low-copy-number DNA markers (discussed later).

It seems likely that the type and amount of highly repeated sequences may have contributed to the divergence within homoeologous chromosome groups. Certain classes of repeated sequences are highly amplified in some genomes and their chromosomal distribution as revealed by *in situ* hybridization has been used in chromosome identification (Appels, Driscoll & Peacock, 1978; Dennis, Gerlach & Peacock, 1980; Flavell *et al.*, 1987; see Fig. 8.2).

Strictly homologous chromosome pairing is controlled by specific sets of genes

Despite the genetic similarity of homoeologous chromosomes, only bivalents are formed in a normal meiotic cycle. The bivalents result from strictly *homologous* pairing of 21 chromosomes comprising seven pairs from each of the A, B and D genomes. The absence of homoeologous pairing is due to a major gene(s) (*Ph1* locus) on the long arm of chromosome 5B. In genetic stocks nullisomic for chromosome 5B, homoeologous pairing was observed (Okamoto, 1957; Riley & Chapman, 1958). Genetic stocks nullisomic for other chromosomes also display homoeologous pairing, but at a reduced level compared to those nullisomic for chromosome 5B, viz. those lacking chromosomes 5BS (Feldman & Mello-Sampayo, 1967; Riley & Chapman, 1967), 5AL and 5DS (Feldman, 1968) and 5DL (Feldman, 1966; Riley, Young & Belfield, 1966*b*). Chromosomes with a suppressing effect on homoeologous pairing include 3DS (*Ph2* locus; Mello-Sampayo & Canas, 1973), 3AS (Upadhya & Swaminathan, 1967; Driscoll, 1973) and 4D (Driscoll, 1973). Regular meiotic behaviour in wheat may thus be considered to be a balance, or an interaction, between enhancers and suppressors of chromosome pairing genes. Various chromosomal associations occur at different stages of meiosis in wheat. The diploid behaviour of allohexaploid wheat, controlled by the *Ph* locus, may operate by controlling the distribution of recombination events. This is because multivalents are frequently observed in zygotene but by the early pachytene stage only bivalents are found (Holm & Rasmussen, 1984).

The *Ph* locus has been exploited in breeding work where wild relatives of wheat have provided a diverse gene pool to broaden the genetic base of wheat. Transfer of traits from closely related diploid progenitor species of the A and D genomes of wheat, *T. urartu* or *T. monococcum* and *T. tauschii*, respectively, can be achieved by normal homologous recombination and do not require the *Ph* locus to be inactivated. Direct crosses of these species into wheat have been used in the transfer of disease resistance (The, 1973; Gill & Raup, 1987; E. Lagudah, R. Eastwood & R. Appels, unpublished results).

Fig. 8.2. A single rye chromatin containing chromosome (marked by arrow) was detected using a H³-cRNA probe synthesized from a DNA fraction specific to rye (May & Appels, 1980). The technique of *in situ* hybridization was used to produce the result shown.

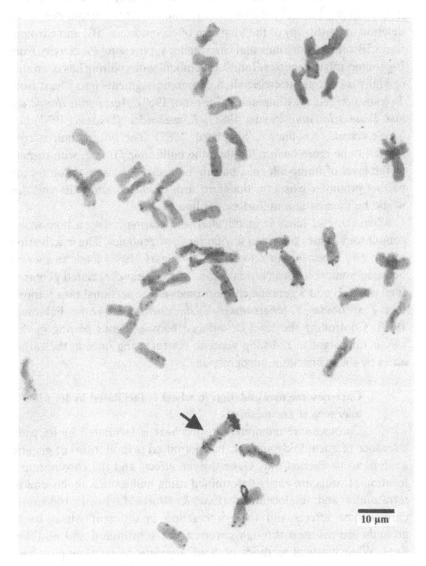

In contrast, introgressions from more diverged species require chromosome manipulation techniques to induce homoeologous pairing and recombination. The induction of homoeologous pairing in wheat and its wild relatives can be achieved by *inactivating* the *Ph* locus. This can be accomplished by the use of lines nullisomic for chromosome 5B, deletion mutants (*ph*) of the long arm of chromosome 5B, and chromosome 5B substitution lines that carry genes suppressing *Ph* activity from B-genome-related species. Induced homoeologous pairing has been successfully used in transferring alien chromatin segments into wheat from *T. comosum* (Riley, Chapman & Johnson, 1968), *Agropyron elongatum* and *T. umbellulatum* (Sears, 1981), *T. speltoides*, (Dvorak, 1977) and *Secale cereale* (Koebner & Shepherd, 1985). The *ph1* mutant is considered to be more favourable than the nullisomic 5B line, with respect to the level of homoeologous pairing induced, because a chromosome pairing promoter exists on the short arm of chromosome 5B and this would be missing in a nullisomic 5B line.

When the *Ph1* locus is inactivated, the pairing between homoeologous chromosomes can cover a wide range of genomes. The inactivation of the *Ph2* locus on chromosome 3DS (Sears, 1982) leads to pairing between homoeologous chromosomes of more closely related genomes such as the B and S genome chromosomes of the sectional taxa Sitopsis (e.g. *T. speltoides*, *T. longissimum*, *T. sharonense*, *T. bicorne*; Feldman, 1988). Controlling the level of induced homoeologous pairing in this way is important in reducing genome restructuring outside the target areas of alien chromatin introgression.

Chromosome manipulation in wheat is facilitated by its wide tolerance of aneuploidy

Chromosome manipulation in wheat is facilitated by its wide tolerance of aneuploidy and this has permitted several forms of genetic analysis to be carried out. Gene dosage effects and the chromosomal location of traits are easily determined using nullisomics, monosomics, tetrasomics and ditelocentrics (Law & Worland, 1973). Individual chromosome effects and their interaction in different wheat backgrounds are assessed through chromosome substitution and addition lines. While classical methods of cloning genetic material employ bacteria, viruses and yeast as cloning vehicles, the use of cytogenetic methods in generating stable chromosome substitution and addition lines provides a *source for 'cloning' large pieces* of DNA (i.e. whole chromosomes, Figs 8.2 and 8.3).

**Fig. 8.3. A pair of rye chromosomes (marked 'A') was detected in the
same way as the rye chromatin shown in Fig. 8.2 (Appels *et al.*,
1978).**

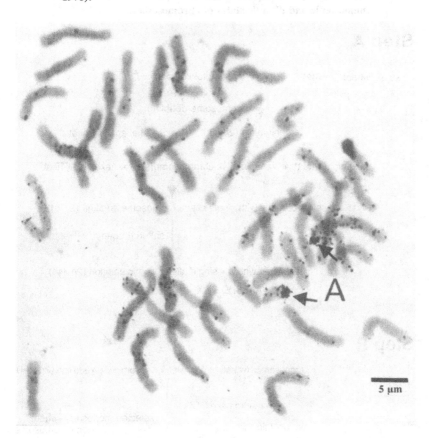

Chromosome substitution lines either from wheat or alien species are
produced in a variety of ways. Intervarietal wheat substitutions are
produced by first crossing the recipient monosomic and euploid donor
and following this by repeated backcrosses of the hybrid-derived
monosomic to the recurrent (recipient) monosomic parent. The final
step in the procedure is to self-fertilize the backcrossed hybrid line,
monosomic for a donor chromosome, and thus recover the respective
donor chromosome in a disomic form. In the development of alien
substitution lines, various modifications of the scheme used by O'Mara
(1940; see Fig. 8.4) are employed. By inducing chromosome doubling in
a hybrid between *T. aestivum* and a diploid wild relative to produce the
amphidiploid, the hybrid can be backcrossed to wheat and the progeny

Fig. 8.4. Standard procedure used in the production of single alien chromosome addition lines (step A) and disomic alien substitution lines (step B) in hexaploid wheat. The symbols 'I' and 'II' refer to monosomic and disomic states of chromosomes.

Step A

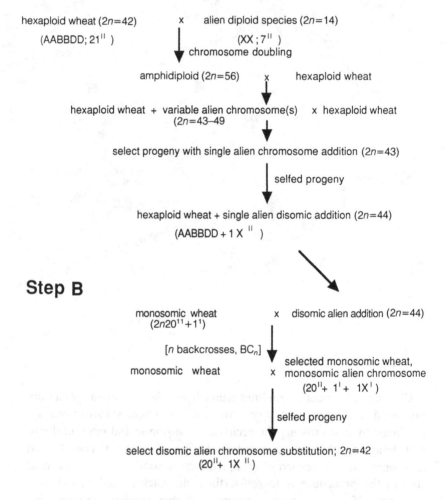

Step B

selected for various chromosome additions of the wild species. Backcrosses of a selected chromosome addition line to the recurrent monosomic wheat parent can then be carried out (see Fig. 8.4, step B) to allow selection of alien chromosome substitutions compensating for its respective homoeologue in wheat. Several alien chromosome substitutions into wheat have been produced from *T. umbellulatum* (Riley, Chapman & Miller, 1973), *T. dichasians* (Muramatsu, 1973), *T. comosum* (Riley, Chapman & Macer, 1966a), *T. longissimum* (Netzle & Zeller, 1984), *Thinopyrum ponticum*, *Th. intermedium*, *Th. elongatum* (Dvorak & Knott, 1974; Dvorak, 1980; for a review, see Pienaar, 1990), and *Secale cereale* (see review of Miller, 1984).

Identification of the alien chromosome addition line is important for the selection of the appropriate aneuploid homoeologue in wheat to facilitate the production of well-characterized disomic substitution lines. Karyotypic analysis is often employed in the identification of alien chromosomes. In addition, similarities in gene content among homoeologous sets of chromosomes have been exploited to locate homoeo-loci for isozymes (Hart, 1987) and restriction fragment length polymorphisms (RFLPs) in DNA (Sharp *et al.*, 1989; Fig. 8.5). Alien disomic addition and substitution lines provide genetic resources for chromosome manipulation; irradiation, tissue culture and inactivation of the *Ph* locus may be used to enable the transfer of alien segments carrying useful agronomic traits (some of which are listed in Table 8.3).

Intercrosses between the hexaploid and tetraploid wheat genomes, introduced another dimension of chromosome manipulation. Pentaploid backcross derivatives using *T. aestivum* as the recurrent parent have been used to disassemble hexaploid wheat by extracting its tetraploid component (Kerber, 1964). Tetraploid genome components of bread wheat cultivars such as Canthatch, Thatcher, Prelude and Chinese Spring are available as Tetra-Canthatch, Tetra-Thatcher etc. Derivation of tetraploids from *T. aestivum* provided a system with which to analyse the relative contribution of the D and AB genomes to gene expression in the hexaploid genome (Kerber & Dyck, 1969; Kerber & Tipples, 1969; Orth, Dronzek & Bushuk, 1973; Rowland & Kerber, 1974; Watanabe, 1981).

Joppa & Williams (1977, 1988) adopted an approach that was complementary to that of Kerber (1964) to extract individual D-genome chromosomes into tetraploid wheat. The technique also exploited the homoeology and aneuploidy of the hexaploid wheat genome to generate monosomics in tetraploid wheat (Fig. 8.6). Disomic substitutions of individual D-genome chromosomes that compensate for their corresponding A and B genome homoeologues made it possible to develop a complete set of 14 D-genome nulli-A or nulli-B chromosomes in the tetraploid wheat Langdon. This chromosome manipulation system constitutes another example of 'cloning' individual chromosomes of wheat.

Table 8.3. *Disease resistance from alien wheat chromatin transferred into cultivated wheat backgrounds*

Source	Disease resistance	Chromosome location involved	Cultivar	Region of cultivation
Agropyron elongatum	*Lr19*	7DL	Agrus	USA
	Lr24/Sr24	3DL	Agent, Blueboy II, Cloud	USA
	Sr25	7DL	Marquis	Canada
	Sr26	6AL	Thatcher Avocet, Flinders, Jabiru, Kite, Takari	Australia
Triticum comosum	*Sr34/Yr8*	2A/2M	Compair	Europe
Triticum timopheevi	*SrTt1*	2BS	Arthur, Arthur 71	USA
			Cook, Mengavi, Timson	Australia
			Kenya Leopard	Kenya
			Dipka, Flamink, Gourtiz	South Africa
			Maris Envoy, M. Templar, M. Huntsman, M. Nimrod	UK
Triticum umbellulatum	*Lr9*	6B	Abe, Arthur 71, McNair 2203	USA
Triticum ventricosum	*Pch1*	7D	Roazon	Europe
Secale cereale	*Sr31/Yr9/Lr26*	1BL.1RS or 1R(1B)	Aurora, Burges 2, Clement, Disponent, Lovrin 10, 12, 13, Saladin, Winnetou, Zorba	Europe
			Bezostaya II, Kavkaz, Skorospelka, Veery 'S'	USSR
			Salmon	Mexico
				Japan

Data on disease transfers and chromosome location are taken partly from Porceddu *et al.* (1988) and McIntosh (1988).

Fig. 8.5. Alien chromosome identification using DNA probes. The sequence pAcc2 from *Thinopyrum elongatum* (E genome) was recovered in a study by McIntyre, Clarke & Appels (1988) and with the restriction endonuclease *Eco*RV (Z-Y. Xin, unpublished observations) reveals an RFLP that is specifically associated with the presence of *Th. elongatum* DNA in wheat. In the X ray illustrated, 10 µg aliquots of DNA from wheat lines suspected of containing an *Th. elongatum* DNA segment conferring stem rust resistance were digested with the restriction endonuclease *Eco*RV. Probing the digests with the pAcc2 sequence, using standard procedures, demonstrated the presence of the *Th. elongatum* DNA segment in only one of the lines (indicated as +). The lower two bands marked by arrows are diagnostic for *Th. elongatum* DNA.

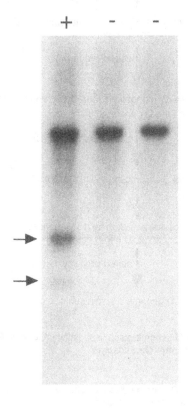

Fig. 8.6. Schematic representation of the sexual crosses and cytogenetic selection employed by Joppa & Williams (1977, 1988) in the production of D genome disomic substitutions for their corresponding chromosome homoeologues in the tetraploid wheat, cv. Langdon. The symbols 'I', 'II' and 'IIII' refer to the monosomic, disomic and tetrasomic states of a chromosome.

Hexaploid wheat , cv. Chinese Spring   X   Tetraploid wheat cv. Langdon
(nulli-A or -B tetra D genome chromosome   (2n=28)
    $2n=42$, 20^{II} + $1D^{IIII}$)

pentaploid derivatives $2n=35$, either monosomic A or B
and $6D^{I}$ + $1D^{II}$

 selfed progeny

select F$_2$ plants with $2n=28$, monosomic for A or B chromosomes
and disomic for a D genome chromosome

backcross to Langdon

BC$_1$, select monosomic-A or B chromosome + monosomic D chromosome

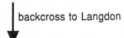

BC$_n$, repeated backcrosses, selecting for 13^{II}+ monosomic-A or B
    +   monosomic D chromosome

 selfed progeny

select for D genome disomic substitution

either $6A^{II}$ + $7B^{II}$ + $1D^{II}$ = D genome disomic substitution for nulli-A
genome chromosome

or $7A^{II}$ + $6B^{II}$ + $1D^{II}$= D genome disomic substitution for nulli-B
genome chromosome

The identification of D genome amplified DNA sequences (Lagudah *et al.*, 1991) combined with the elegant 'cloning' system of Langdon–D-genome substitution lines provide a means whereby DNA fragments enriched for each D-genome chromosome may be isolated as originally suggested by May & Appels (1987).

While alien species have been well documented as a resource for broadening the genetic variation of wheat (Gale & Miller, 1987), the primary and secondary gene pool species of wheat, e.g. *T. spelta, T. macha, T. vavilovii, T. dicoccoides*, provide the major sources for increasing variation at homologous loci. The production of a set of chromosome substitution lines in which *T. dicoccoides* chromosomes replace their corresponding homologues in tetraploid and hexaploid wheat (Joppa, 1988; E. R. Sears, unpublished results) provides useful parental lines for generating recombinant disomic substitution lines. Recombinant disomic substitution lines allow for the fixation of divergent homozygous alleles, a 'clonal' source of variation usually obtained from F_2 segregation, and consequently are considered ideal material for detailed genetic mapping studies.

The addition of alien chromosomes to wheat has revealed unusual chromosome behaviour

There are reports of unusual chromatin behaviour occurring in the course of production alien chromosome addition and substitution lines either from the donor or the recipient genome. Certain alien chromatin introductions into wheat from *T. longissimum* (Maan, 1975), *T. triunciale* (Endo & Tsunewaki, 1975), *T. cylindricum* (Endo, 1979) and *T. sharonense* (Miller, Hutchinson & Chapman, 1982), in their monosomic states, are preferentially transmitted. The preferential transmission is often associated with sterility in other gametes lacking the alien chromosome. The reduction in fertility has been attributed to a gametocidal effect of the preferentially transmitted alien chromosome (referred to as a gametocidal chromosome). Although the alien disomic state of the gametocidal chromosomes rarely causes sterility, this state can lead to a high frequency of non-random chromosomal changes involving translocations, deletions and ring chromosomes (Fig. 8.7). Endo (1988) detected the frequency of chromosome mutations induced by *T. cylindricum* to be much higher in the B than in the A and D genomes of wheat. In some wheat backgrounds, suppression of chromosome mutation induced by gametocidal chromosomes from *T. longissimum* and *T. sharonense* is associated with chromosome 4B (current

Fig. 8.7. Some examples of chromosome mutations induced by gametocidal chromosomes as detected by C-banding (kindly supplied by T. Endo & B. S. Gill). From left to right, normal chromosome 2B (far left) and a series of deletions of the short arm of 2B (top line) and a series of deletions of the long arm of 2B (bottom line).

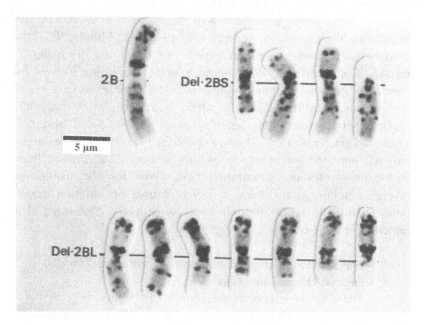

designation), while those from *T. triunciale* are modulated by a gene(s) on chromosome 3B (Tsujimoto & Noda, 1988; Endo, 1988; Tsujimoto & Tsunewaki, 1985). The isolation of cytotypes of wheat with deletion series of their respective chromosomes provides a useful source of genetic stocks for physical mapping of genes (see Fig. 8.7, for an example of chromosome 2B).

The short arm of chromosome 4D carries the semi-dwarf gene *Rht2*, which has been found to segregate for aneuploidy in some wheat varieties, resulting in undesirable tall plants. In an attempt to ensure regular transmission of *Rht2*, the high frequency of transmission of gametocidal chromosomes from *T. sharonense* (4S[1]) has been exploited by King, Reader & Miller (1988) by producing translocations involving chromosome 4D and 4S[1].

The manipulation of chromosomes in wheat has been widely exploited for agronomic improvement

Several traits with major gene effects on plant morphology, development, reproduction and grain quality have been manipulated by selection and breeding practices in wheat domestication, (see McIntosh (1988) for a listing of all major genetic loci studied in wheat). The genetics of these traits have been reported as Mendelian or quantitative and their chromosomal location deduced from aneuploid stocks.

Species relationships in the wheat gene pool have been well studied. Consequently, numerous examples of chromosome manipulation involving either part of, or a whole, chromosome to the complete genome of wild/alien species have been produced in bread wheat backgrounds (for a review, see Gale & Miller, 1987; Appels & Lagudah, 1990). A number of these alien chromosome transfers carry useful agronomic traits such as high grain protein content, disease resistance and tolerance to abiotic stress. The agronomic significance of some of the alien chromosomes added to wheat has been exploited to develop commercial cultivars. The major impact of cultivars carrying alien chromatin on the domestication of wheat arises from the fact that they provide novel sources of disease resistance (see Table 8.3). The effectiveness of these alien sources of resistance in overcoming the diseases listed in Table 8.3 has often been marred by accompanying deleterious effects on wheat yields and/or grain processing quality. The smaller the amount of alien chromatin carrying the desirable trait, the less its deleterious effects are on agronomic performance. Consequently a major objective in alien chromosome manipulation is to ensure that the minimum chromatin containing, the target loci, is introgressed without major restructuring in the recipient wheat genome.

The genome structure of wheat includes a large number of repetitive DNA sequence families

The DNA content of the haploid wheat genome is estimated at 18.1 pg, with an average nuclear volume ranging from 240 to 160 000 μm^3 due to the occurrence of monoploid to endopolyploid nuclei in different tissues (Bennett *et al.*, 1973; Flavell *et al.*, 1987). Studies of denaturation and renaturation kinetics of DNA have shown that various classes of DNA exist in the wheat genome. Repeated sequences account for approximately 80% while 10–20% are of low copy numbers (Mitra & Bhatia, 1973; Flavell *et al.*, 1974; Flavell & Smith, 1976; Ranjekar,

Pallota & Lafontaine, 1976; Rimpau, Smith & Flavell, 1978). An estimate of the proportion of sequences in wheat with a coding function can be estimated from the observation that, in a plant such as *Arabidopsis*, an array of genes essential for plant life is present in a genome of approximately 7×10^7 bp of DNA. Assuming a similar array of genes would be present in each of the diploid genomes of wheat containing approximately 5.5×10^9 bp of DNA, we estimate that at most 1.3% of the genome has a normal coding function. This estimate is consistent with biophysical studies on the wheat genome (references cited above in this paragraph).

Certain classes of the highly repeated nucleotide sequences have been cloned and sequenced. Their chromosomal distribution has been shown to be either terminal, or centromeric, or dispersed. Some of these sequences have proved useful in identifying chromosomes belonging to a specific genome as well as in identifying individual chromosomes within and between genomes. Among the 1.3% of nucleotide sequences which have a coding function, an increasing number of genes have been isolated using molecular cloning techniques. These include genes for alcohol dehydrogenase, rDNA, 5 S DNA, storage proteins, early methionine polypeptide, germination-specific shoot polypeptide, 7 S globulin, chlorophyll a/b binding proteins, α-amylases, wheat endosperm ATP-dependent glucose-1-phosphate adenosyltransferase and the small subunit of carboxylase. The chromosomal location of most of these sequences has been identified and incorporated into a genetic linkage map (Chao *et al.*, 1989*a,b*; Futers *et al.*, 1990; Lagudah *et al.*, 1991; Gill *et al.*, 1991; Baum & Appels, 1991).

A study of the evolution of the repetitive units suggests that amplification/deletion events may be relatively common. Major deletion events involving the tandemly arranged ribosomal RNA genes (*Nor* locus) and repetitive sequence array comprising the 350-family at the terminal regions of the 1R chromosome have been reported (Brettell *et al.*, 1986; Koebner, Appels & Shepherd, 1986). Amplification events, arising during the course of evolution, involving a restricted set of units belonging to a repetitive sequence family have been observed (Appels *et al.*, 1989; Baum & Appels, 1991; Guidet *et al.*, 1991; D. McNeil, E. S. Lagudah & R. Appels, unpublished results) and clearly represent major forces in the modification of Triticeae genomes. The full impact of genome restructuring of this type on gene expression remains an area of active research.

The genetic mapping of hexaploid wheat

In recent years there has been a world-wide move toward genetic mapping of the major food crops. This has been made possible by advances in molecular biology, computer technology and the realization by plant breeders that closely linked genetic markers can provide significant gains in the process of cultivar development. Some of the gains include the tracking of characters which are normally difficult to assay, the pyramiding of genes with the same phenotype, an increased precision in manipulation of chromosome segments and the fingerprinting (identification) of plant varieties.

A prerequisite for producing genetic maps from the progeny of a species is the availability of markers capable of differentiating parental genotypes. Traditionally, markers have been obtained as naturally occurring or induced morphological variants. The available range was extended by the use of biochemical markers such as isozymes and storage proteins that differ in electrophoretic mobility because of variation in the size and charge of their constituent polypeptides. In recent years the use of restriction endonucleases that cleave DNA at specific regions to generate fragments with different sizes has widened the scope for detecting variability in different organisms (Botstein *et al.*, 1980). The fragments generated by restriction endonuclease digests of DNA can be fractionated by electrophoresis, transferred onto membrane filters, and specific regions can be detected by hybridization to cloned DNA sequences (Southern, 1975) to reveal RFLPs between genotypes.

In bread wheat, the large genome size and low level of restriction endonuclease site differences between cultivars creates a formidable challenge in developing a molecular genetic linkage map. From a range of restriction endonuclease digests of genomic DNA, bread wheat shows less polymorphism than do other crops, such as barley and rye, in the Triticeae (Gale, Chao & Sharp, 1990). The polyploid nature of bread wheat, which accounts for its large genome size, means that triplicate loci corresponding to the three genomes will frequently be present. Furthermore each homoeologous locus needs to be clearly distinguished to enable unequivocal identification of segregating markers for all 21 chromosomes (seven homoeologous groups from three genomes) to be mapped. The task of producing a wheat genetic map has brought together several laboratories from the USA, Australia and Europe to form the International Triticeae Mapping Initiative (ITMI). Laboratories involved with ITMI have each been assigned specific chromosome groups, to locate, map and exchange DNA clones.

Despite the difficulty of mapping the wheat genome, the availability of a wide range of wheat aneuploids and alien chromosome addition and substitution lines has facilitated the assignment of cloned DNA sequences to homoeologous chromosome groups (see Fig. 8.8).

Chromosome banding procedures provide landmarks that can be incorporated into genetic maps

Karyotype analyses have provided a valuable method for identifying the constituent chromosomes of the wheat genome. Karyotypic analysis using chromosome banding relies on differences in the distribution, type, size and number of heterochromatic sequences. Techniques that permit the detection of differences in condensation in neighbouring regions of the chromatin produce regions with different banding patterns that can be observed cytologically. The C- and N-banding methods are based on alkali and acid pretreatments, respectively, prior to the application of Giemsa stain to locate heterochromatic chromosomal regions (Gill & Kimber 1974; Gerlach, 1977). In hexaploid wheat, identification of all chromosomes is possible with C-banding (for a review, see Gill, 1987), whereas 16 chromosomes are detected using N-banding (Endo & Gill, 1984). The level of homologous chromosome differentiation between wheat varieties as determined by chromosome banding can, in a limited number of cases, be used to incorporate specific bands into a chromosome map (group 1 chromosomes shown in Fig. 8.9).

The genetic distances in chromosome maps such as is shown in Fig. 8.9 are based on the frequency of recombination events between adjacent markers and do not consistently correspond with their physical distances, which in this figure are determined by C-banding. For example, the locus *XcsIH69* on the short arm of chromosome 1B shows a recombination frequency of 4.2% from the centromere, but is located physically in the distal end of the chromosome.

Mapping the large genome of wheat is made simpler by mapping diploid progenitor species

An approach to overcoming some of the constraints in mapping the wheat genome (due its large size) is to use their diploid progenitors and relatives. The D genome progenitor, *T. tauschii*, by reason of its diploid genome obviates the need to identify homoeologous loci, it contains wide genetic variation and is the least diverged of the putative progenitor species relative to their corresponding wheat genome. Consequently *T. tauschii* has been exploited to produce a genetic

Fig. 8.8. Chromosome location of a genomic clone revealed by autoradiography of DNA from wheat, barley and barley chromosome addition lines to wheat. DNA (10 µg) from leaf samples was digested with the restriction endonuclease *Hind*III, subjected to electrophoresis, transferred onto membrane filters and probed with a CT³²P-labelled genomic clone from *T. tauschii* (Lagudah *et al.*, 1991). Lane B, barley cv. Betzes, lane CS, 'Chinese Spring' wheat; lanes 1–7, single disomic barley (cv. Betzes) chromosome 1H–7H additions into wheat (Islam, Shepherd & Sparrow, 1981). The fragments of the barley genome are assigned to chromosomes 2H and 4H addition lines (lanes 2 and 4) which show additional fragments which are additional to those of the euploid wheat genome (lane CS).

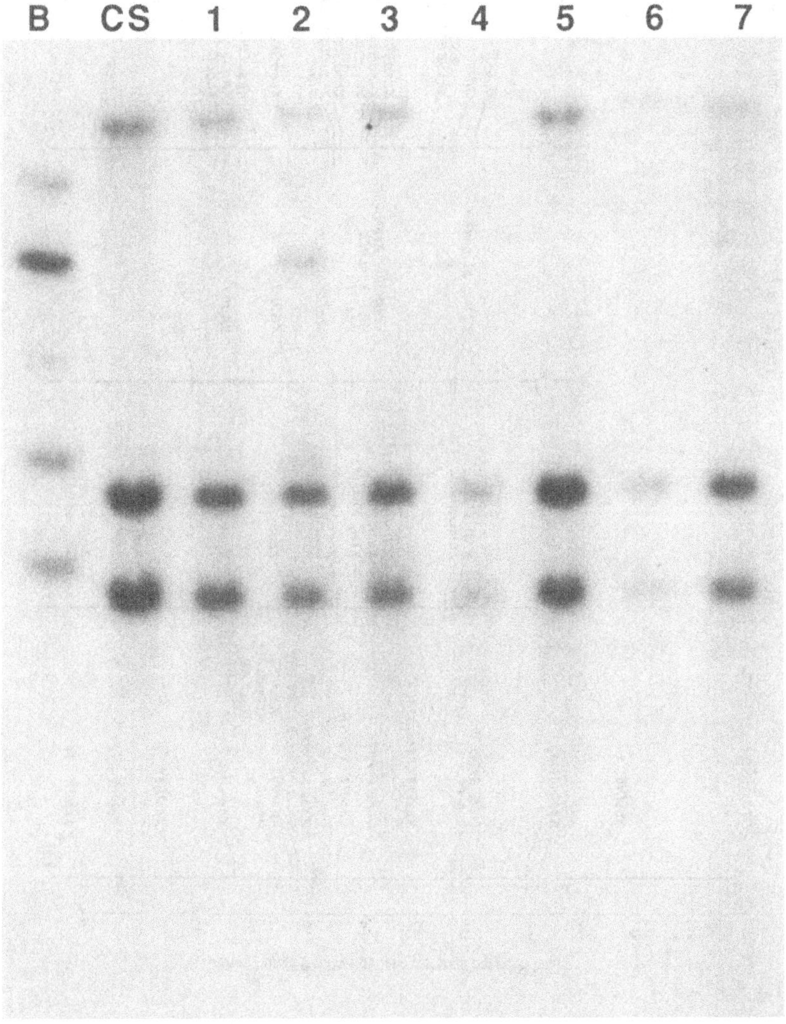

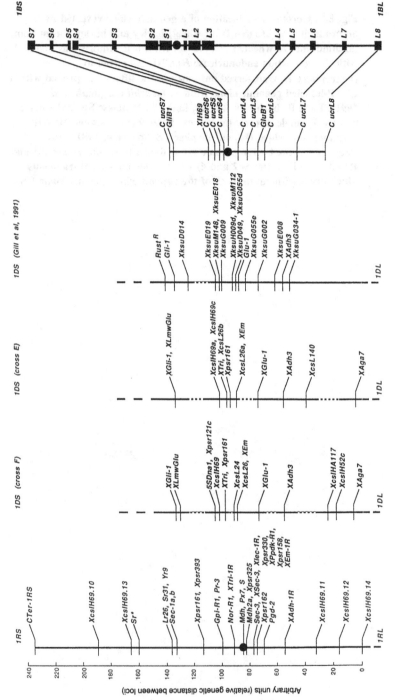

linkage map equivalent to the D genome of wheat (Gill *et al.*, 1991; Lagudah *et al.*, 1991). Genetic maps available for chromosome 1 from *T. tauschii*, wheat and rye and for chromosome 7 from *T. tauschii* and wheat have revealed a degree of conserved gene synteny (Payne, 1987; Chao *et al.*, 1989*b*; Gill *et al.*, 1991; Lagudah *et al.*, 1991; Curtis & Lukaszewski, 1991). Parental genotypes used in generating the mapping population have an effect on the actual level of gene synteny that occurs. Some wheat varieties contain translocations that are not present in their diploid progenitor species or relatives (Vega & Lacadena, 1982; Naranjo *et al.*, 1987) and the converse has also been detected for chromosomes 1D and 7D in certain *T. tauschii* genotypes.

The evolution of wheat is closely linked to its domestication

Polyploidy constitutes a major feature of wheat evolution and domestication. Wheat comprises a polyploid series of $2n=14$, 28 and 42, with $x=7$. In the diploid wheats, *T. urartu* and *T. boeticum* constitute the wild forms, while the cultivated types are *T. monococcum* and *T. sinskajae*. In the tetraploids, *T. dicoccoides* and *T. araraticum* represent the wild forms of the AABB and AAGG genomes, respectively. Major gene effects involving three loci subsequent to amphidiploidization of the AABBDD genome have contributed substantially to morphological changes in hexaploid wheat. Major gene effects of Q factor on chromosome 5A determine the presence of free-threshing grains and a tough rachis (MacKey, 1966) and exist in different strengths among the free-threshing and non-free-threshing *Triticum* species (Swaminathan, 1966). Spike compactness, is controlled by the C locus in chromosome 2D, while spherical grains result from a recessive gene *sl* on chromosome 3D. On the basis of these three loci, hexaploid wheats can be classified as follows (Swaminathan & Rao, 1961; Morris & Sears, 1967):

Fig. 8.9. Genetic linkage maps of chromosome 1R of rye (Baum & Appels, 1991), 1D from three different *T. tauschii* populations (Gill *et al.*, 1991; Lagudah *et al.*, 1991) and the genetic and physical maps (based on C-banding) of chromosome 1B of tetraploid wheat (Curtis & Lukaszewski, 1991). The markers used in producing the maps are based on isozymes (e.g. *Mdh*, *Gpi*), storage proteins (e.g. *Sec1*, *Glu-1*), RFLPs (prefixed with the symbol '*X*'), and C-banding (prefixed with '*C*'). The corresponding physical map positions of the C-bands are designated as S1–S7 and L1–L8 on the short and long arms, respectively, of chromosome 1B.

T. aestivum (common wheat)	*Q c Sl*
T. compactum (club wheat)	*Q C Sl*
T. sphaerococcum (shot wheat)	*Q c sl*
T. macha	*q C Sl*
T. spelta (spelt wheat) and *T. vavilovii*	*q c Sl*

Increasing ploidy level shows parallels with the domestication of wheat, from einkorns (diploids) through to emmers (tetraploids) and to the dinkels (hexaploids). Phylogenetic studies on the evolution and domestication of wheat must take into account complementary evidence from palaeoethnobotanical studies. Most of the early-dated archaeological findings of wheat samples have been wild einkorn and cultivated emmer (Harlan, 1981), while naked wheats and cultivated einkorn and emmer have been found in the Near East in archaeological remains dating to Neolithic times (Zohary, 1973). There is uncertainty about the ploidy level of carbonized grains of wheat from archaeological sites, because moisture content at the time of carbonization can cause considerable differences in grain size (Stewart & Robertson, 1971). Genetic studies that often implicate *T. spelta* as the putative primitive form of hexaploid wheat (McFadden & Sears, 1946) do not accord with the late appearance of *T. spelta* in archaeological findings (Helbaek, 1966; Hjelmquist, 1966; Janushevich, 1978). However, Listina (1978) provided archaeological evidence for the presence of *T. spelta* in the Transcaucasus as early as the sixth to seventh millenium BC, predating the earlier findings of European *T. spelta*.

The relatively recent evolution of wheat means that extant diploid progenitors can be identified

The polyploid composition of wheat has intrigued several workers and stimulated them to study the identity, structure, organization and evolution of its constituent genomes (for a review, see Lilienfield, 1951; Flavell *et al.*, 1987; Kimber & Sears, 1987). Emerging from the studies on constituent genomic analyses is the availability of a wide array of techniques for inferring phylogenetic relationships. These include meiotic chromosome pairing in hybrids, electrophoresis of proteins and isozymes, immunochemical analyses, patterns of heterochromatin distribution, DNA hybridization of repeated sequences, karyotype analysis and plant morphology. From these studies the A- and D-genome progenitors of hexaploid wheat have generally been accepted as *T. monococcum*, or more likely *T. urartu*, and *T. tauschii*, respectively. In the case of the B-genome donor a group of species belonging to the sectional taxa, Sitopsis – *T. speltoides*, *T. bicorne*, *T. sharonense*, *T. searsii* and *T. longissimum* – have been identified as potential candidates. The ambiguity surrounding the precise B-genome donor is highlighted by assumptions that are often made in inductive derivations of genome donors. They presuppose no major modification in the diploid progenitor and the corresponding genome in the polyploid. In most of the methods employed in deducing genome donors, the conflicting results can be attributed to the special evolutionary characteristics of the particular portion of the genome analysed in nominated progenitor species.

Analysis of meiotic chromosome pairing of several hexaploid wheat intervarietal hybrids reveals varying degrees of differentiation between homologous sets of chromosomes (Dvorak & McGuire, 1981). Chromosome differentiation was very high in the B compared to the A genomes, while differentiation in the D genome was rare. These differential changes within and between genomic components of hexaploid wheat are also reflected in the relatively high level of polymorphism present in the B genome for storage proteins (Payne & Lawrence, 1983; Metakovsky *et al.*, 1984) and DNA markers (Gale *et al.*, 1990). An alternative explanation for the non-random evolutionary changes in the hexaploid wheat genome is the suggestion that the B genome is polyphyletic in origin. A polyphyletic origin may arise from mixed genomes where genetic interchanges between tetraploid (AABB) amphidiploids carrying different putative B-genome donor species, but a common A genome, would result in a high level of differentiation in the B genome. Furthermore, inadequate sampling of the level of variation within a putative diploid donor species when phylogenetic relationships for a polyploid genome are studied can result in conflicting results. For example, variable patterns of meiotic chromosome pairing may be observed in hybrids between the B genome of wheat and *T. speltoides* genotypes with or without genes that suppress the *Ph* locus.

A locus that has been studied in detail in both wheat and its possible progenitors is the *5SDna* locus (Gerlach & Dyer, 1980; Scoles *et al.*, 1988; Reddy & Appels, 1989; Lagudah, Clarke & Appels, 1989; Vakhitov, Gimatov & Shumyatskii, 1989; Kolchinsky *et al.*, 1990; Appels, Baum & Clarke, 1992). The units that contain the 5 S RNA genes range in size from 300 to 500 (bp) and have a *Bam*HI (restriction endonuclease) site located 30 bp from the 5′ start of the gene. The units are tandemly arranged at one or two *5SDna* loci (Appels *et al.*, 1992) that are often distinguishable by the length of the particular unit at the respective locus. The units are of interest because their small size allows the entire structure to be readily solved at the DNA sequence level. The 5 S DNA units contain regions that evolve relatively quickly and allow the sequence information to be used to examine relationships between species.

Appels *et al.* (1992) isolated 44 5 S DNA units from wheat for sequencing, with the aim of investigating the evolution of 5 S DNA in a polyploid. A comparison of the available data is shown in Fig. 8.10. The 5 S DNA entries *T. aestivum* S4 and *T. aestivum* S6 group into a cluster of 5 S DNA entries from *T. tauschii*, *T. comosum*, *T. bicorne*, *T. longissimum*, *T. sharonense*, *T. tripsacoides*, *T. dichasians*, *T. umbellulatum* and *T. searsii*. The *T. aestivum* S4 and *T. aestivum* S6 entries are closest in sequence to *T. tauschii*, an observation that is consistent with earlier indications (Lagudah *et al.*, 1989) suggesting a very close sequence relationship between the short 5 S DNA units from *T. tauschii* var. *strangulata* and the short 5 S DNA units located on chromosome 1D of *T. aestivum* cv. Chinese Spring. A direct sequence comparison indicates that *T. aestivum* S4 and *T. tauschii* S differ at only seven positions, two in the gene region and five in the spacer region.

The *T. aestivum* S3 entry groups with the available *T. monococcum* entries while the *T. aestivum* L1 5 S DNA entry groups with 5 S DNA entries from *T. speltoides* L and *T. aestivum* L-G&D. Entries *T. aestivum* L2 and *T. aestivum* L3 group with *T. tauschii* L and entry *T. aestivum* L4 groups with *T. monococcum* L. The ability to find relationships at the DNA sequence level in this way suggests that DNA sequence changes at this repetitive locus are relatively minor, at least within the period of time in which modern wheat has been evolving. Equally important is the observation that two clear lineages of sequences are evident in Fig. 8.10, namely one set of units marked with an S and another set marked with an L. The S and L correspond to the *5SDna* loci on group 1 and group 5 chromosomes, respectively (in those cases where it has been possible to analyse chromosomal locations), and their clear independence from each other suggests that free exchange between these loci is not possible even though they have regions of very similar sequences (the conserved 5 S RNA gene regions).

In the analysis described in Fig. 8.10 comparisons with progenitor species allowed the 5 S DNA units from *T. aestivum* to be grouped into units that, it could reasonably be argued, originated from the *5SDna*

Fig. 8.10. Comparisons of 5 S DNA sequences. Consensus sequences for the 5 S DNAs from a given species were determined using CLUSTAL adapted to run on the VAX/VMS (Appels *et al.*, 1992). Cladistic analyses were carried out on an aligned matrix of the consensus sequences using PAUP version V.2.4 and PHYLIP V.3.3 (kindly carried out by B. Baum). The S or L designation of many of the entries relates to the length of the 5 S DNA unit (S=320–469 bp and L=470–499 bp). The Rus designation of some of the clones refers to the sequences determined by Vakhitov *et al.* (1989) and Kolchinsky *et al.* (1990) and the G&D designation identifies the clones sequenced by Gerlach & Dyer (1980).

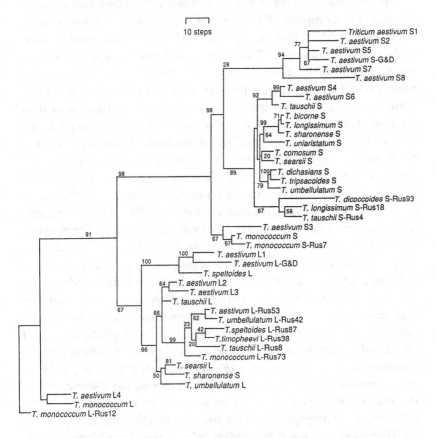

locus on chromosome 1D (*T. aestivum* S4 and S6), chromosome 1A (*T. aestivum* S3), chromosome 5B (*T. aestivum* L1), chromosome 5D (*T. aestivum* L2 and L3) and chromosome 5A (*T. aestivum* L4). A major group that could not be 'accounted' for included the units identified as *T. aestivum* S1, S2, S5, S7, S8 and *T. aestivum* S-G&D (a 5 S DNA from

Chinese Spring wheat sequenced by Gerlach & Dyer (1980)). It is possible that they originate from chromosome 1B because this *5SDna* locus remains to be accounted for in the above list. The existence of this separate group of 5 S DNA units suggests that there may exist a progenitor that has not yet been assessed in the analysis and that may be significant in the evolution of wheat.

The evolution of hexaploid wheat has also been studied at the level of DNA molecules present in the cytoplasm. Studies on the large subunit of ribulose bisphosphate carboxylase (Chen, Gray & Wildman, 1975; for a review, see Wildman, 1980), as well as the restriction endonuclease site map of chloroplast DNA locating specific genes relative to each other, (reviewed in May & Appels, 1987), have been used to infer evolutionary relationships. The work showed that *T. turgidum* was the female parent in the cross leading to hexaploid wheat. Studies using mitochondrial DNA came to the same conclusion (Vedel *et al.*, 1978). More recent studies have led to the detailed characterization of both chloroplast and mitochondrial DNA from wheat (for reviews, see Bonen, 1992; Ogihara, 1992). Transcripts from wheat mitochondrial DNA were shown to have been subject to RNA editing (mainly the conversion of some cytosine to uracil residues) and it is of particular interest that a defect in this editing process may be related to cytoplasmic male sterility (review in Walbot, 1991). These molecular studies are providing new markers that will help to resolve the problem of characterizing proposed progenitors in the evolution of wheat.

The discovery of the putative diploid progenitors of polyploid wheats and other species in the subtribe Triticineae (Kihara, 1954) led to attempts to resynthesize several polyploids. Production of amphidiploids either by inducing chromosome doubling through colchicine treatment or through the phenomenon of cytologically functional unreduced gamete formation in interspecific hybrids, results in a stable disomic genomic composition. The amphidiploids are characterized by regular meiotic chromosome pairing resulting from the disomic state of the constituent genomes, which enables stable reproduction of the polyploid. In a comprehensive appraisal of synthetic polyploid species in the Triticineae, Riley & Bell (1958) concluded that it was highly unlikely that the direct use of synthetic species as cultivated crops would reach a level comparable with that for cultivated wheat. However, they may provide a valuable resource for crop improvement as bridges for the transfer of genes to wheat from species that do not cross easily with wheat. So far, the triticales are the only group of synthetic hybrids in the Triticineae that has achieved a commercial place in agriculture.

The adaptive value of polyploid wheats

The widespread distribution of polyploid wheat may partly be due to the combined effects of the differently adapted populations of their parental genomes. Clearly domestication by humans has contributed greatly to the extensive distribution of wheat across different ecological zones. The significance of the adaptive value of polyploidy is evident from the localized distribution of several diploid progenitor species (such as *T. umbellulatum*, *T. comosum*, *T. dichasians*) in contrast to the wider geographical spread of allopolyploids (e.g. *T. ovatum*, *T. triaristatum*, *T. triunciale*, *T. cylindricum*) in the Triticineae (Miller, 1987). However, polyploids and closely related diploids of the same genus very often do not occupy the same habitat with equal success (de Wet, 1986). This uneven distribution is striking with the D-genome-containing species. The diploid progenitor *T. tauschii* (D) occurs over a wide ecological zone in contrast to the relatively localized distribution of the polyploids *T. crassum* (DM^{cr} and DD*M^{cr}), *T. syriacum* (D*M^{cr}S^{p}) and *T. juvenalis* (D*M^{j}U); the D* genome is considered to be a modified form of the D genome present in wheat. The relationship between the diploid and the corresponding D genome in polyploids shows that some of the polyploid species have very little homology to the present day diploid genome. Recent evidence from DNA hybridization studies of a highly amplified D-genome sequence isolated from the diploid *T. tauschii*, and present on all seven hexaploid wheat D-genome chromosomes, revealed its absence in the polyploid species *T. juvenalis* and *T. syriacum* (D. McNeil, E. Lagudah & R. Appels, unpublished results).

The adaptive value of newly evolved polyploids may be due not only to new genome combinations, but also to their ability to compete with the prevalent parents for existing and new habitats. Watanabe (1981) examined the competitive abilities of the hexaploid wheat cultivar Canthatch and five resynthesized hexaploids obtained by crossing the extracted tetraploid component of Canthatch with five *T. tauschii* genotypes. This assessment was based on differences in whole plant weight and spike number and weight when planted as mixtures with the wheat cultivar Chinese Spring as a control. In the experiments the competitive ability of Canthatch was greatest while only minor differences were found among the resynthesized hexaploids. It is likely that the poor adaptability of the synthetic hexaploid may limit its potential for direct use as a cultivated crop. Successful polyploids may have inherited the adaptive attributes of their parental genomes; in the case of the resynthesized hexaploids, tetra-Canthatch cannot be considered as an adapted tetraploid.

Expansion of gene pools by mutation and secondary hybridization of newly evolved polyploids may further enhance the colonization capacities of these polyploids (de Wet, 1986). Reduced fertility is often found in newly synthesized polyploids, but with further selection for improved fertility their adaptive capacity may equal or exceed their progenitor species. The cytologically diploid behaviour of allohexaploid wheat regulated by the *Ph* locus may be a major factor contributing to the fertility and stability of the polyploid genome.

The *Rht* genes were important in the more recent domestication of wheat

Under primitive agricultural conditions, factors influencing both the amount of grain harvested per unit of time and the energy expended may have constituted the major attributes looked for during domestication. Large grain size, increased grain number and non-brittle rachis must have been favoured selection targets and are associated with wheat evolution and domestication. Increases in grain size under wheat domestication is partly due to its polyploidy, but has not been accompanied by an increase in leaf photosynthetic rates (Evans & Dunstone, 1970); the fall in the maximum rate of flag leaf photosynthesis in cultivated wheats has been associated with an increase in the size of the leaf and of the mesophyl cells (Dunstone & Evans, 1974).

Until the early part of the nineteenth century most of the wheats used in agriculture were landraces. They constituted a heterogeneous mixture from which individual plant selections were developed to produce homozygous pure lines. The observations made from Mendel's genetic studies reshaped the direction of plant breeding by suggesting to breeders that selection for useful recombinants from the segregating progeny of hybrids could be achieved (for a review, see Lupton, 1987). Selection and breeding schemes under improved agricultural practices have resulted in plants with increased assimilate supply and conversion to grain yield. In a study of winter wheat yields developed between the beginning of this century and 1980 in the UK, Austin *et al.* (1980) showed that under optimum growing conditions newly bred varieties outyielded older varieties. Comparable trends for wheats grown in the USA, Mexico and Canada have been observed (Evans, 1981). When the breeding objective is for adaptive or quality characteristics, however, the yield trends have been found to be variable with different types of wheat. For example, the yield increases from modern varieties show that the high quality wheats tend to lag behind partly owing to the negative correlation between yield and increased grain protein.

One of the most striking features in yield trends has been the sharp rise in yield that was associated with the impact of the semi-dwarf wheat varieties in the 1970s. Semi-dwarf wheats contain *Rht* genes (reduced height) and are characterized by a high conversion of assimilates from the straw to the grain, reduction in yield losses from lodging and stem elongation which is either gibberellic acid insensitive (*Rht1*, *Rht2* and *Rht3*) or sensitive (*Rht8* and *Rht9*). Gale, Marshall & Rao (1981) have identified different types of *Rht* genes in a world collection of wheat varieties. *Rht1* and *Rht2* genes located on chromosomes 4B (current designation) and 4D, respectively, are prevalent in commercial semi-dwarf wheats. The wheat variety 'Norin 10' was selected from crosses between the Japanese variety 'Daruma' (the source of the *Rht1* gene) and American varieties, and was subsequently used in many breeding programmes around the world (for a review, see Lupton, 1987).

Pleiotropic effects of the *Rht* genes that confer increased yields are well documented with winter wheats, whereas both yield losses and gains have been observed with spring wheats. Some of the variable responses of *Rht* genes in spring wheats are associated with the duration of crop growth. Hoogendoorn *et al.* (1988) studied 22 pairs of nearly isogenic *Rht* and tall (*rht*) wheats and found that yield gains associated with *Rht* genes were greater in wheat backgrounds with relatively long days to flowering. It has been suggested that increased yields in spring wheats with *Rht* genes are maximized under conditions suited for high biomass production. These observations highlight the assertion by Pugsley (1983) that major genes controlling characteristics of the wheat plant such as photoperiodic response (*Ppd*) and vernalization requirement (*Vrn*) and their interaction with *Rht* genes must be managed in concert in breeding programmes for optimum yields to be achieved.

Wheat domestication – prospects

The presence of the chromosome 1RS translocation in wheat backgrounds has provided new sources of resistance to stripe, leaf and stem rust and in certain environments significant gains in yields have been observed (Zeller, 1973; for a review, see Baum & Appels, 1991). As a result, wheat lines with 1RS–1BL translocations or 1R(1B) chromosome substitutions are being utilized in breeding programmes in several countries. Most of the present 1R chromosomes in commercial wheats are derived from very few genotypes of rye, hence the need to widen the genetic base of the 1R chromatin by exploiting new genotypic sources. Considering the impact of chromosome 1R on the agronomic performance of wheat, it is likely that selection for other alien chromatin in an appropriate combination with the wheat genome could result in new genotypes with agronomic performances similar to or even better than those based on 1R.

One trait where possibilities for genome restructuring could be exploited is a high photosynthetic rate. The occurrence, at saturating light, of a higher net photosynthetic rate in the wild diploid A genomes than in bread wheat is thought to be associated with quantitative differences in the distribution of chlorophyll and chlorophyll–protein complexes, which may thus influence the photosynthetic electron transport capacity (Austin et al., 1988). Restructuring of the A genome of bread wheat may lead to new genomic combinations derived from T. urartu, such as translocations, single or multiple chromosome substitutions or homologous recombinant genotypes, which can be examined for photosynthetic efficiency under various climatic conditions. Preliminary studies by Austin et al. (1988) have shown that certain plant selections of T. urartu introgressions in bread wheat have consistently produced higher biomass and grain yields under field conditions.

In recent years yields of tetraploid wheats have been shown to equal or sometimes exceed that of bread wheat. The prospects for further improvement of tetraploid wheat are quite promising given that less effort has gone into its breeding than into that of bread wheat. The cooking quality of tetraploid wheats with high gluten strength for a range of pasta products is good, although a major limitation to the range of end use attributes is the comparatively low baking quality of these wheats. The poor baking quality of flour from tetraploid wheat has been attributed to the absence of the D-genome encoding storage proteins. Translocations involving either the long arm or the entire chromosome 1D addition to tetraploid wheats have been found to confer improved dough properties (Kaltsikes, Evans & Larter, 1968; Joppa, Bietz & McDonald, 1975). Chromosome manipulation techniques aimed at transferring segments of chromosome 1D which code for endosperm storage proteins into appropriate regions of the tetraploid genome without major losses of gluten proteins located on chromosomes 1A and 1B could be exploited for quality improvement in tetraploid wheat.

The endosperm storage proteins of hexaploid wheat, particularly the high molecular weight (HMW) subunits of glutenin, have been studied in detail in attempts to relate their structure to their function (Thompson, Bartels & Harberd, 1985; Payne, 1987; Greene *et al.*, 1988; Halford *et al.*, 1988; Shewry, Halford & Tatham, 1989). Some of these findings point to certain classes of repetitive sequences, involving hexa-peptides and nonapeptides, that have structures predicted to be involved in the elasticity of the aggregated protein. Similar studies on these classes of storage proteins are required in tetraploid wheats, to examine where structural changes may relate to potential differences in function between the hexaploid and tetraploid genomes. In the event that stable genetic transformation systems become available for wheat, modifications of the tetraploid wheat endosperm proteins could be achieved by inserting sequences characteristic of the hexaploid wheat genome. Alternatively, the number of glutenin gene copies, particularly those with sequences identical with those derived from the D genome, could be increased and incorporated into the tetraploid genome. It can be anticipated that modification of the expression of the HMW glutenin genes will be achieved by manipulating the sequences that have been identified, using reporter genes in transgenic plants, as being endosperm specific, in that they direct, as well as enhance, the expression of storage protein genes (Colot *et al.*, 1987; Thomas & Flavell, 1990).

Acknowledgement
The authors are grateful to Dr U. Hohmann for critical comments on this manuscript.

References
Appels, R., Baum, B. & Clarke, B. C. (1992). The 5 S DNA units of *Triticum* species. *Plant Systematics and Evolution* (in press).
Appels, R., Driscoll, C. & Peacock, W. J. (1978). Heterochromatin and highly repeated DNA sequences in rye (*Secale cereale*) chromosomes. *Chromosoma*, **70**, 265–277.
Appels, R. & Lagudah, E S. (1990). Manipulation of chromosomal segments from wild wheat for the improvement of bread wheat. *Australian Journal of Plant Physiology*, **17**, 253–266.
Appels, R., Reddy, P., McIntyre, C. L., Moran, L. B., Frankel, O. H. & Clarke, B. C. (1989). The molecular-cytogenetic analysis of grasses and its application to studying relationships among species of the Triticeae. *Genome*, **31**, 122–133.

Austin, R. B., Bingham, J., Blackwell, R. D., Evans, L. T., Ford, M. A., Morgan, C. L. & Taylor, M. (1980). Genetic improvements in winter wheat yields since 1900 and associated physiological changes. *Journal of Agricultural Science*, **94**, 675–689.

Austin, R. B., Morgan, C. L., Ford, M. A., Roscoe, T. J. & Miller, T. E. (1988). Increasing the photosynthetic capacity of wheat by incorporating genes from A genome diploid species. In *Proceedings of the seventh international wheat genetics symposium*, ed. T. E. Miller & R. M. D. Koebner, pp. 203–208. Cambridge: Institute of Plant Science Research, Cambridge Laboratory.

Baum, M. B. & Appels, R. (1991). The cytogenetic and molecular architecture of chromosome 1R – one of the most widely used sources of alien chromatin in wheat varieties. *Chromosoma*, **100**, 1–10.

Bennett, M. D., Rao, M. K., Smith, J. B. & Bayliss, M. W. (1973). Cell development in the anther the ovule and the young seed of *Triticum aestivum* L. var. Chinese Spring. *Philosophical Transactions of the Royal Society of London*, B, **266**, 39–81.

Bonen, L. (1992). Wheat mitochondrial protein-coding gene structure and expression. *Seiken Ziho*, (in press).

Botstein, D., White, R. L., Skolnick, M. & Davis, R. W. (1980). Construction of a genetic map in man using restriction fragment length polymorphisms. *American Journal of Human Genetics*, **32**, 314–331.

Brettell, R. I. S., Pallota, M. A., Gustafson, J. P. & Appels, R. (1986). Variation at the NOR loci in triticale derived from tissue culture. *Theoretical and Applied Genetics*, **71**, 637–643.

Chao, S., Raines, C., Longstaff, M., Sharp, P. J., Gale, M. D. & Dyer, T. A. (1989*a*). Chromosomal location and copy number in wheat and some of its close relatives of genes for enzymes involved in photosynthesis. *Molecular and General Genetics*, **218**, 423–430.

Chao, S., Sharp, P. J., Worland, A. J., Warham, E. J., Koebner, R. M. D. & Gale, M. D. (1989*b*). RFLP-based genetic maps of wheat homoeologous group 7 chromosomes. *Theoretical and Applied Genetics*, **78**, 495–504.

Chen, K., Gray, J. C., Wildman, S. G. (1975). Fraction 1 protein and the origin of polyploid wheats. *Science*, **190**, 1304–1306.

Chen, P. D. & Gill, B. S. (1984). The origin of chromosome 4A and genomes B and G of tetraploid wheats. In *Proceedings of the sixth international wheat genetics symposium*, ed. B. Sakamoto, pp. 39–48. Tokyo: Faculty of Agriculture, Kyoto University.

Colot, V., Robert, L. S., Kavanagh, T. A., Bevan, M. W. & Thompson, R. D. (1987). Localization of sequences in wheat endosperm protein genes which confer tissue-specific expression in tobacco. *EMBO Journal*, **6**, 3559–3564.

Curtis, C. A. & Lukaszewski, A. J. (1991). Genetic linkage between C-bands and storage protein genes in chromosome 1B of tetraploid wheat. *Theoretical and Applied Genetics*, **81**, 245–252.

de Wet, J. M. J. (1986). Hybridization and polyploidy in the Poaceae. In *Grass systematics and evolution*, ed. T. R. Soderstrom, K. W. Hilu, C. S. Campbell & M. Barkworth, pp. 188–194. Washington, DC: Smithsonian Institution Press.

Dennis, E. S., Gerlach, W. L. & Peacock, W. J. (1980). Identical polypyrimidine-polypurine satellite DNA's in wheat and barley. *Heredity*, **44**, 349–366.

Driscoll, C. J. (1973). Minor genes affecting homoeologous pairing in hybrids between wheat and related genera. *Genetics*, **74**, s66.

Dunstone, R. L. & Evans, L. T. (1974). Role of changes in cell size in the evolution of wheat. *Australian Journal of Plant Physiology*, **1**, 157–165.

Dvorak, J. (1977). Transfer of leaf rust resistance from *Aegilops speltoides* to *Triticum aestivum*. *Canadian Journal of Genetics and Cytology*, **19**, 133–141.

Dvorak, J. (1980). Homoeology between *Agropyron elongatum* chromosomes and *T. aestivum* chromosomes. *Canadian Journal of Genetics and Cytology*, **22**, 237–259.

Dvorak, J. (1983). The origin of wheat chromosomes 4A and 4B and their genome reallocation. *Canadian Journal of Genetics and Cytology*, **25**, 210–214.

Dvorak, J. & Knott, D. R. (1974). Disomic and ditelosomic addition of diploid *Agropyron elongatum* chromosomes to *Triticum aestivum*. *Canadian Journal of Genetics and Cytology*, **16**, 399–417.

Dvorak, J. & McGuire, P. E. (1981). Nonstructural chromosome differentiation among wheat cultivars with special reference to differentiation of chromosomes in related species. *Genetics*, **97**, 391–414.

Dvorak, J., Resta, P. & Kota, R. S. (1989). Molecular evidence on the origin of chromosomes 4A and 4B. *Genome*, **33**, 30–39.

Endo, T. R. (1979). Selective gametocidal action of a chromosome of *Aegilops cylindrica* in a cultivar of common wheat. *Wheat Information Service*, **50**, 24–28.

Endo, T. R. (1988). Chromosome mutations induced by gametocidal chromosomes in common wheat. In *Proceedings of the seventh international wheat genetics symposium*, ed. T. E. Miller & R. M. D. Koebner, pp. 259–266. Cambridge: Institute of Plant Science Research, Cambridge Laboratory.

Endo, T. R. & Gill, B. S. (1984). A somatic karyotype, heterochromatin distribution, and nature of chromosome differentiation in common wheat, *Triticum aestivum* L. *Chromosoma*, **89**, 361–369.

Endo, T. R. & Tsunewaki, K. (1975). Sterility of common wheat with *Aegilops triuncialis* cytoplasm. *Journal of Heredity*, **66**, 13–18.

Evans, L. T. (1981). Yield improvement in wheat: empirical or analytical? In *Wheat science – today and tomorrow*, ed. L. T. Evans & W. J. Peacock, pp. 203–222. Cambridge: Cambridge University Press.

Evans, L. T. & Dunstone, R. L. (1970). Some physiological aspects of evolution in wheat. *Australian Journal of Biological Science*, **23**, 725–741.

Feldman, M. (1966). The effect of chromosomes 5B, 5D, and 5A on chromosome pairing in *Triticum aestivum*. *Proceedings of the National Academy of Sciences, USA*, **55**, 1447–1453.

Feldman, M. (1968). Regulation of somatic association and meiotic pairing in common wheat. *Proceedings of the third international wheat genetics symposium*, ed. K. W. Finlay & K. Shepherd, pp. 31–40. Canberra: Australian Academy of Science.

258   E. S. Lagudah and R. Appels

Feldman, M. (1988). Cytogenetic and molecular approaches to alien gene transfer in wheat. pp. In *Proceedings of the seventh international wheat genetics symposium*, ed. T. E. Miller & R. M. D. Koebner, pp. 23–32. Cambridge: Institute of Plant Science Research, Cambridge Laboratory.

Feldman, M. & Mello-Sampayo, T. (1967). Suppression of homoeologous pairing in hybrids of polyploid wheats × *Triticum speltoides*. *Canadian Journal of Genetics and Cytology*, **9**, 307–313.

Flavell, R. B., Bennett, M. D., Seal, A. G. & Hutchinson, J. (1987). Chromosome structure and organisation. In *Wheat breeding, its scientific basis*, ed. F. G. H. Lupton, pp. 211–268. London: Chapman & Hall.

Flavell, R. B., Bennett, M. D., Smith, J. B. & Smith, D. B. (1974). Genome size and the proportion of repeated sequence of DNA in plants. *Biochemical Genetics*, **12**, 257–269.

Flavell, R. B. & Smith, D. B. (1976). Nucleotide sequence organisation in the wheat genome. *Heredity*, **37**, 231–252.

Futers, T. S., Vaughan, T. J., Sharp, P. J. & Cuming, A. C. (1990). Molecular cloning and chromosomal location of genes encoding the "early-methionine-labelled" (Em) polypeptide of *Triticum aestivum* L. var. Chinese Spring. *Theoretical and Applied Genetics*, **80**, 43–48.

Gale, M. D. & Miller, T. E. (1987). The introduction of alien genetic variation in wheat. In *Wheat breeding its scientific basis*, ed. F. G. H. Lupton, pp. 173–210. London: Chapman & Hall.

Gale, M. D., Chao, S. & Sharp, P. J. (1990). RFLP mapping in wheat – progress and problems. In *Gene manipulation in plant improvement, II*, ed. J. P. Gustafson, pp. 353–364. New York: Plenum Press.

Gale, M. D., Marshall, G. A. & Rao, M. V. (1981). A classification of the Norin 10 and Tom Thumb dwarfing genes in British, Mexican, Indian and other hexaploid bread wheat varieties. *Euphytica*, **30**, 355–361.

Gerlach, W. L. (1977). N-banded karyotypes of wheat species. *Chromosoma*, **62**, 49–56.

Gerlach, W. L. & Dyer, T. A. (1980). Sequence organization of the repeating units in the nucleus of wheat which contain r5S RNA genes. *Nucleic Acids Research*, **8**, 4851–4865.

Gill, B. S. (1987). Chromosome banding methods, standard chromosome band nomenclature, and applications in cytogenetic analysis. In *Wheat and wheat improvement*, ed. E. G. Heyne, pp. 243–254. Madison, WI: American Society of Agronomy.

Gill, B. S. & Kimber, G. (1974). Giemsa C-banding and the evolution of wheat. *Proceedings of the National Academy of Sciences, USA*, **71**, 4086–4090.

Gill, B. S. & Raup, W. J. (1987). Direct genetic transfers from *Aegilops squarrosa* L. to hexaploid wheat. *Crop Science*, **27**, 445–450.

Gill, K. S., Lubbers, E. L., Gill, B. S., Raupp, W. J. & Cox, T. S. (1991). A genetic linkage map of *Triticum tauschii* (DD) and its relationship to the D genome of bread wheat (AABBDD). *Genome*, **34**, 362–374.

Greene, F. C., Anderson, O. D., Yip, R. E., Halford, N. G., Malpica Romero, J.-M. & Shewry, P. R. (1988). Analysis of possible quality-related sequence variations in the 1D glutenin high molecular weight subunit genes of wheat.

In *Proceedings of the seventh international wheat genetics symposium*, ed. T. E. Miller & R. M. D. Koebner, pp. 735–740. Cambridge: Institute of Plant Science Research, Cambridge Laboratory.

Guidet, F., Rogowsky, P., Taylor, C., Song, W. & Langridge, P. (1991). Cloning and characterization of a new repeated sequence. *Genome*, **34**, 81–87.

Halford, N. G., Forde, J., Anderson, O. D., Greene, F. C. & Shewry, P. R. (1988). The structure and expression of genes encoding the high molecular weight (HMW) subunits of wheat glutenin. In *Proceedings of the seventh international wheat genetics symposium*, ed. T. E. Miller & R. M. D. Koebner, pp. 745–750. Cambridge: Institute of Plant Science Research, Cambridge Laboratory.

Harlan, J. R. (1981). The early history of wheat: earliest traces to the sack of Rome. In *Wheat science – today and tomorrow*, ed. L. T. Evans & J. W. Peacock, pp. 1–19. Cambridge: Cambridge University Press.

Hart, G. E. (1987). Genetic and biochemical studies of enzymes. In *Wheat and wheat improvement*, ed. E. G. Heyne, pp. 199–214. Madison, WI: American Society of Agronomy.

Helbaek, H. (1966). Commentary on the phylogenesis of *Triticum* and *Hordeum. Economic Botany*, **20**, 350–360.

Hjelmquist, H. (1966). Some notes on the old wheat species of Gotland. *Hereditas*, **56**, 382–393.

Holm, P. B. & Rasmussen, S. W. (1984). In *Chromosomes today*, ed. M. D. Bennett, A. Gropp & U. Wolf, pp. 104–116. London: Allen & Unwin.

Hoogendoorn, J., Pfeiffer, W. H., Rajaram, S. & Gale, M. D. (1988). Adaptive aspects of dwarfing genes in CIMMYT germplasm. In *Proceedings of the seventh international wheat genetics symposium*, ed. T. E. Miller & R. M. D. Koebner, pp. 1093–1100. Cambridge: Institute of Plant Science Research, Cambridge Laboratory.

Islam, A. K. M. R., Shepherd, K. W. & Sparrow, D. H. B. (1981). Isolation and characterisation of euplasmic wheat–barley chromosome addition lines. *Heredity*, **46**, 161–174.

Janushevich, Z. V. (1978). Prehistoric food plants in south-west of the Soviet Union. *Bericht Deutscher Botanisscher Gesellschaft*, **91**, 59–66.

Joppa, L. R. (1988). Cytogenetics of tetraploid wheat. In *Proceedings of the seventh international wheat genetics symposium*, ed. T. E. Miller & R. M. D. Koebner, pp. 197–202. Cambridge: Institute of Plant Science Research, Cambridge Laboratory.

Joppa, L. R., Bietz, J. A. & McDonald, C. (1975). Development and characteristics of a disomic-1D addition line of durum wheat. *Canadian Journal of Genetics and Cytology*, **17**, 355–463.

Joppa, L. R. & Williams, N. D. (1977). D-genome substitution monosomics of durum wheat. *Crop Science*, **17**, 772–776.

Joppa, L. R. & Williams, N. D. (1988). Langdon durum disomic substitution lines and aneuploid analysis of tetraploid wheat. *Genome*, **30**, 222–228.

Kaltsikes, P. J., Evans, L. T. & Larter, E. N. (1968). Identification of a chromosome segment controlling bread-making quality in common wheat. *Canadian Journal of Genetics and Cytology*, **10**, 763.

Kerber, E. R. (1964). Wheat. Reconstitution of tetraploid component (AABB) of hexaploids. *Science*, **143**, 253–255.

Kerber, E. R. & Dyck, P. L. (1969). Inheritance in hexaploid wheat of leaf rust resistance and other characters derived from *Aegilops squarrosa*. *Canadian Journal of Genetics and Cytology*, **11**, 639–647.

Kerber, E. R. & Tipples, K. H. (1969). Effects of the D genome on milling and baking properties of wheat. *Canadian Journal of Plant Science*, **49**, 225–263.

Kihara, H. (1944). Discovery of the DD analyser, one of the ancestors of *T. vulgare*. *Agric. Hort.*, **19**, 889–890.

Kihara, H. (1954). Consideration on the evolution and distribution of *Aegilops* species based on the analyser method. *Cytologia*, **19**, 336–357.

Kimber, G. & Sears, E. R. (1987). Evolution in the genus *Triticum* and the origin of cultivated wheat. In *Wheat and wheat improvement*, ed. E. G. Heyne, pp. 154–164. Madison, WI: American Society of Agronomy.

King, I. P., Reader, S. M & Miller, T. (1988). Exploitation of the 'cuckoo' chromosome (4S^1) of *Aegilops sharonensis* for eliminating segregation for height in semi-dwarf *Rht2* bread wheat cultivars. In *Proceedings of the seventh international wheat genetics symposium*, ed. T. E. Miller & R. M. D. Koebner, pp. 337–342. Cambridge: Institute of Plant Science Research, Cambridge Laboratory.

Koebner, R. M. D., Appels, R. & Shepherd, K. W. (1986). Rye heterochromatin II. Characterization of a derivative of chromosome 1DS. 1RL with a reduced amount of the major repeating sequence. *Canadian Journal of Genetics and Cytology*, **28**, 658–664.

Koebner, R. M. D. & Shepherd, K. W. (1985). Induction of recombination between rye chromosome IRL and wheat chromosomes. *Theoretical and Applied Genetics*, **71**, 208–215.

Kolchinsky, A., Kanazin, V., Yakovleva, E., Gazumyan, A., Kole, C. & Ananiev, E. (1990). 5S-RNA genes of barley are located on the second chromosome. *Theoretical and Applied Genetics*, **80**, 333–336.

Lagudah, E. S., Appels, R., Brown, A. H. D. & McNeil, D. (1991). The molecular-genetic analysis of *Triticum tauschii* – the D genome donor to hexaploid wheat. *Genome*, **34**, 375–386.

Lagudah, E. S., Clarke, B. C. & Appels, R. (1989). Phylogenetic relationships of *Triticum tauschii*, the D genome donor to hexaploid wheat. 4. Variation and chromosomal location of 5 S DNA. *Genome*, **32**, 1017–1025.

Law, C. N. & Worland, A. J. (1973). Aneuploidy in wheat and its uses in genetic analysis. In *Annual Report, 1972, of the Plant Breeding Institute*, pp. 25–65. Cambridge: Plant Breeding Institute.

Lilienfield, F. A. (1951). H. Kihara: genome analysis in *Triticum* and *Aegilops*. Concluding review. *Cytologia*, **16**, 101–123.

Listina, G. N. (1978). Main types of ancient farming on the Caucasus – on the basis of palaeo-ethnobotanical research. *Bericht Deutscher Botanisscher Gesellschaft*, **91**, 47–57.

Lupton, F. G. H. (1987). History of wheat breeding. In *Wheat breeding, its scientific basis*, ed. F. G. H. Lupton, pp. 51–70. London: Chapman & Hall.

Maan, S. S. (1975). Exclusive preferential transmission of an alien chromosome in common wheat. *Crop Science*, **15**, 287–292.

MacKey, J. (1966). Species relationships in *Triticum*. *Proceedings of the second international wheat genetics symposium. Hereditas* (suppl.), **2**, 237–276.

May, C. E., & Appels, R. (1980). Rye chromosome translocation in hexaploid wheat: a reevaluation of the loss of heterochromatin from rye chromosomes. *Theoretical and Applied Genetics*, **56**, 17–23.

May, C. E. & Appels, R. (1987). The molecular genetics of wheat: toward an understanding of 16 billion base pairs of DNA. In *Wheat and wheat improvement*, ed. E. G. Heyne, pp. 165–198. Madison, WI: American Society of Agronomy.

McFadden, E. S. & Sears, E. R. (1946). The origin of *Triticum spelta* and its free-threshing hexaploid relatives. *Journal of Heredity*, **37**, 81–89; 107–116.

McIntosh, R. A. (1988). Catalogue of gene symbols for wheat. In *Proceedings of the seventh international wheat genetics symposium*, ed. T. E. Miller & R. M. D. Koebner, pp. 1225–1324. Cambridge: Institute of Plant Science Research, Cambridge Laboratory.

McIntyre, C. L., Clarke B. C. & Appels, R. (1988). Amplification and dispersion of repeated DNA sequences in the Triticeae. *Plant Systematics and Evolution*, **160**, 39–59.

Mello-Sampayo, T. & Canas, A. P. (1973). Suppressors of meiotic chromosome pairing in common wheat. *Proceedings of the fourth international wheat genetics symposium*, ed. E. R. Sears & L. M. S. Sears, pp. 709–713. Columbia: University of Missouri.

Metakovsky, E. V., Novoselskaya, A. Yu., Kopus, M. M., Sobko, T. A. & Sosinov, A. A. (1984). Blocks of gliadin components in winter wheat detection by one-dimensional polyacrylamide gel electrophoresis. *Theoretical and Applied Genetics*, **67**, 559–568.

Miller T. E. (1984). The homoeologous relationship between the chromosomes of rye and wheat. Current status. *Canadian Journal of Genetics and Cytology*, **26**, 578–589.

Miller T. E. (1987). Systematics and evolution. In *Wheat breeding: its scientific basis*, ed. F. G. H. Lupton, pp. 1–30. London: Chapman & Hall.

Miller T. E., Hutchinson, J. & Chapman, V. (1982). Investigation of a preferentially transmitted *Aegilops sharonensis* chromosome in wheat. *Theoretical and Applied Genetics*, **61**, 27–33.

Mitra, R. & Bhatia, C. R. (1973). Repeated and non-repeated sequences in diploid and polyploid wheat species. *Heredity*, **31**, 251–262.

Morris, R. & Sears, E. R. (1967). The cytogenetics of wheat and its relatives. In *Wheat and wheat improvement*, ed. K. S. Quisenberry & L. P. Reitz, pp. 19–87. Madison, WI: American Society of Agronomy.

Muramatsu, M. (1973). Genic homology and cytological differentiation of the homoeologous-group-5 chromosomes of wheat and related species. In *Proceedings of the fourth international wheat genetics symposium*, ed. E. R. Sears & L. M. S. Sears, pp. 719–724. Columbia: University of Missouri.

Naranjo, T., Roca, A., Goicoechea, P. G. & Giraldez R. (1987). Arm homoeology of wheat and rye chromosomes. *Genome*, **29**, 873–882.

Naranjo, T., Roca, A., Goicoechea, P. G. & Giraldez, R. (1988). Chromosome structure of common wheat: genome reassignment of chromosome 4A and 4B. In *Proceedings of the seventh international wheat genetics symposium*, ed. T. E. Miller & R. M. D. Koebner, pp. 115–120. Cambridge: Institute of Plant Science Research, Cambridge Laboratory.

Netzle, S. & Zeller, F. J. (1984). Cytogenetic relationship of *Aegilops longissima* chromosomes with common wheat chromosomes. *Plant Systematics and Evolution*, **145**, 1–13.

Ogihara, Y. (1992). Diversity and evolution of chloroplast DNA. *Seiken Ziho* (in press).

Okamoto, M. (1957). Asynaptic effect of chromosome V. *Wheat Information Service*, **5**, 6.

O'Mara, J. G. (1940). Cytogenetic studies on triticale. I. A method for determining the effects of individual *Secale* chromosomes on *Triticum*. *Genetics*, **25**, 401–408.

Orth, R. A., Dronzek, B. L. & Bushuk, W. (1973). Studies of glutenin IV. Microscopic structure and its relation to bread-making quality. *Cereal Chemistry*, **50**, 688–701.

Payne, P. I. (1987). Genetics of wheat storage proteins and the effect of allelic variation on bread-making quality. *Annual Review of Plant Physiology*, **38**, 141–153.

Payne, P. I. & Lawrence, G. J. (1983). Catalogue of alleles for the complex gene loci, *Glu-A1*, *Glu-B1* and *Glu-D1* which code for the high-molecular-weight subunits of glutenin in hexaploid wheat. *Cereal Research Communications*, **11**, 29–35.

Pienaar, R. de V. (1990). Wheat × *Thinopyrum* hybrids. In *Biotechnology in agriculture and forestry 13: wheat*, ed. Y. P. S. Bajaj, pp. 167–217. Berlin: Springer-Verlag.

Pugsley, A. T. (1983). The impact of plant physiology on Australian wheat breeding. *Euphytica*, **32**, 743–748.

Porceddu, E., Ceoloni, C., Lafiandra, D., Tanzarella, O. A. & Scarascia Mugnozza, G. T. (1988). Genetic resources and plant breeding: problems and prospects. In *Proceedings of the seventh international wheat genetics symposium*, ed. T. E. Miller & R. M. D. Koebner, pp. 7–22. Cambridge: Institute of Plant Science Research, Cambridge Laboratory.

Ranjekar, P. K., Pallota, D. & Lafontaine, J. G. (1976). Characterisation of repetitive DNA in barley and wheat. *Biochemical et Biophysica Acta*, **425**, 30–40.

Reddy, P. & Appels, R. (1989). A second locus for the 5S multigene family in *Secale* L.: sequence divergence in two lineages of the family. *Genome*, **32**, 456–467.

Riley, R. & Bell, G. D. H. (1958). The evaluation of synthetic species. *Proceedings of the first international wheat genetics symposium*, pp. 161–180. Manitoba, Canada.

Riley, R. & Chapman, V. (1958). Genetic control of the cytologically diploid behaviour of hexaploid wheat. *Nature*, **182**, 713–715.

Riley, R. & Chapman, V. (1967). Effect of 5BS in suppressing the expression of altered dosage of 5BL on meiotic chromosome pairing in *Triticum aestivum*. *Nature*, **216**, 60–62.

Riley, R., Chapman, V. & Johnson, R. (1968). The incorporation of alien disease resistance in wheat by genetic interference with the regulation of meiotic chromosome synapsis. *Genetic Research Cambridge*, **12**, 199–219.

Riley, R., Chapman, V. & Macer, R. C. F. (1966*a*). The homoeology of an *Aegilops* chromosome causing stripe rust resistance. *Canadian Journal of Genetics and Cytology*, **8**, 616–630.

Riley, R., Chapman, V. & Miller, T. E. (1973). The determination of meiotic chromosome pairing. In *Proceedings of the fourth international wheat genetics symposium*, ed. E. R. Sears & L. M. S. Sears, pp. 731–738. Columbia: University of Missouri.

Riley, R., Young, R. M. & Belfield, A. M. (1966*b*). Control of meiotic chromosome pairing by the chromosomes of homoeologous group 5 of *Triticum aestivum*. *Nature*, **212**, 1475–1477.

Rimpau, J., Smith, D. B. & Flavell, R. B. (1978). Sequence organization analysis of the wheat and rye genomes by interspecies DNA/DNA hybridization. *Journal of Molecular Biology*, **123**, 327–359.

Rowland, G. G. & Kerber, E. R. (1974). Telocentric mapping in hexaploid wheat of genes for leaf rust resistance and other characters derived from *Aegilops squarrosa*. *Canadian Journal of Genetics and Cytology*, **16**, 137–144.

Scoles, G. J., Gill, B. S., Xin, Z.-Y., Clarke, B. C., McIntyre, C. L., Chapman, C. & Appels, R. (1988). Frequent duplication and deletion events in the 5 S RNA genes and the associated spacer regions of the *Triticeae*. *Plant Systematics and Evolution*, **160**, 105–122.

Sears, E. R. (1954). *The aneuploids of common wheat*. Missouri Agricultural Experimental Station Research Bulletin, no. 572. Columbia: University of Missouri.

Sears, E. R. (1966). Nullisomic–tetrasomic combinations in hexaploid wheat. In *Chromosome manipulations and plant genetics*, ed. R. Riley & K. R. Lewis, pp. 29–45. Edinburgh: Oliver & Boyd.

Sears, E. R. (1981). Transfer of alien genetic material to wheat. In *Wheat science – today and tomorrow*, ed. L. T. Evans & W. J. Peacock, pp. 75–89. Cambridge: Cambridge University Press.

Sears, E. R. (1982). A wheat mutation conditioning an intermediate level of homoeologous chromosome pairing. *Canadian Journal of Genetics and Cytology*, **24**, 715–719.

Sharp, P. J., Chao, S., Desai, S. & Gale, M. D. (1989). The isolation, characterization and application in the Triticeae of a set of wheat RFLP probes identifying each homoeologous chromosome arm. *Theoretical and Applied Genetics*, **78**, 342–348.

Shewry, P. R., Halford, N. G. & Tatham, A. S. (1989). The high molecular weight subunits of wheat, barley and rye: genetics, molecular biology, chemistry and role in wheat gluten structure and functionality. In *Oxford surveys of plant molecular and cell biology*, vol. 6, ed. B. Miflin, pp. 163–220. Oxford: Oxford University Press.

264   E. S. Lagudah and R. Appels

Southern, E. M. (1975). Detection of specific sequences among DNA fragments separated by gel electrophoresis. *Journal of Molecular Biology*, **98**, 503–517.

Stewart, R. B. & Robertson III, W. (1971). Moisture and seed carbonisation. *Economic Botany*, **25**, 381.

Swaminathan, M. S. (1966). Mutational analysis of the hexaploid *Triticum* complex. *Proceedings of the second international wheat genetics symposium. Hereditas* (suppl.), **2**, 418–437.

Swaminathan, M. S. & Rao, M. V. P. (1961). Macro-mutations and sub-specific differentiation in *Triticum*. *Wheat Information Service*, **13**, 9–11.

The, T. T. (1973). Chromosome location of genes conditioning stem rust resistance transferred from diploid to hexaploid wheat. *Nature New Biology*, **241**, 256.

Thomas, M. S. & Flavell, R. B. (1990). Identification of an enhancer element for the endosperm-specific expression of high molecular weight glutenin. *Plant Cell*, **2**, 1171–1180.

Thompson, R. D., Bartels, D. & Harberd, N. P. (1985). Nucleotide sequence of a gene from chromosome 1D of wheat encoding a HMW glutenin subunit. *Nucleic Acids Research*, **13**, 6833–6846.

Tsujimoto, H. & Noda, K. (1988). Chromosome breakage in wheat induced by the gametocidal gene of *Aegilops triuncialis* L.: its utilization for wheat genetics and breeding. In *Proceedings of the seventh international wheat genetics symposium*, ed. T. E. Miller & R. M. D. Koebner, pp. 455–460. Cambridge: Institute of Plant Science Research, Cambridge Laboratory.

Tsujimoto, H. & Tsunewaki, K. (1985). Gametocidal genes in wheat and its relatives. II. Suppressor of the chromosome 3C gametocidal gene of *Aegilops triuncialis*. *Canadian Journal of Genetics and Cytology*, **27**, 178–185.

Upadhya, M. D. & Swaminathan, M. S. (1967). Mechanisms regulating chromosome pairing in *Triticum*. *Biol. Zentralbl*, **87** (Suppl.), 239–255.

Vakhitov, V. A., Gimalov, F. R. & Shumyatskii, G. P. (1989). Nucleotide sequences of 5 S rRNA genes of polyploid wheat species and *Aegilops* species. *Molekulyarnaya Biologiya*, **23**, 431–440.

Vedel, F., Lebacq, P., Dosba, F., Doussinault, G. (1978). Study of wheat phylogeny by *Eco*R1 analysis of chloroplastic and mitochondrial DNAs. *Plant Science Letters*, **13**, 97–102.

Vega, C. & Lacadena, J. R. (1982). Cytogenetic structure of common wheat cultivars from or introduced into Spain. *Theoretical and Applied Genetics*, **61**, 129–133.

Walbot, V. (1991). RNA editing fixes problems in plant mitochondrial transcripts. *Trends in Genetics*, **7**, 37–39.

Watanabe, N. (1981). Competitive abilities of the resynthesized hexaploid wheats. *Wheat Information Service*, **52**, 11–13.

Wildman, S. G. (1980). Molecular aspects of wheat evolution: RUBISCO composition. In *Wheat science – today and tomorrow*, ed. L. T. Evans & W. J. Peacock. Sydney: Cambridge University Press.

Zeller, F. J. (1973). 1B/1R wheat–rye chromosome substitutions and translocation. In *Proceedings of the fourth international wheat genetics symposium*, ed. E. R. Sears & L. M. S. Sears, pp. 209–221. Columbia: University of Missouri.

Zohary, D. (1973). The origin of cultivated cereals and pulses in the Near East. *Chromosomes Today*, **4**, 307–320.

9

Maize as a model system

David Hoisington

Introduction

Maize (*Zea mays* L.) is the leading grain crop in the United
States, and the second leading crop in the world after wheat. In Mexico,
Central America, parts of South America, China and Africa it
represents one of the main food-producing crops. In the United States
and Europe, its primary use is as feed for livestock, although it has
several industrial uses as well. Because of its importance to commercial
agriculture and its use as a food source, and for several important
properties, maize has received widespread attention by scientists in
many disciplines. The research has led to a better understanding of the
biology of maize, and of plants in general. This, in turn, has led to the
enhanced ability to create adapted varieties of maize and is the basis for
future efforts in using biotechnology to modify and improve the maize
plant. This chapter describes the features of the maize plant which make
it particularly amenable to modification using certain aspects of biotechnology and discusses the application of molecular techniques to the
improvement of maize.

One beneficiary of all research into improving maize will be breeders,
whose task it is to develop improved maize varieties and hybrids for
farmers and consumers world-wide. Maize breeders have been quite
successful in producing adapted and improved varieties over the past
100 years. The early breeding efforts directed to improving open-pollinated varieties were not very successful in improving yield, but were able
to develop varieties better adapted to particular growing conditions.
One of the largest contributions to increasing maize production has
arisen from the demonstration of hybrid seed production in the early
1900s (Shull, 1909). Since then, breeders have developed new and
improved inbred lines and, as a result, improved hybrids resulting in a
1% gain in yield per year (Duvick, 1979, 1983). Additional gains in yield

were achieved through the introduction and use of artificial fertilizers, herbicides and insecticides. The increased reliance on chemicals will not be advisable in the future and increases in genetic gains in yield will not go on forever. Thus, additional approaches will be necessary to continue the improvement of maize.

Not only has maize received an immense amount of attention from breeders, it has also been the focus of research by many geneticists, cytogeneticists, biochemists and molecular biologists. The results of these efforts have established maize as one of the best genetically characterized plants. The current genetic data base for maize consists of over 600 individual loci identified by their visible phenotype or by the analysis of their biochemical nature (Coe, Neuffer & Hoisington, 1988). An additional 500 or more loci have been identified but lack rigorous proof regarding allelism and genetic nature. Factors are known which affect any tissue type in the plant, at any developmental stage, and in almost any manner possible. Almost all of these 600 loci have been localized at the chromosome level and nearly 400 mapped relative to one another. The linkage maps for the 10 chromosomes of maize contain physical features such as translocation break-points and knob positions in addition to single genetic loci.

Over 1000 probes for molecular loci (restriction fragment length polymorphisms (RFLPs)) have been isolated and mapped in maize by several researchers (Burr *et al.*, 1988; Helentjaris, 1987; Hoisington & Coe, 1990). This work has been aided by the high level of polymorphism found in maize, coupled with the potential importance of such markers in increasing the efficiency of breeding programmes. The developed sets of probes have been shared among the various workers as quickly as possible, adding to the rapidity with which new markers can be utilized and resulting in much less duplication of effort and better correlation of mapping results.

Features of maize important to applied molecular genetics

Besides its importance as a food and feed crop, maize has several features which make it particularly amenable to genetic and molecular analysis. The monoecious nature of maize, in which male and female flowers are borne on separate parts, allows for easy and rapid pollinations. Both self-pollination and outcross pollinations to many other individuals can be performed with the same individual. The large size of the maize ear permits multiple pollinations to occur on the same ear (Sheridan & Clark, 1987). For example, one half of the ear may be selfed to detect the presence of a recessive gene and the other half crossed by a suspected allelic gene. If the phenotype is observed on both halves of the ear, the two genes can be considered allelic. The considerable number of seeds produced on each ear also facilitates the production of large population sizes from a single cross for use in mapping studies.

Maize is also rich in cytogenetic features, which have been important tools for many genetic analyses and for the development of molecular maps. The chromosome number is small ($2n=20$) and the chromosomes are sufficiently large to permit cytological examination. Distinctive knobs on some chromosomes along with their relative length permit cytological identification of each chromosome. *In situ* hybridization with single-copy sequences has been reported (Shen, Wang & Wu, 1987), although not applied routinely.

There also exists a large collection of translocations, both between members of the A chromosome set (Longley, 1961) and between the A chromosomes and the supernumerary B chromosome (Roman, 1947; Beckett, 1978). Subsets of the over 900 reciprocal A–A translocations have been useful in various mapping studies involving dominant markers and in the determination of chromosomal segments involved in several quantitative traits (Burnham, 1982). A method has been proposed that would allow the physical dissection of each chromosome for accurate placement of molecular markers relative to one another and to the physical position on the chromosome (Hoisington & Coe, 1990).

Individuals containing a translocation between an A and a B chromosome, termed B–A translocations, undergo non-disjunction during microsporogenesis (Roman, 1947). The resulting pollen lacks the segment of the A chromosome that was translocated with the B chromosome. The progeny which are produced following pollination onto normal female plants will be hypoploid for the chromosome segment missing in the pollen grain. These hypoploid individuals can be identified by their phenotype or by the expression of a recessive gene 'uncovered' by the translocation.

A set of B–A translocations is available which will produce hypoploid progeny for 18 of the 20 chromosome arms in maize (Beckett, 1978). This set has been extremely useful in locating to chromosome arm many genetic factors (Roman & Ullstrup, 1951), including molecular markers (Weber & Helentjaris, 1989). There is also a report of the potential use of hypoploid and hyperploid individuals to localize quantitative trait loci to chromosome arm (Lee *et al.*, 1991). While limited in their application to breeding, B–A translocations are very useful in localizing many types of genes to chromosome segment and in studying the effects of gene dosage on the expression of complex phenotypes.

Monosomic individuals in maize can be produced through the use of the *r-x1* deficiency (Weber, 1986). This small deletion on the long arm of chromosome 10 results in non-disjunction in the female gametophyte giving eggs which lack individual chromosomes or chromosome segments (Simcox, Shadley & Weber, 1987). Following fertilization, a random set of monosomic progeny is produced and can be identified by their particular phenotypes or by cytological examination. Monosomic individuals can be used to locate visible phenotypes and molecular markers (Helentjaris, Weber & Wright, 1986) in a manner similar to that of B–A translocations.

The various cytogenetic features have aided in the development of the current conventional and molecular linkage maps for maize. Molecular characterization of several of the genes contained on these maps has also progressed rapidly. The presence of numerous transposable element systems and their application in gene tagging has resulted in the isolation and sequencing of many of the pigment, starch and protein biosynthesis genes (for a review, Walbot & Messing, 1988). Several additional genes have been cloned by using heterologous sequences from other organisms, by virtue of abundant mRNAs, and by using antibodies to screen expression libraries. All of these studies have led to a greater understanding of gene structure and function in plants.

Molecular markers and isozymes have received very substantial attention, both because of their potential value in manipulating segments of the genome, and because of their ease of use in maize. Maize is one of the most, if not the most, polymorphic species studied to date (Burr *et al.*, 1983). The ability readily to determine differences between even closely related lines means that maize will serve as an important model species for the application of molecular markers in plant breeding. If the location of quantitative genetic factors and their successful manipulation in a breeding programme are not feasible in a highly polymorphic species such as maize, it is difficult to imagine the successful application of these factors in less polymorphic species.

Molecular approaches to maize improvement

While the efforts of breeders have been quite successful in developing improved maize varieties, it is highly probable that they will need additional techniques in the future if new varieties are to be developed rapidly. Not only will increases in yield be necessary, but the new varieties will also require improved levels of resistance to biotic and abiotic stresses. Often the gene or genes conferring resistance will be available in other varieties and will need only to be transferred to new germplasm. In other cases, appropriate levels of resistance will not be known and will need to be engineered.

Many of the techniques of molecular biology are being used currently to develop new tools for plant breeding in the future. In general, these techniques can be classified under two major headings: genome manipulation and genome modification. Techniques under the heading of genome manipulation involve the analysis of the *in situ* genome of an organism in order to identify specific regions important for a particular phenotype. Once these regions have been identified, they can be transferred (manipulated) from germplasm to germplasm by conventional crossing methods. The markers which are the result of the molecular analysis serve as the selection criteria. Genome modification techniques involve the direct modification of the existing genome of a plant, usually by the insertion of a novel gene or genes. These novel genes range from ones which encode for increased or modified protein levels in the developing kernel or plant to ones which can provide resistance to pathogens or herbicides. Approaches involving techniques under both of these headings, along with traditional breeding methods, will be important if improved varieties of any plant are to be produced in the future.

GENOME MANIPULATION TECHNIQUES

In order to identify the specific regions of the genome involved in a particular phenotype, some type of marker is required. Four types have been tried or are being tried in maize: the above mentioned reciprocal A–A translocations, isozymes, RFLPs and the newest marker class, randomly amplified polymorphic DNAs (RAPDs). The first three of these marker systems have been successful in initially locating the traits of interest, although the use of each marker system in detecting linkages and in breeding has certain disadvantages.

Translocations

Reciprocal translocations are not normally found in germplasm used in breeding programmes and have to be crossed into the desired germplasm before analysis of the trait or traits can begin. Translocations have been used to locate to chromosome or chromosome arm genes involved in resistance to fungal diseases (Saboe & Hayes, 1941; Burnham & Cartledge, 1939; Jenkins, Robert & Findley, 1957; El-Rouby & Russell, 1966); and to insects (Ibrahim, 1954; Scott, Dickie & Pesho, 1966; Onukoga *et al.*, 1978). Very few additional studies have utilized translocations as markers owing to the difficulty in handling such markers and to the fact that only a low resolution of the genome is possible.

Isozymes

Almost 30 isozyme systems have been identified for maize; they allow for the detection of nearly 80 loci (Coe *et al.*, 1988). Isozymes offer several advantages over translocations in locating genetic traits. They already exist in each plant and do not need to be introduced into the material under study; they are neutral in nature and do not affect the expression of the trait; they are usually co-dominant, allowing the determination of both parental alleles in the heterozygous state; and, they provide fairly good coverage of the genome, with at least one or two isozyme loci present on each chromosome of maize. Unfortunately, a major disadvantage is the level of isozyme variants one detects between maize varieties. Of the 80 isozyme loci known, usually fewer than 10 are polymorphic in crosses involving elite materials (C. W. Stuber, personal communication). This severely limits the resolution possible in mapping traits of interest.

Several studies have utilized isozymes in the analysis of germplasm diversity (Goodman & Stuber, 1983; Smith, 1984; Smith & Smith, 1987, 1989), in assessing the value of isozymes for predicting hybrid yield (Frei, Stuber & Goodman, 1986*a*), and in locating quantitative traits to chromosome segments (Edwards, Stuber & Wendel, 1987; Stuber, Edwards & Wendel, 1987). The studies of Stuber *et al.* (1987) demonstrated the feasibility of locating genetic factors involved in the expression of complex traits. Associations for yield, 24 yield-related traits and 25 morphological traits were detected using two F_2 populations and 17–20 isozyme loci. Multiple types of gene action and variation in the phenotypic effect were found at many of the isozyme loci. The authors pointed out the necessity for further evaluation of these associations using additional crosses before using linked isozymes as selection criteria. Changes in allele frequency at isozyme loci following selection for yield have been reported (Stuber & Moll, 1972); and isozyme loci have been used as selectable markers for improving yield and ear number (Stuber, Goodman & Moll, 1982; Frei, Stuber & Goodman, 1986*b*).

These studies developed the framework for the current efforts with RFLPs in locating and manipulating quantitative trait loci (QTLs).

Restriction fragment length polymorphisms

The third type of marker, RFLPs, has been studied in maize and found to be quite useful in both genetic studies and in locating more complex traits in the genome. RFLPs are almost perfect markers for a genome: they exhibit Mendelian co-dominant behaviour; they are already present in the plants under investigation; they can be detected in almost all stages of development and in almost all tissues of the plant; they are neutral and, therefore, do not interfere with the trait under study; they are not affected by environmental conditions or stresses that an individual might be placed in; they are detected by a uniform method; and, perhaps most important, they are potentially unlimited in number and, thus, offer extensive coverage of the genome. The limitations in the use of RFLPs for genome analysis are: the number of steps required to detect an RFLP; the number of individuals and speed at which they may be analysed (although this is improving); and the difficulty in detecting polymorphisms in some species. Even with these difficulties, RFLPs are receiving significant attention, particularly as applied to locating the genetic loci involved in more complex traits.

RFLP mapping was first described in 1975 using adenovirus (Grodzicker et al., 1975). In 1980, Botstein and colleagues described the construction of a linkage map for humans based on RFLPs (Botstein et al., 1980). Since that publication, there has been impressive progress in the development of RFLP linkage maps for many different plant species. Recently, it has been demonstrated that RFLPs can be used to locate genes involved in complex traits in tomato (Patterson et al., 1988).

An RFLP is simply a difference in the DNA between two individuals. This difference may be due to any number of reasons, the most common being, the insertion or deletion of a small segment of DNA, or a change in one or two bases in the DNA sequence. To detect an RFLP, the DNA from an individual is isolated (in maize, dried leaf material or cob tissue is preferred) and digested with one or more restriction enzymes – enzymes which cleave the DNA sequence at specific sites. The digested DNA is then separated by gel electrophoresis, resulting in a separation based on the length of each fragment. Millions of fragments are generated following digestion of genomic DNA in higher organisms and additional steps are required to detect only a single fragment.

For ease of use in the detection steps, the DNA separated in the gel is transferred to a membrane, giving an exact replica of the DNA pattern in the gel. This membrane is then used as the template for the detection of individual fragment lengths by a process termed hybridization. Hybridization consists of soaking the membrane in a solution containing a probe which is a single-stranded sequence of DNA that matches a part of the sequence of the fragment to be detected. Once the probe has been given sufficient time to attach to its corresponding sequence on the membrane, non-hybridized probes are removed and the remaining, hybridized probe located.

For detection of the probes, the probe is labelled with a substance that can be detected on the membrane. Probes can be labelled by incorporating a molecule of the radionuclide, P^{32} into the DNA strand. Detection then consists of placing the washed membrane containing radio-labelled probe next to X-ray film and developing the film after one to several days. Detected fragments appear as dark bands (or spots) on the film in the exact position that they occurred on the gel. A typical autoradiograph from an RFLP survey of tropical maize germplasm is presented in Fig. 9.1. Seven RFLP morphs are visible across the germplasm in the figure. Digestion of the DNA with a different restriction enzyme or hybridization with a different probe would result in a totally different RFLP pattern.

The use of radioactivity involves several precautions and non-radioactive detection of DNA fragments would reduce this risk involved in RFLP analyses. Most of the methods described for non-radioactive detection involve labelling the DNA probe by incorporating a chemical compound which will serve as a hapten. Hybridization and removal of non-hybridized probes is the same as for radioactively labelled probes. Instead of the washed membrane being immediately placed on film, several additional steps are performed to detect the presence of the hapten molecule. First, an antibody–enzyme complex is allowed to attach to the hapten. Then a chemical is used which, when acted on by the enzyme, results in a visible product. Insoluble dyes have been used, but are generally not sensitive enough for single-copy sequences and do not allow the membrane to be rehybridized with additional probes.

More recently, a protocol based on light emission has been developed (Kreike, de Koning & Krens, 1990). In this method, the hapten is

Fig. 9.1. Germplasm survey using RFLPs. Autoradiograph depicting
the RFLP profile of 54 maize inbreds for probe *UMC103*. Each DNA
sample was digested with the restriction enzyme *Hind*III.

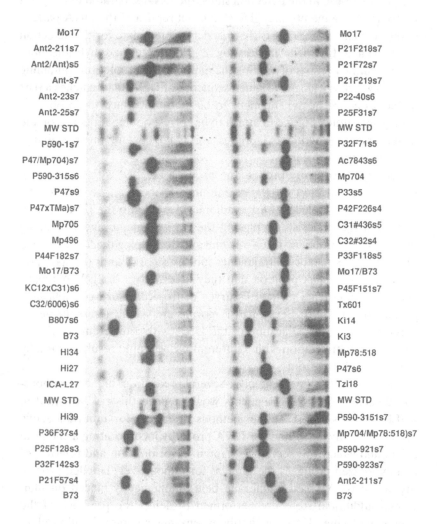

Mo17	Mo17
Ant2-211s7	P21F218s7
Ant2/Ant)s5	P21F72s7
Ant-s7	P21F219s7
Ant2-23s7	P22-40s6
Ant2-25s7	P25F31s7
MW STD	MW STD
P590-1s7	P32F71s5
P47/Mp704)s7	Ac7843s6
P590-315s6	Mp704
P47s9	P33s5
P47xTMa)s7	P42F226s4
Mp705	C31#436s5
Mp496	C32#32s4
P44F182s7	P33F118s5
Mo17/B73	Mo17/B73
KC12xC31)s6	P45F151s7
C32/6006)s6	Tx601
B807s6	Ki14
B73	Ki3
Hi34	Mp78:518
Hi27	P47s6
ICA-L27	Tzi18
MW STD	MW STD
Hi39	P590-3151s7
P36F37s4	Mp704/Mp78:518)s7
P25F128s3	P590-921s7
P32F142s3	P590-923s7
P21F57s4	Ant2-211s7
B73	B73

digoxigenin and the detection substrate is AMPPD (3-(2'-spiroadaman-
tane)-4-methoxy-4-(3''-phosphoryloxy)phenyl-1,2-dioxetane). When an
AMPPD molecule is dephosphorylated by the antibody–enzyme complex, an unstable intermediary is produced which gives off visible light
as its relaxes to its more stable form. Although the light given off is not
intense enough to be seen by the naked eye, it is possible to detect it
with the same X-ray film (or other sensitive films) used in detecting

radioactivity. The resulting luminograph is indistinguishable from that of the autoradiograph. This chemiluminescent procedure should allow RFLP technology to be used in many laboratories not currently equipped for, or experienced in, the handling of radioactivity.

Maize has been quite amenable to RFLP analysis owing to the large amount of tissue available for DNA extraction, its diploid nature, the small number of chromosomes, and, most importantly, the extremely high amount of polymorphism present. Maize is one of the most polymorphic plants found so far. From sequence analysis, it is estimated that five of every 100 bases are modified in the maize genome (Shattucks-Eidens *et al.*, 1990). Studies of germplasm diversity have detected as many as five different morphs on average across temperate germplasm (Melchinger *et al.*, 1990*a*, *b*) and over six morphs across tropical and subtropical germplasm (D. A. Hoisington, unpublished results). This high level of polymorphism allows one rapidly to detect differences between maize lines for use in mapping quantitative and qualitative trait loci.

Several applications for RFLPs have been proposed and are being investigated. The major applications are listed below.

> Mapping of qualitative traits (single genetic factors and translocation break-points).
> Determination of genetic diversity and relationships in the germplasm base.
> Enhancement of the introgression of desired traits in backcrossing programmes.
> Mapping of quantitative trait loci (complex genetic factors involving several genes).

As mentioned above, maize is particularly rich in the number and types of single genetic loci already identified. Not included in the current genetic database is a large collection of mutants produced by chemical mutagenesis in pollen by Neuffer at the University of Missouri (Neuffer, 1978). Future genetic, biochemical, physiological and developmental studies of maize will depend on having precise positional information for all of the loci relative to each other and to molecular markers. Mapping of single loci is relatively easy; however, the complex nature of many of the loci and the large number available for mapping make the task very difficult. RFLPs provide an alternative approach to developing an integrated linkage map for an organism. Through the use of interval mapping, a method which relies on the use of homozygous individuals for mapping, any gene can be defined to a genetic interval with equal efficiency (Hoisington & Coe, 1990). As part of a large project at the University of Missouri to produce an integrated linkage map for maize, several genes have already been mapped into a core map of RFLP markers (Howell *et al.*, 1991). Further work will result in the refinement of the current working linkage maps and lead to the placement of each gene to a 'box' defined by molecular markers.

As mentioned, maize is extremely polymorphic and this, coupled with the commercial importance of maize, has led to the development of many RFLP maps (Helentjaris, 1987; Burr *et al.*, 1988; Murray *et al.*, 1988; Hoisington & Coe, 1990; Beavis & Grant, 1991). Currently, over 1000 probes have been isolated and mapped. The mapping project at the University of Missouri also involves the integration of as many of these maps as possible and the production of a core set of probes for future integration of mapping data.

The ability to develop a large amount of segregation information in one population has led to the development of more efficient methods for linkage analysis. Algorithms involving the maximum likelihood determination of multipoint maps (Lander & Green, 1987) have been incorporated into computer programs such as MAPMAKER (Lander *et al.*, 1987) and allowed very accurate linkage maps of the maize genome to be developed. Since the number of molecular loci that could be mapped in any one population is virtually unlimited, maps based on eternal F_2s and recombinant inbreds (Burr *et al.*, 1988) were developed. The latter has allowed new molecular probes to be mapped into the existing linkage map with only one or two hybridizations.

The applications of RFLP analysis to increasing the efficiency of breeding programmes are perhaps the major reason for the large efforts in maize. From surveys of germplasm using RFLPs, molecular diversity measures can be calculated and used to classify the germplasm into groups (González de León *et al.*, 1989; Melchinger *et al.*, 1990*b*). RFLPs, along with isozymes and protein profiles, appear to be the best method for the classification of new or existing germplasm (J. S. C. Smith, personal communication). RFLP profiles also provide fingerprints of each line for use in varietal protection.

Some correlation between the number of differences detected by RFLPs between two lines (molecular heterozygosity) and the heterosis observed in the hybrid between them has been reported (Lee *et al.*, 1989; Melchinger *et al.*, 1990*a*). Previous studies using isozymes showed little or no correlation (Price *et al.*, 1986; Lamkey, Hallauer & Kahler,

1987), perhaps owing to the limited number of isozyme loci analysed. If molecular heterozygosity is found to be a dependable indicator of the level of hybrid vigour, molecular markers will be extremely useful in the development of new inbred lines.

Once the RFLP pattern has been determined for a line, RFLPs can be used in backcrossing programmes involving the line, to increase the rate at which the undesired germplasm is eliminated (Tanksley *et al.*, 1989; Young & Tanksley, 1989). It has been estimated that through the use of RFLP-assisted backcrossing, introgression of a new gene(s) can be achieved in as little as two to three backcrosses, compared to six to seven with traditional methods. This would mean that new varieties could be developed relatively quickly with considerable savings in time.

Most of the attention in the application of RFLP analysis to plant breeding is directed at their use as markers for the location of complex traits of interest to breeders. These traits range from relatively simple ones, such as disease resistances, to very complex ones, such as yield and tolerances to abiotic stresses. The genetic loci responsible for the complex phenotypes are QTLs.

The locating of a QTL is not very different from mapping a single locus, but the precision with which one can map a QTL is much less. A population segregating for the trait or traits of interest is developed (either F_2 backcross, or recombinant inbred) and scored in the field for the phenotypic value of the trait and in the laboratory for the genotype at each segregating RFLP locus. These two data sets are then analysed to detect correlations between the presence of a particular RFLP allele and a phenotypic value. Efficient maximum likelihood methods have been developed to increase the resolution of the analysis and to take advantage of the multipoint linkage data available from RFLPs (Lander & Botstein, 1989). Computer programs, such as MAPMAKER-QTL, have also been developed that implement these algorithms into usable packages.

Although there is a tremendous amount of research being conducted in mapping QTLs in maize, very little has reached publication in journals. Several presentations and posters have been presented at meetings which have indicated that it is feasible to map complex traits in maize (Bubeck *et al.*, 1990; Fincher, Beavis & Grant, 1990; Grant *et al.*, 1990). A number of these have also pointed out several of the problems associated with QTLs. Many of the traits which have been located using RFLPs appear to be very complex and demonstrate variability across different populations and environments (Fincher, Beavis & Grant,

1990). Before the use of RFLPs as selectable markers becomes routine it will be necessary to study each trait–marker association in several different populations, in several different environments and across several seasons. Marker–trait associations which are found to be consistent across all of these factors will be good candidates for use in following the trait in a breeding programme. Associations which are inconsistent across one or more of these factors will be useful in that particular environment or cross, but will not be useful as a general indicator of the trait.

Less complex traits such as viral resistance have also been mapped using RFLPs. A single locus on chromosome 6 which is required for resistance to maize dwarf mosaic virus has been identified (McMullen & Louie, 1989). Some variation in resistance was observed in different genotypes indicating that additional loci may be involved in viral resistance. The use of isogenic lines may lead to further refinement of the location of the resistance locus and associated markers of less complex traits (Muehlbauer et al., 1988; Weck et al., 1990).

Randomly amplified polymorphic DNAs

One of the major limitations of RFLP analysis as applied to plant breeding is the limited number of samples which can be analysed. In the laboratory at the International Center for Maize and Wheat Improvement (CIMMYT), it has been estimated that using current protocols for DNA isolation, digestion, electrophoresis and blotting, and the chemiluminescent protocol for probe detection, one person could perform 20 000 lane-hybridizations in 3 months. If the entire genome of maize were to be tested, requiring 50 probes and one enzyme, this number of data points would represent the analysis of only 400 individuals. This figure is substantially lower than the population sizes used in most breeding programmes and points to the large discrepancy between sample sizes that are handled in the laboratory and in the field. If molecular markers are to be applied routinely as selection criteria, modifications must be made allowing large sample sizes to be analysed.

Recently, a second type of molecular marker has been described (Welsh & McClelland, 1990; Williams et al., 1990). These markers are termed randomly amplified polymorphic DNAs (RAPDs) and are based on the amplification of DNA by the polymerase chain reaction (PCR). PCR is a simple process in which a specific segment of DNA is synthesized repeatedly, resulting in the production of large amounts of a single DNA sequence starting from a minute quantity of template (Saiki *et al.*, 1985). The process depends on primer sequences of DNA which match flanking sequences at both ends of the targeted sequence. Through repeated denaturing, annealing and synthesis steps, the intervening sequence is synthesized in a 2^n amplification. PCR has revolutionized molecular biology, allowing for rapid production of probes by direct synthesis from bacteria, isolation of single genes, DNA sequencing without the isolation of the targeted gene sequence, and development of diagnostic assays for the presence of specific alleles (Erlich, Gibbs & Kazazian, 1989).

For the detection of RAPDs, the primer sequences used to initiate the targeted DNA synthesis are chosen somewhat at random from 10-base sequences. Only a few criteria must be met for a 10-mer to qualify as a potential primer (Williams *et al.*, 1990), so that thousands could be synthesized for testing. The typical reaction consists of a simple buffer, the four dinucleotide triphosphates (dATP, dCTP, dGTP and dTTP), a single primer, DNA polymerase, and a few nanograms of DNA from the organism under study. Following 4–5 hours of amplification, the products are analysed by gel electrophoresis. Instead of blotting, and hybridizing the resulting DNA pattern, the gel is simply stained for DNA and photographed. Since the probability of any 10-base sequence being present in the genome in a position to result in amplification of an intervening sequence is very low, the pattern seen on the gel usually consists of a few bands. Often, only one to two major bands are observed and polymorphisms are detected as the presence or absence of each band. Figure 9.2 shows the results of using a primer sequence to amplify the DNA of 20 individuals from an open-pollinated population of maize. Extreme care must be taken in order to assure that the results obtained are accurate and repeatable.

RAPDs have most of the advantages of RFLPs with the primary exception of being dominant, instead of co-dominant. The following features are unique to RAPDs.

> Primer sequences are synthesized at random and do not require complicated probe development.
> Primers are universal and function in any species.
> The amount of template DNA required is extremely small and, thus, requires simple DNA extraction procedures starting with small amounts of tissue.
> No gel transfer or hybridization is required, greatly shortening the time required for analysis.

Fig. 9.2. RAPD analysis of an open-pollinated maize variety.
Photograph of the ethidium-stained gel following PCR amplification
of genomic DNA from 16 individuals of the open-pollinated maize
variety V25. The primer sequence used was ACAACGCCTC. MW,
molecular weight.

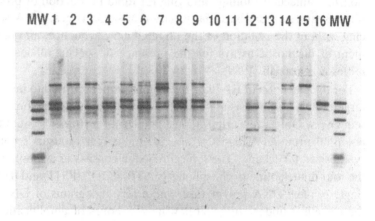

Large numbers of samples can be analysed in a relative short
period of time.

Given all of the advantages one might think that RAPDs are the
perfect marker system. Unfortunately, there are a few disadvantages
which limit the utility of RAPDs as genetic markers. Because the typical
polymorphism observed is a presence or an absence of a band, RAPDs
detect dominant loci (not co-dominant). Therefore, it is difficult or
impossible to determine the heterozygous condition in an individual.
This severely limits the amount of genetic information derived from
each individual in a selfed population, although RAPDs are quite useful
in backcross and recombinant inbred populations. In maize, the number
of morphs detected with each primer appears to be lower than that
detected with each RFLP probe (D. A. Hoisington, unpublished
results). This observation coupled with the care necessary for obtaining
accurate results may preclude the use of RAPDs for general germplasm
surveys. A perhaps more serious problem is the fact that the same
primer may not detect the same locus in different populations. Since the
RAPD system relies on a competitive reaction of primer to template
and some mismatching of the two sequences can occur, the resulting
amplified DNA may originate from different starting sequences located
in different regions of the genome. Thus, once a primer–QTL association has been found in one population, direct use of the primer in

another population may not be possible. This would severely limit the use of RAPDs as a universal marker for a QTL.

By using RAPDs in a backcrossing program, larger numbers of individuals in each generation could be analysed resulting in a more rapid progression towards the recipient parent. One would simply have to determine which primers detect polymorphisms between the donor and recipient lines and follow those that do in each backcross generation. A protocol could be developed which couples RFLP analysis of specific genomic regions containing QTLs with RAPD analysis for the remainder of the genome.

RAPDs can also provide a mechanism to acquire RFLP probes for a specific region of the genome, since the sequence generated by a primer can be labelled and used as an RFLP probe (Williams *et al.*, 1990). In this manner, RAPDs could be used initially to locate a QTL, and then in the production of an RFLP probe for the region.

GENOME MANIPULATION

The use of molecular markers to locate and trace specific genomic segments depends on the presence of sufficient variation of the trait of interest. Again, maize is somewhat unique in that there exists remarkable variation for almost any phenotype, including resistance to disease and environmental stresses. The success of maize breeders in adapting maize to many different environments is testimony to this variation in the maize genome. Unfortunately, even with the potential use of RFLPs and RAPDs, the time required to develop new varieties is often too slow to adapt existing varieties to new situations. If the breeders are to be able rapidly to modify new germplasm to withstand new pressures from diseases and stresses, methods for the rapid modification of the maize genome will need to be available. Genome manipulation through genetic transformation offers a method to modify quickly the existing genome in a specific way without interfering with the remainder of the genome.

Genetic transformation relies on the ability to insert a novel piece of DNA into a cell, on the incorporation of the new DNA into the nuclear genome, and on the cell being part of the germline of the plant. Many methods have been employed in attempting to transform maize, but only recently has successful transformation been reported (Fromm *et al.*, 1990; Gordon-Kamm *et al.*, 1990). Most of the lack of success in producing transgenic maize resides in the fact that the method for insertion of DNA requires some form of tissue culture. While culturing and regeneration of dicotyledons has proved relatively easy, mono-cotyledons and, in particular, cereals have proved to be quite recalcitrant to most culture methods (Phillips, Somers & Hibberd, 1988). Numerous lines of maize must be screened before ones which result in type II callus formation and subsequent regeneration can be found.

Any method of genetic transformation also requires that there be a gene or genes whose introduction into the crop is desired. Most of the research has focused on the use of herbicide, insect or virus resistance genes. While these are potentially important traits, future efforts will require new genes to be identified and isolated. Efforts underway to clone genes in other plant species such as *Arabidopsis* are potentially a source of supply for initial gene sequences for use in maize. Because of the level of synteny often found between species (Bonierbale, Plaisted & Tanksley, 1988), gene sequences from even distantly related species may be useful in identifying similar genes in maize. Other potential genes may even include those involved in QTLs. Robertson (1985) proposed that QTLs are typical genetic loci for which severe mutations may already be known (i.e. dwarf loci may be involved in plant height). If correlations between QTLs and qualitative loci are found, it may be possible to devise methods for cloning QTLs. These cloned genes could then be useful in improving specific traits through transformation.

From the numerous approaches to gene transfer in plant cells, the following three methods have been reported as resulting in transgenic maize plants.

1 Protoplast electroporation.
2 *Agrobacterium*-mediated transfer.
3 Particle bombardment.

One of the first reports of maize transformation was by the electro-poration of protoplasts (Fromm, Taylor & Walbot, 1986). The first reported regenerated plants following protoplast electroporation were unfortunately sterile (Rhodes *et al.*, 1988). Despite additional reports on the successful regeneration of fertile plants from protoplasts (Prioli & Sondahl, 1989), the regeneration of fertile transformed plants has not been reported.

Agrobacterium-mediated transformation has been reported only recently in maize (Gould *et al.*, 1991). This method has great promise in that it seems to be less genotype specific than other methods requiring some form of tissue culture, although additional work remains to be done before it can be applied routinely to maize.

Two reports on the transformation and regeneration of fertile plants of maize using particle bombardment have been published (Fromm *et al.*, 1990; Gordon-Kamm *et al.*, 1990). Transformation using the particle gun initially involves the development of a starting callus that demonstrates type II callus formation and regeneration capacity. Embryogenic suspension culture cells are bombarded with microscopic metal particles coated with a DNA construct (containing the gene(s) to be inserted and appropriate promoters). Following bombardment, the cultures are allowed to recover and are then placed on a selective medium for 1–2 weeks. During this phase, cells which have received the DNA construct will continue to grow and can be screened for the presence of the introduced construct by DNA hybridization or PCR analysis. The selected calli will then be transferred to embryogenic media in order to initiate root and shoot primordia. Plantlets will form in 2–3 months and can be carefully transferred to soil media and finally to greenhouse conditions. The entire process can take as long as 6–7 months. The regenerated plants are tested for the correct phenotype and crosses made to determine the inheritance of the newly acquired trait. If successful, stable transformed progeny will be obtained which can be analysed for the new phenotype in field conditions.

All of the studies on maize transformation reported to date have involved gene constructs useful for studying the transformation process. Efforts in the future will focus on the use of gene sequences for insect, herbicide and virus resistance. Although difficult to predict, the first transgenic maize varieties containing agriculturally important genes may be available by the mid-1990s.

Conclusion

It is impossible to reference all of the current biotechnological research in maize; however, it should be apparent that the maize plant offers an almost unlimited number of possibilities for study. Maize is particularly suited for molecular studies in that it is easily manipulated both in the field and in the laboratory. The great interest in understanding and manipulating complex traits (QTLs) has led to the development of a vast body of research using maize. All of these efforts will lead to a rapid development of knowledge about the maize genome. If QTLs are to be mapped and the information utilized in large-scale breeding programmes, it is likely that maize will prove to be the candidate for success. If molecular markers cannot be employed successfully in maize breeding, the probability of their successful use in other, less polymorphic, species is low.

Maize will also continue to be one of the major species for genetic investigation, given its rich collection of already identified genetic loci, the large collection of new mutations, the wealth of cytological tools available, and the presence of numerous transposable element systems to allow for the cloning of new genes. Future improvement in our knowledge about the maize genome and our ability to manipulate it will depend on the application of all of these areas. Breeders, geneticists and molecular biologists will have to take advantage of the wealth of tools available, both traditional and modern, in order to continue to provide new insights into the maize plant.

References

Beavis, W. D. & Grant, D. (1991). A linkage map based on information from four F_2 populations. *Maize Genetics NewsLetter*, **65**, 70–71.

Beckett, J. B. (1978). B–A translocations in maize. *Journal of Heredity*, **69**, 27–36.

Bonierbale, M. W., Plaisted, R. L. & Tanksley, S. D. (1988). RFLP maps based on a common set of clones reveal modes of chromosomal evolution in potato and tomato. *Molecular and General Genetics*, **203**, 8–14.

Botstein, D., White, R. L., Skolnick, M. & Davis, R. W. (1980). Construction of a genetic linkage map in man using restriction fragment length polymorphism. *American Journal of Human Genetics*, **32**, 314–331.

Bubeck, D. M., Goodman, M. M., Beavis, W. D. & Grant, D. (1990). Quantitative trait loci controlling resistance to gray leaf spot in maize. *Agronomy Abstracts*, p. 194.

Burnham, C. R. (1982). The locating of genes to chromosome by the use of chromosomal interchanges. In *Maize for biological research*, ed. W. F. Sheridan, pp. 65–70. Charlottesville, VA: Plant Molecular Biological Association.

Burnham, C. R. & Cartledge, J. L. (1939). Linkage relations between smut resistance and semisterility in maize. *Journal of the American Society of Agronomy*, **31**, 924–928.

Burr, B., Burr, F. A., Thompson, K. H., Albertsen, M. C. & Stuber, C. W. (1988). Gene mapping with recombinant inbreds of maize. *Genetics*, **188**, 519–526.

Burr, B., Evola, S. V., Burr, F. & Beckmann, J. S. (1983). The application of restriction fragment length polymorphisms to plant breeding. In *Genetic engineering: principles and methods*, vol. 5, ed. J. K. Setlow & A. Hollaender, pp. 45–59. New York: Plenum Press.

Coe, E. H. Jr, Neuffer, M. G. & Hoisington, D. A. (1988). The genetics of corn. In *Corn and corn improvement*, 3rd edn, ed. G. F. Sprague & J. W. Dudley, pp. 81–257. Madison, Wisconsin: American Society of Agronomy.

Duvick, D. N. (1983). Genetic contributions to yield gains of U.S. hybrid maize 1900–1980. *ASA Special Publication*, **7**, 15–20.

Edwards, M. D., Stuber, C. W. & Wendel, J. F. (1987). Molecular marker-facilitated investigations of quantitative trait loci in maize. I. Numbers, distribution, and type of gene action. *Genetics*, **116**, 113–125.

El-Rouby, M. M. & Russell, W. A. (1966). Locating genes determining resistance to *Diplodia maydis* in maize by using chromosomal transloca-tions. *Canadian Journal of Genetics and Cytology*, **8**, 233–236.

Erlich, H. A., Gibbs, R. & Kazazian, H. H. Jr (1989). *Polymerase chain reaction*. New York: Cold Spring Harbor Laboratory Press.

Fincher, R., Beavis, B. & Grant, D. (1990). Genotype by environment interac-tion associated with molecular markers identifying quantitative trait loci effects. *Agronomy Abstracts*, p. 196.

Frei, O. M., Stuber, C. W. & Goodman, M. M. (1986*a*). Use of allozymes as genetic markers for predicting performance in maize single cross hybrids. *Crop Science*, **26**, 37–42.

Frei, O. M., Stuber, C. W. & Goodman, M. M. (1986*b*). Yield manipulation from selection of allozyme genotypes in a composite of elite corn lines. *Crop Science*, **26**, 917–921.

Fromm, M. E., Morrish, F., Armstrong, C., Williams, R., Thomas, J. & Klein, T. M. (1990). Inheritance and expression of chimeric genes in the progeny of transgenic maize plants. *Bio/Technology*, **8**, 833–839.

Fromm, M. E., Taylor, L. P. & Walbot, V. (1986). Stable transformation of maize after gene transfer by electroporation. *Nature*, **319**, 791–793.

González de León, D., Hoisington, D. A., Jewell, D. & Deutsch, J. (1989). Genetic evaluation of inbred tropical maize germplasm using RFLPs. *Agronomy Abstracts*, p. 82.

Goodman, M. M. & Stuber, C. W. (1983). Races of maize. VI. Isozyme vari-ation among races of maize in Bolivia. *Maydica*, **28**, 169–187.

Gordan-Kamm, W. J., Spencer, T. M., Mangano, M. L., Adams, T. R., Daines, R. J., Start, W. G., O'Brien, J. V., Chambers, S. A., Adams, W. R. Jr, Willets, N. G., Rice, T. B., Mackey, C. J., Krueger, R. W., Kausch, A. P. & Lemaux, P. G. (1990). Transformation of maize cells and regenera-tion of fertile transgenic plants. *Plant Cell*, **2**, 603–618.

Gould, J., Devey, M., Hasegawa, O., Ulian, E. C., Peterson, G. & Smith, R. H. (1991). Transformation of *Zea mays* L. using *Agrobacterium tumefa-ciens* and the shoot apex. *Plant Physiology*, **95**, 426–434.

Grant, D., Beavis, B., Fincher, R. & Albertsen, M. (1990). Correspondence between qualitative genes and quantitative trait loci in maize. *Agronomy Abstracts*, p. 196.

Grodzicker, T., Williams, J., Sharp, P. & Sambrook, J. (1975). Physical map-ping of temperature-sensitive mutations of adenovirus. *Cold Spring Habor Symposium on Quantative Biology*, **39**, 439–445.

Helentjaris, T. (1987). A genetic linkage map for maize based on RFLPs. *Trends in Genetics*, **3**, 217–222.

Helentjaris, T., Weber, D. F. & Wright, S. (1986). Use of monosomics to map cloned DNA fragments in maize. *Proceedings of the National Academy of Sciences, USA*, **83**, 6035–6039.

Hoisington, D. A. & Coe, E. H. Jr (1990). Mapping in maize using RFLPs. In *Stadler genetics symposium – gene manipulation in plant improvement II*, ed. J. P. Gustafson, pp. 331–352. New York: Plenum Press.

Howell, C., Coe, E., Gardiner, J., Melia-Hancock, S. & Chao, S. (1991). The integrated mapping project: chromosome nine. *Maize Genetics NewsLetter*, **65**, 52.

Ibrahim, M. A. (1954). Association tests between chromosomal interchanges in maize and resistance to the European corn borer. *Agonomy Journal*, **49**, 293–299.

Jenkins, M. T., Robert, A. L. & Findley, W. R., Jr (1957). Genetic studies of resistance to *Helminthosporium turcicum* in maize by means of chromosomal translocations. *Agronomy Journal*, **49**, 197–201.

Kreike, C. M., de Koning, J. R. A. & Krens, F. A. (1990). Non-radioactive detection of single-copy DNA–DNA hybrids. *Plant Molecular Biology Reporter*, **8**, 172–179.

Lamkey, K. R., Hallauer, A. R. & Kahler, A. L. (1987). Allelic differences at enzyme loci and hybrid performance in maize. *Journal of Heredity*, **78**, 231–234.

Lander, E. S. & Botstein, D. (1989). Mapping Mendelian factors underlying quantitative traits using RFLP linkage maps. *Genetics*, **121**, 185–199.

Lander, E. S. & Green, P. (1987). Construction of multilocus genetic linkage maps in humans. *Proceedings of the National Academy of Science, USA*, **84**, 2363–2367.

Lander, E. S., Green, P., Abrahamson, J., Barlow, A., Daly, M., Lincoln, S. & Newburg, L. (1987). MAPMAKER: an interactive computer program for constructing genetic linkage maps of experimental and natural populations. *Genomics*, **1**, 174–181.

Lee, E. A., Baxter, D. R., Darrah, L. L. & Coe, E. H. Jr (1991). Chromosome arm dosage analysis — identification of potential QTLs on the short arm of chromosome 5. *Maize Genetics NewsLetter*, **65**, 58.

Lee, M., Godshalk, E. B., Lamkey, K. R. & Woodman, W. W. (1989). Association of restriction fragment length polymorphisms among maize inbreds with agronomic performance of their crosses. *Crop Science*, **29**, 1067–1071.

Longley, A. E. (1961). Breakage points for four translocation series and other corn chromosome aberrations maintained at the California Institute of Technology. *US Department of Agriculture – Agriculture Research Service*, 34–16.

McMullen, M. D. & Louie, R. (1989). The linkage of molecular markers to a gene controlling the symptom response in maize to maize dwarf mosaic virus. *Molecular Plant–Microbe Interactions*, **2**, 309–314.

Melchinger, A. E., Lee, M., Lamkey, K. R., Hallauer, A. R. & Woodman, W. L. (1990*a*). Genetic diversity for restriction fragment length polymorphisms and heterosis for 2 diallele sets of maize inbreds. *Theoretical and Applied Genetics*, **80**, 488–496.

Melchinger, A. E., Lee, M., Lamkey, K. R. & Woodman, W. L. (1990*b*). Genetic diversity for restriction fragment length polymorphisms – relation to estimated genetic effects in maize inbreds. *Crop Science*, **30**, 1033–1040.

Muehlbauer, G. J., Specht, J. E., Thomas-Compton, M. A., Staswick, P. E. & Bernard, R. L. (1988). Near-isogenic lines – a potential resource in the integration of conventional and molecular marker linkage maps. *Crop Science*, **28**, 729–735.

Murray, M., Cramer, J., Ma, Y., West, D., Romero-Severson, J., Pitas, J., DeMars, S., Vilbrandt, L., Kirshman, J., McLeester, R., Schilz, J. & Lotzer, J. (1988). Agrigenetics maize RFLP linkage map. *Maize Genetics NewsLetter*, **63**, 89–91.

Neuffer, M. G. (1978). Induction of genetic variability. In *Maize breeding and genetics*, ed. D. B. Walden, pp. 579–600. New York: Wiley-Interscience.

Onukoga, F. A., Guthrie, W. D., Russell, W. A., Reed, G. L. & Robbins, J. C. (1978). Location of genes that condition resistance in maize to sheath-collar feeding by second-generation European corn borers. *Journal of Economic Entomology*, **71**, 1–5.

Patterson, A. H., Lander, E. S., Hewitt, J. D., Peterson, S., Lincoln, S. E. & Tanksley, S. D. (1988). Resolution of quantitative traits into Mendelian factors by using a complete RFLP linkage map. *Nature*, **335**, 721–726.

Phillips, R. L., Somers, D. A. & Hibberd, K. A. (1988). Cell/tissue culture and *in vitro* manipulation. In *Corn and corn improvement*, 3rd edn, ed. G. F. Sprague & J. W. Dudley, pp. 345–387. Madison, Wisconsin: American Society of Agronomy.

Price, S. C., Kahler, A. L., Hallauer, A. R., Charmley, P. & Giegel, D. A. (1986). Relationships between performance and multilocus heterozygosity at enzyme loci in single-cross hybrids of maize. *Journal of Heredity*, **77**, 341–344.

Prioli, L. M. & Sondahl, M. R. (1989). Plant regeneration of recovery of fertile plants from protoplasts of maize (*Zea mays* L.). *Bio/Technology*, **7**, 589–594.

Rhodes, C. A., Pierce, D. A., Mettler, I. J., Mascarenhas, D. & Detmer, J. J. (1988). Genetically transformed maize plants from protoplasts. *Science*, **240**, 204–207.

Robertson, D. S. (1985). A possible technique for isolating genic DNA for quantitative traits in plants. *Journal of Theoretical Biology*, **117**, 1–10.

Roman, H. (1947). Mitotic nondisjunction in the case of interchanges involving the B-type chromosome in maize. *Genetics*, **32**, 391–409.

Roman, H. & Ullstrup, A. J. (1951). The use of B–A translocations to locate genes in maize. *Agronomy Journal*, **43**, 450–454.

Saboe, L. C. & Hayes, H. K. (1941). Genetic studies of reactions to smut and of firing in maize by means of chromosomal translocations. *Journal of the American Society of Agronomy*, **33**, 463–465.

Saiki, R. K., Scharf, S., Faloona, F., Mullis, K. B., Horn, G. T., Erlich, H. A. & Arnheim, N. (1985). Enzymatic amplification of β-globin genomic sequences and restriction site analysis for diagnosis of sickle cell anemia. *Science*, **230**, 1350–1354.

Scott, G. E., Dicke, F. F. & Pesho, G. R. (1966). Location of genes conditioning resistance in corn to leaf feeding of the European corn borer. *Crop Science*, **6**, 444–448.

Shattucks-Eidens, D. M., Bel, R. N., Neuhausen, S. L. & Helentjaris, T. (1990). DNA sequence variation within maize and melon – observations from polymerase chain reaction amplification and direct sequencing. *Genetics*, **126**, 207–217.

Shen, D., Wang, Z. & Wu, M. (1987). Gene mapping on maize pachytene chromosomes by *in situ* hybridization. *Chromosoma*, **95**, 311–314.

Sheridan, W. F. & Clark, J. K. (1987). Allelism testing by double pollination of lethal maize *dek* mutants. *Journal of Heredity*, **78**, 49–50.

Shull, G. H. (1909). A pure line method of corn breeding. *Report of the American Breeders' Association*, **5**, 51–54.

Simcox, K. D., Shadley, J. D. & Weber D. F. (1987). Detection of the time of occurrence of nondisjunction induced by the *r-X1* deficiency in *Zea mays* L. *Genome*, **29**, 782–785.

Smith, J. S. C. (1984). Genetic variability within U.S. hybrid maize: multivariate analysis of isozyme data. *Crop Science*, **24**, 1041–1046.

Smith, J. S. C. & Smith, O. S. (1987). Associations among inbred lines of maize using electrophoretic, chromatographic, and pedigree data. I. Multivariate and cluster analysis of data for 'Lancaster Sure Crop' derived lines. *Theoretical and Applied Genetics*, **73**, 654–664.

Smith, J. S. C. & Smith, O. S. (1989). The description and assessment of distances between inbred lines of maize. II. The utility of morphological, biochemical, and genetic descriptors and a scheme for the testing of distinctiveness between inbred lines. *Maydica*, **34**, 151–161.

Stuber, C. W., Edwards, M. D. & Wendel, J. F. (1987). Molecular marker-facilitated investigations of quantitative trait loci in maize. II. Factors influencing yield and its component traits. *Crop Science*, **27**, 639–647.

Stuber, C. W., Goodman, M. M. & Moll, R. H. (1982). Improvement of yield and ear number resulting from selection at allozyme loci in a maize population. *Crop Science*, **22**, 737–740.

Stuber, C. W. & Moll, R. H. (1972). Frequency changes of isozyme alleles in a selection experiment for grain yield in maize (*Zea mays* L.). *Crop Science*, **12**, 337–340.

Tanksley, S. D., Young, N. D, Pateerson, A. H. & Bonierbale, M. W. (1989). RFLP mapping in plant breeding: new tools for an old science. *Bio/Technology*, **7**, 257–264.

Walbot, V. & Messing, J. (1988). Molecular genetics of corn. In *Corn and corn improvement*, 3rd edn, ed. G. F. Sprague & J. W. Dudley, pp. 389–429. Madison, WI: American Society of Agronomy.

Weber, D. F. (1986). The production and utilization of monosomic *Zea mays* in cytogenetic studies. In *Gene structure and function in higher plants*, ed. G. Reddy & E. Coe, pp. 191–204. New Dehli: Oxford and IBH Publishing Co.

Weber, D. F. & Helentjaris, T. (1989). Mapping RFLP Loci in maize using B–A translocations. *Genetics*, **121**, 583–590.

Weck, E., Beckman, D., Maed, D., Bredenkamp, C. & Trainor, M. (1990). Near-isogenic line localization of MDMV resistance to chromosome 6S. *Maize Genetics NewsLetter*, **62**, 99–101.

Welsh, J. & McClelland, M. (1990). Fingerprinting genomes using PCR with arbitrary primers. *Nucleic Acids Research*, **18**, 7213–7218.

Williams, J. G., Kubelik, K., Livak, A. R., Rafalski, J. & Tingey, S. V. (1990). DNA polymorphism amplified by arbitrary primers are useful as genetic markers. *Nucleic Acids Research*, **18**, 6531–6535.

Young, N. D. & Tanksley, S. D. (1989). Restriction fragment length polymorphism maps and the concept of graphical genotypes. *Theoretical and Applied Genetics*, **77**, 95–101.

10

Rice as a model system
G. Kochert

Introduction

Rice stands in the unique position among major crop plants of being eminently suitable as a model system: for studies of plant molecular biology, for systematic and evolutionary studies, to investigate the dispersal and evolution of plants under domestication, and for the application of molecular genetics to plant breeding. For no other plant is there the combination of favourable characters and background knowledge. In this review, those features which make rice suitable as a model system are reviewed, and then some areas of research for which rice is especially well suited or that would be of great benefit to rice production are described.

SYSTEMATICS

The genus *Oryza* contains more than 20 wild species and two species which have been brought under cultivation, *O. sativa* and *O. glaberrima* (Chang, 1976a). Its distribution is pan-tropical. Of the two cultivated forms, *O. sativa* has been the much more widely grown. *Oryza glaberrima* was apparently domesticated in Africa, where it was grown only in some small areas of West Africa, and its cultivation is now being displaced by cultivars of *O. sativa* (Chang, 1984). Extensive morphological and cytogenetic analyses have been carried out on the various species of *Oryza* (Nayar, 1973; Kurata, 1986). The base chromosome number is $x=12$, and extensive studies on chromosome pairing in interspecific crosses have classified the various accessions into several genome types (Chang, 1976b). Most species are diploids, including cultivated forms of rice, which have an AA genome. BBCC and CCDD genome allotetraploids also occur (Table 10.1).

The relationship of rice to other members of the grass family has been interpreted in various ways. Early treatments, such as that of Hitchcock

Table 10.1. *The species of* Oryza

Species	Genome type	Chromosome number	Distribution
O. barthii	AA	24	West Africa
O. glaberrima	AA	24	West Africa
O. glumaepatula	AA	24	South America, West Indies
O. longistaminata	AA	24	Africa
O. nivara	AA	24	India
O. rufipogon	AA	24	South and Southeast Asia
O. sativa	AA	24	Asia, cultivated world-wide
O. eichingeri	BBCC	48	East and Central Africa
	CC	24	
O. punctata	BBCC	48	Africa
	BB	24	
O. minuta	BBCC	48	Southeast Asia
O. officinalis	CC	24	South and Southeast Asia
O. alta	CCDD	48	Central and South America
O. grandiglumis	CCDD	48	South America
O. latifolia	CCDD	48	Central and South America
O. australiensis	EE	24	Australia
O. brachyantha	FF	24	West and Central Africa
O. granulata	?	24	South and Southeast Asia
O. longiglumis	?	48	New Guinea
O. meridionalis	?	24	Australia
O. meyeriana	?	24	Southeast Asia
O. ridleyi	?	48	Southeast Asia
O. schlechteri	?	?	New Guinea

(1971), which considered mainly the characteristics of the spikelets, placed rice in a tribe, Oryzeae, of the subfamily Festucoideae. The main attribute of this tribe was laterally compressed florets with rudimentary or no glumes. Later treatments, which examined many more characters including anatomy of the embryo and leaf, have placed rice and closely related plants in a separate subfamily (Gould & Shaw, 1983; Watson, Clifford & Dallwitz, 1985). Affinities with the bamboos were noted early, and certain features of leaf anatomy have tended to confirm this view (Ellis, 1986). Accordingly, rices and bamboos are grouped in the same subfamily in some treatments, but additional research is needed using molecular methods.

GERMPLASM

Rice was domesticated very early in Asia, and it has been continuously cultivated for as long as 7000 years in some areas. Its use as a food crop has spread to all continents, although it is primarily a crop of Third World tropical countries (Chang, 1976a,b). Rice production is second to that of maize and its acreage is second only to wheat (FAO, 1985). Most rice is directly consumed by humans close to its place of production, and half of the earth's people consume rice as their staple food. There is not much international trade in rice compared with that in corn or wheat, and the use of rice for animal food of industrial purposes is limited. Thus rice is the most important human food crop. Since most rice production is in poor countries with large and rapidly growing populations, research findings which are of practical value can be immediately applied in a place where help is urgently needed. The importance of rice as a food crop will continue to grow because of continued population growth and increasing dependence on vegetable sources of calories, as fisheries are depleted and the use of animals for food becomes prohibitively expensive.

Rice has always been produced primarily in small plots with a large input of human labour and very little mechanization. Its long history of cultivation over a large geographical area has lead to the production of a very large number of cultivars, often adapted to a very specific microclimate (Chang, 1984). Fortunately, much of this germplasm has been conserved. Most areas where rice is grown were not suitable, from socio-economic or geographical considerations, for large-scale mechanization, which brings a pressure for the use of the same cultivars over large areas. Local varieties were thus widely grown until very recently. Rice germplasm centres have been established in many countries, and the largest of these is the International Rice Research Institute (IRRI) in the Philippines, which maintains a collection of more than 70 000 cultivars of rice. This represents an 'enormous and remarkable' array of genetic diversity, unmatched by any other crop plant (Chang, 1976a). Wild species of *Oryza* have also been extensively collected, and multiple accessions of all the species are available from germplasm collections. Extensive evaluation of all this germplasm has been made, making rice germplasm the most extensively conserved, the best evaluated, and the most used (Chang, 1976a). A wide array of valuable traits has been discovered in rice germplasm, including resistance to insects and diseases and valuable agronomic characters. The most easily accessible of these traits have been introduced into elite cultivars by conventional breeding, and many more await future technological advances (Khush, 1987).

The huge rice germplasm resource is also being studied, using modern methods of classification, to a greater extent than that of other crops. Second (1982, 1985) and Glaszmann (1988, 1989, 1990) reported extensive studies using isozymes to study diversity and phytogeography. The huge *O. sativa* complex can be separated into seven groups by isozyme analysis (Glaszmann, 1988, 1989). Higher resolution analyses using nuclear restriction fragment length polymorphism (RFLP) techniques have tended to confirm this breakdown (Wang & Tanksley, 1989). Useful classification of wild species germplasm has also been undertaken with RFLP analysis (Jena & Kochert, 1991; S. D. Tanksley, personal communication).

GENETICS
Rice is a very favourable organism for conventional genetics and molecular genetics. It is a diploid with a very small genome. Using microspectrophotometric analysis of vegetative tissue, it was determined that the genome size was 0.6 pg or 580 Mbp, 10^6 base points (Bennett & Smith, 1976), but more recent analyses based on separation of nuclei and flow cytometry give an estimate of about 0.45 pg or 4.35 Mbp, (E. Earle, unpublished data). This is much smaller than the genome of maize 7200 Mbp, and is 40 times smaller than that of wheat and 10 times smaller than that of barley. In fact the rice genome is only about six times larger than that of *Arabidopsis*. A small genome has major advantages for molecular genetics. The entire genome can be represented in genomic libraries by a relatively small number of clones. To produce a complete yeast artificial chromosome (YAC) library of rice with an average insert size of 250 kb would require only about 1800 clones. A comparable maize or wheat YAC library would require tens of thousands of clones. Screening of such libraries for valuable genes is thus much easier, and techniques such as map-based gene cloning are greatly simplified.

Rice genetics have been studied assiduously since the 1920s, and rice is one of the genetically best-characterized organisms. A conventional genetic map containing more than 100 markers is available through the efforts of scientists from many countries (Khush, 1986). Most of the markers are for morphological characters, but some valuable agronomic traits such as resistance to rice blast, brown planthoppers and seed traits have also been mapped.

Rice is an inbreeding plant and homozygous lines can be constructed by repeated selfing or backcrossing. This makes it possible to construct very valuable breeding lines, such as recombinant inbreds and near-isogenic lines. A complete set of primary trisomics is available (Khush *et al.*, 1984), and alien addition lines are available for some combinations (Jena & Khush, 1989). There is a wealth of breeding material because of the establishment of IRRI and its dedication to detailed record keeping and storage of breeding materials such as remnant seed. Molecular genetics studies are greatly facilitated by the availability of such material; witness the use of McClintock's material by maize molecular geneticists and the use of Sears' material by wheat researchers.

In the mid-1980s the Rockefeller Foundation, which had supported the founding of IRRI in the early 1960s, realized that the study of the molecular biology of rice was lagging behind that of other major crop plants such as maize and wheat. This was partly caused by the fact that the United States and Western Europe, which has most of the research infrastructure and were leaders in the emerging science of plant molecular genetics, were not major rice producers and had very little incentive to extend their studies to rice. Accordingly, the Foundation decided to recruit leading plant molecular biologists for rice research by making research grants and setting up an international network to promote training of Third World scientists and to facilitate cooperation and dissemination of information. This programme has greatly stimulated research into the molecular genetics of rice, and ongoing research rivals that of any other crop plant. It should serve as a model for the production of valuable research results with minimum waste caused by overlap of projects and poor communication.

One of the projects funded by the Rockefeller Foundation was the construction of an RFLP map. The first map utilized an F_2 population derived from remnant seed from crosses made at IRRI (McCouch *et al.*, 1988). The map was extended using a backcross population from a cross which had been made in the Ivory Coast (McCouch *et al.*, 1991). The result is that rice has a high-density RFLP map with more than 400 markers mapped thus far. Another map has been developed by Japanese researchers and has more than 100 markers (A. Saito, unpublished data).

Although rice has a small genome, the RFLP map is large when measured in centimorgans (cM; McCouch *et al.*, 1988). This has an important practical consideration for the cloning of valuable genes by reverse genetics. The rice genome contains only about 450×10^6 bases, but the recombination map is 1800 cM in length. This means that each map unit is only about 250 kb. Thus it will be relatively easy to get RFLP markers that are close to genes of interest, and cloning projects that

Table 10.2. *Cloned and characterized rice genes*

Gene	Reference
Chloroplast (entire genome)	Hiratasuka *et al.*, 1989
Prolamin	Kim & Okita, 1988
	Barbier & Ishihama, 1990
Glutelin	Masumura *et al.*, 1989
	Takaiwa, Kikuchi & Oono, 1989
Basic chitinase	Zhu & Lamb, 1991
Endochitinase	Huang *et al.*, 1991
Protein kinase	Feng & Kung, 1991
Alpha-amylase	Huang *et al.*, 1990
Phytochrome	Kay *et al.*, 1989
Cytochrome *c*	Kemmerer, Lei & Wu, 1991
Cytochrome oxidase	Kao, Moon & Wu, 1984
Glycine-rice cell wall protein	Lei & Wu, 1991
Phenylalanine ammonia lyase	Minami *et al.*, 1989
Actin	Reece, McElroy & Wu, 1990
Chlorophyll a/b binding protein	Sakamoto *et al.*, 1991
rRNA-17 S	Takaiwa, Oono & Sugiura, 1984
-25 S-17 S intergenic spacer	Takaiwa, Kikuchi & Oono, 1990
Waxy	Wang *et al.*, 1990
Histone-H3	Wu *et al.*, 1989
Alcohol dehydrogenase	Xie & Wu, 1989

utilize chromosome walking from linked RFLP markers will be greatly facilitated. The size of a map unit in rice compares favourably with *Arabidopsis*, a plant of no commercial utility, but which is touted as a model system for plant molecular biology. One map unit in rice is only about four times that in *Arabidopsis*.

Characterization of individual genes has also progressed rapidly in recent years. Many rice genes have been cloned (Table 10.2), and a considerable amount of sequence information is available. This rapid progress has, in many cases, been made possible by advances in other organisms. For example, many rice genes such as actin, alcohol dehydrogenase and histones were first cloned in other organisms, and thus heterologous probes were available for use in selecting rice genes from genomic libraries. However, the next step is now being taken, and rice is beginning to serve as the model for isolation of completely new genes. Several genes or gene families of potential agronomic signifi-cance have been characterized in rice. These include genes for seed storage proteins and enzymes important in starch metabolism such as alpha-amylase, the starch-granule-bound isomerase encoded by the waxy gene, and the phytochrome gene (Table 10.2). The chloroplast genome of rice has also been completely sequenced, and it is the only major crop plant for which this information is available (Hiratsuka *et al.*, 1989).

TRANSFORMATION AND REGENERATION

Much recent research has concentrated on tissue culture and regeneration of rice, and it is the model system among cereal crops (Toriyama *et al.*, 1988; Shimamoto *et al.*, 1989; Datta *et al.*, 1990; Hodges *et al.*, 1990; Li, Chen & Chen, 1990; Suh *et al.*, 1990; Wu, Kemmerer & McElroy, 1990; Kothari *et al.*, 1991; Meijer *et al.*, 1991). Anther culture systems have been extensively used for the production of homozygous lines in one step, and this procedure has been integrated with conventional breeding to produce cultivars which are grown over considerable acreages. Unfortunately, grasses are not naturally suscep- tible to *Agrobacterium tumefaciens*, so this most useful transformation system cannot be directly used. A great deal of effort has gone into finding alternative systems which will work in rice, and considerable success has been achieved. Protoplast culture and regeneration is now possible with a few genotypes, and transformation by several methods has been achieved, but in not all cases has integration and transfer to the next generation been demonstrated (Potrykus, 1990; Kothari *et al.*, 1991). However, rice is closer than any other cereal crop to being routinely transformable and progress is very rapid.

Current research and future prospects

This background of material from decades of germplasm collec- tion and conventional genetic and plant breeding coupled with the newly developed tools of plant molecular biology make rice a leading candidate for use in plant molecular genetics. Thanks to the fact that rice research is largely supported by international groups and not by private companies, the great majority of the germplasm, breeding material, RFLP markers and cloned genes are freely available to any scientist. In the immediate future, fundamental advances are likely in several major areas.

GENOME MAPS
Genetic maps

The rice RFLP map currently has several hundred markers, and more are being steadily added (McCouch *et al.*, 1988; S. D. Tanksley, unpublished data). The ultimate goal is a map containing about 1000 markers, which would make any region of the genome accessible from RFLP markers located within a few centimorgans. This mapping effort can be made much more efficient if the various groups engaged in mapping the genome can use a common mapping population. Otherwise separate maps are developed which later must be reconciled by exchange of probes and extra mapping effort. Common mapping populations can only be developed using plants which can be asexually propagated or those that are true-breeding (completely homozygous) and can be propagated by seed. The first rice map was developed from an F_2 population, which was a finite resource and not suitable for a common mapping population. The rice map currently being sponsored by the Rockefeller Foundation is being developed from an interspecific cross between *O. longistaminata* and *O. sativa* and is being used for mapping by the two main US research groups. The backcross plants developed from this cross are easily propagated asexually, and can be distributed to all interested research groups (S. D. Tanksley, personal communication). Since the population was developed from an interspecific cross, the level of RFLP polymorphism is very high, and most randomly selected probes can be mapped with a small number of enzymes. Two other kinds of mapping population can be shared among many laboratories: (1) a set of recombinant inbred lines derived by single seed descent from the F_2s or (2) a set of progeny developed by anther culture from the F_1 of a mapping cross. Both these types of population will breed true and seed can be distributed along with accumulated RFLP data. Any group that maps further probes can then add them to the existing map, rather than having to start a new one.

Mapped RFLP probes should be stored in a central location, and duplicate sets should be maintained to prevent loss. An alternative, which is in some ways better, is to sequence enough of each mapped probe to enable its amplification by the polymerase chain reaction (PCR). Information on the PCR primers can then be published, giving access to that region of the genome to anyone and minimizing the change of loss or mislabelling of clones (Beckmann & Soller, 1990).

Physical maps

The ultimate map of the rice genome will, of course, be the complete nucleotide sequence. Lack of sufficiently efficient technology prevents a direct approach to this goal for the present, but some of the preliminary steps are being undertaken. Foremost among these is the development of a set of overlapping clones (contigs) covering the entire genome. Early efforts to accomplish this in *Arabidopsis* and *Caenorhabditis* used cosmid vectors, which are capable of accepting inserts up to about 40 kb. Although a great deal of useful information was gained, the resultant contigs seldom contained more than a few clones and could not be joined to produce complete maps (Coulson *et al.*, 1988). Moreover, because of the large size of eukaryotic genomes, a great number of cosmid clones would be required for full coverage. Failure of the contigs to join is not completely understood, but some of the problem certainly is caused by sequences in the eukaryotic genome which are unclonable in prokaryotic vectors. Such sequences have been found in most studies where any appreciable contiguous portions of the genome have been cloned.

Cloning vectors based on yeast artificial chromosomes (YACs) seem to offer a better alternative than cosmid vetors for producing complete genomic libraries (Burke, 1990; Schlessinger, 1990). Larger segments can be cloned, with 250 kb average size inserts a realistic goal. Many fewer YACs would thus be required for a contig library. It may be also that sequences which are unclonable in prokaryotic vectors will be more stable in yeast, which is a eukaryote. The development of YAC libraries can go hand-in-hand with RFLP mapping; each will be a useful adjunct of the other. Ordering of YAC clones will be facilitated by probing the libraries with mapped RFLP markers, and absolute distances between linked RFLP markers can be determined with reference to YAC contigs.

MAP USES

RFLP maps can be used for comparative genome analysis, where they provide a level of resolution superior to previously utilized methods. In rice, the different genomes (A to F) which have been described in the various species have largely been differentiated on the basis of chromosome pairing in interspecific hybrids (Nayar, 1973). If the chromosomes pair, they are considered to represent the same genome. This sort of analysis is limited, of course, to those plants where interspecific hybrids can be produced. Also, the genetic basis of chromosome pairing is not clear. In wheat, pairing can be greatly affected by a single gene (Sears, 1976). Thus it is not clear just how different are the genomes of rice species whose chromosomes do not pair.

Karyotypic analysis has also been carried out to compare the genomes of *Oryza* species. However, rice chromosomes are small, and, even with the best current staining techniques, it is difficult to gain much resolution (Kurata, 1986).

RFLP maps provide a superior way to compare the order of markers on the chromosomes of different organisms and thus to disover regions of conserved linkages, inversions, deletions or translocations. Comparative RFLP mapping is not limited to those organisms which can be intercrossed as is conventional genome analysis by cytogenetics. As long as the genome retains sufficient homology to be able to hybridize to the mapped markers, the analysis can be done. Tomato RFLP probes have been used to analyse the genomes of two other solanaceous plants, potato (*Solanum tuberosum*) and pepper (*Capsicum annuum*). The pepper genome is extensively scrambled relative to tomato (Tanksley *et al.*, 1988), but the potato genome differs from that of tomato only by four paracentric inversions (Bonierbale, Plaisted & Tanksley, 1988). Such analysis has not been carried out with the various species of *Oryza* or related genera such as *Leersia*, but preliminary results indicate a high degree of conservation of gene order between the A and C genomes of rice (G. Kochert, unpublished data). Such information will give a detailed picture of chromosome evolution in rice and its relatives.

RFLP maps allow researchers to follow the transmission of all segments of all the chromosomes in suitable genetic crosses. Therefore it will be possible to make complete chromosome segment maps of hybrid plants. For example, IR-8 is the first of the improved varieties to have been developed through the breeding programme at IRRI. IR-8 was developed from a cross of Dee-geo-woo-gen, a variety from China, with Peta, a variety from Indonesia. Since this was a fairly wide cross, there is considerable RFLP between the parents. Therefore a segment map of the IR-8 chromosomes can be developed that shows just which segments of each chromosome were derived from each one of the parents. Since both of the parents of IR-8 have been used in other crosses it is of potential interest to determine which segments of Peta, for example, have been repeatedly selected by plant breeders in the various crosses where it has been utilized. Such repeatedly selected segments may be expected to contain valuable agronomic traits. Parentage analysis of any rice pedigree could be carried out using these methods.

Another useful adjunct to RFLP mapping would be the development of RFLP probes for especially polymorphic regions of the rice genome. Such 'hypervariable' probes would be valuable in fingerprinting cultivars and in certain cases where it would be valuable to follow segregation in crosses between closely related lines. Hypervariable regions in other organisms have been shown to have their origin in tandem repeats of various sequence elements (Nakamura, Leppert & O'Connell, 1987). The elements may be very short, such as the dinucleotide repeats found in microsatellites (Vergnaud *et al.*, 1991), or they may be longer and form minisatellites (Jeffreys, Wilson & Thein, 1985). Primer sequences in single-copy regions flanking such satellites provide the means for rapid analysis and mapping of these markers by PCR analysis.

CLONING GENES

One of the goals of rice molecular genetics is to identify and clone interesting or agronomically valuable genes. Such cloned genes can then be subjected to DNA sequence analysis to learn about the gene product, or they can be transformed into rice plants to study their expression and to develop superior rice varieties. Most genes which have been cloned have used the following method: first a genomic library is produced and then the desired gene is selected from the library by hybridization to a probe which is specific for the desired gene. Thus cloning the gene is not the problem; any reasonable complete rice genomic library will contain a wealth of valuable and interesting genes. The problem is the selection of clones containing desired genes from the millions of clones contained in a library. If one has information on the gene product, or if the gene has been cloned in another organism, a specific probe can be developed. Unfortunately, most valuable agronomic genes are known only by their phenotype, and the gene product or even the mode of action is completely unknown.

Genes can be cloned in the absence of product information in one of two ways: insertional mutagenesis or map-based cloning. Insertional mutagenesis is based on the inactivation of a gene by insertion of a DNA sequence in, or immediately adjacent to, it. The inserted element can be either a transposon or an element introduced into the plant by transformation, for example the Ti plasmid in dicotyledons. If the transformed element integrates into the genome, there is some chance that the element will insert into a gene and cause a mutation. No well-characterized transposon systems have yet been discovered in rice, but apparently transposons from other organisms, such as the Ac element from maize, can be transformed into rice and actively transpose (Izawa *et al.*, 1991; Zhang *et al.*, 1991). To clone a specific gene by insertional mutagenesis, the resultant mutation will have to cause a phenotypic change that can be relatively easily determined, because large numbers of plants will have to be screened to find the relatively rare insertion into the gene of interest. Dominant genes or genes controlling easily observable plant traits, such as height, should be the most amenable to cloning by insertional mutagenesis. If plasmids or other elements are to be used in a way analogous to that in which the Ti plasmid system is used, a very efficient transformation system will have to be developed.

Map-based cloning presents a more general approach that should be useful for the cloning of any gene. For this technique, markers which are closely linked to the gene of interest must first be obtained as a starting point for cloning efforts. Since the rice genome is highly recombinogenic and quite small, it should be especially good for this sort of analysis.

Closely linked genetic markers can be obtained by analysis of segregating progeny or by probing of near-isogenic lines (NILs). The ideal NIL contains only a single introgressed chromosome segment containing the gene of interest. Markers which are located in the introgressed segment, and which are by definition closely linked can be found by probing the two parents and a single plant from the NIL with mapped or unmapped markers (Young *et al.*, 1988). Conventional RFLP markers can be used, but PCR analysis using single primer methods can be used to screen a large number of markers in very rapid fashion (Martin, Williams & Tanksley, 1991). The many NILs that have been developed by rice breeders are thus a very valuable resource for the mapping and eventual cloning of valuable genes. In practice, NILs have not proved to be ideal in that they have always been found to contain more than one introgressed chromosome segment, making the problem of false positives a significant one (S. D. Tanksley, personal communication). Putative linked probes have to be confirmed by analysis in segregating populations.

Analysis of segregating progeny has the disadvantage that relatively large numbers of plants have to be analysed to find closely linked markers, the more closely linked the larger the number of probes that must be screened. Also such analysis can be efficient only in fairly wide crosses which exhibit a reasonable degree of RFLP or by the use of hypervariable probes. None the less, if a dense RFLP map is available, it is a method with general applicability.

When genetic markers closely linked to traits of interest are found, they are of direct use in plant breeding in addition to their use as starting points for gene cloning. Plant breeding schemes can be greatly improved in efficiency by the use of the indirect selection made possible by linked RFLP markers (Tanksley *et al.*, 1989). One of the most profitable activities that rice breeders can undertake in the immediate future is the development of mapping populations that allow 'tagging' of valuable agronomic traits with RFLP markers. Since no one cross will be segregating for any sizeable number of valuable genes, many such segregating populations will have to be constructed. Recombinant inbred lines will also be valuable for tagging genes and for the analysis of quantitative trait loci (QTLs).

To clone genes for which closely linked markers have been found, the next step would be to probe the markers onto YAC libraries. The ideal situation would accrue if single YAC clones or single contigs could be found which contain RFLP markers flanking the gene of interest. The only problem remaining would be the identification of the gene in the YAC clones, but this is not a trivial problem. Much of the genomic DNA in rice, or any other eukaryote, is not genes but rather it is the space between genes. Thus most of the clones in a small library formed from one or a few YACs would not be genes. Even if the genes could be separated from non-gene DNA, several genes might be present, and it will be necessary to determine which is the gene of interest.

Several ways of identifying candidate genes in small libraries, such as those developed from YAC clones, have been proposed, and one of these is the screening of transgenic plants containing candidate clones. In this approach, subclones from YAC clones would be transformed into plants that have the requisite genetic background to allow the desired gene to be selected. For example, putative genes for insect resistance could be screened by transformation into plants sensitive to the insect. For this approach a very efficient transformation system would be required. Only relatively small genes could be assayed by this method and the genes would probably have to have a dominant effect. Since this is a functional assay, the clone being transformed would have to have the entire gene including control regions. To ensure that this was the case, overlapping clone libraries would have to be constructed, and the problem would become more complex.

In screening clones derived from YAC libraries, it would obviously be of some help to be able to select clones which contain coding sequences. This would eliminate the need to test, for example by transformation, clones which represent intergenic spacer DNA. One way to do this is to take advantage of the fact that coding regions of a genome are usually more highly conserved than non-coding regions. By hybridizing candidate clones to distantly related plants, those clones that contain coding sequences would hybridize while non-coding regions would not.

Another approach is to hybridize candidate clones to cDNA libraries derived from tissues expressing the gene of interest. Clones which contain coding regions could be selected on the basis of their hybridization to cDNAs. This approach is limited by the fact that in many cases the mRNA for the gene of interest might be present in low amounts or only at certain developmental stages and thus might not be present in a cDNA library. However, anything that can be done which will lessen the number of clones which must be screened by transformation will be of great utility, and both these approaches have been successfully used in map-based cloning of human disease genes (Rommens *et al.*, 1989).

All of the tools discussed above for the cloning of valuable genes in rice are either in place or are being rapidly developed. Important genes to target in future cloning efforts have, in many cases, already been identified. Many of these genes are inherently interesting in terms of the control of plant growth and development, and some of them have direct economic importance.

QUANTITATIVE TRAIT LOCI

The methods discussed above for the cloning of genes by insertional mutagenensis or by map-based cloning are models for the cloning of genes for traits controlled by single major genes. However, many agronomic traits are not controlled by single genes, but involve multiple genes. Such traits are called quantitative traits and are controlled by genes called quantitative trait loci (QTLs). Before the advent of molecular mapping techniques, QTLs could only be investigated in terms of the total effect of all the loci affecting a trait on the phenotype of the plant. It was enormously difficult to determine the number of genes controlling a quantitative trait or the chromosomal location of those genes from the phenotypic data. However, an RFLP map provides a way of analysing QTLs. This approach has not been utilized yet in rice, but it has proved successful in mapping QTLs in tomato (Patterson *et al.*, 1991). Rice is an ideal plant for the study of QTLs because an RFLP map is available, so much information is available about the segregation of such traits, and many assays for the phenotype of various quantitative traits have already been developed.

CHARACTERIZATION OF PATHOGENS

Cultivating rice involves a constant struggle to protect it from insects and pathogens. Such problems have increased with the decrease in biodiversity brought about by large-scale utilization of improved varieties. Rice breeding efforts and research in insecticides and fungicides have managed to stay one step ahead of the problem, but more fundamental knowledge of the insect pests and pathogens is needed if future progress is to be maintained. One area which seems fruitful for future research is characterization of the genetic diversity present in insect pests and rice pathogens. It is clear from traditional breeding studies that a great deal of variability is present, but this has never really been investigated in a quantitative fashion. Bioptyes of insects, such as planthoppers, and strains or races of fungal or bacterial pathogens are examples of variable phenotypes. However, screening methods for the detection of such differences have to utilize inoculation schemes onto tester rice lines, and are time consuming and cumbersome to implement. The long-standing debate over the number of races of the rice blast pathogen in rice is an example of the problems encountered by traditional screening methods (Ou, 1985).

Molecular markers provide a way to quantitate genetic variability in important pests and pathogens. This approach has proved successful in characterizing pathogen variability in certain other systems. Vilgalys & Gonzalez (1990) found a tremendous amount of genetic variability in isolates of *Rhizoctonia solani* from a single field. Isolates of pathogenic bacteria can, in many cases, be genetically fingerprinted by DNA probes (Lazo, Roffey & Gabriel, 1987).

Great benefits will arise from the application of this technique to the study of rice insect pests and diseases. Rapid diagnosis of disease problems would benefit epidemiological studies, and would allow prompt use of the appropriate treatment. Development of resistant varieties would also be aided by knowledge of the spectrum of pests or pathogens that are likely to be encountered upon introduction of a new variety.

EXPANDING THE GENE POOL

Plant breeders carry out their craft by surveying the germplasm available to them for useful genes and then using intercrossing and selection strategies to assemble these genes into cultivars. The basic goals of plant breeding will remain the same, but the tools and germplasm resources available to plant breeders will be greatly expanded by the new techniques described above.

For one thing, the gene pool available to breeders will be greatly expanded. Conventional breeding protocols of necessity can utilize only that portion of the germplasm which will intercross with the plant of interest. In rice, the gene pool most used in breeding consists of land races and local cultivars of cultivated rice and those wild species which have an A genome in common with cultivated rice. This is the gene pool that has accounted for the great majority of the advances which have been made in rice breeding. However, this gene pool does not contain genes for certain traits which are critical for further progress. For example, genes for satisfactory resistance to sheath blight and for stem borer have not been found in the primary gene pool, and these are all serious problems. Much greater areas could be planted with rice if sources of salt resistance, drought tolerance and resistance to adverse soils could be found, but the primary germplasm may be inadequate for these purposes.

Another source of valuable genes in rice consists of wild species of *Oryza* with genomes other than the AA genome of cultivated rice. This gene pool contains many valuable genes, but transferring genes from these species into cultivated rice by wide crossing programme is tedious (Sitch, 1990). A high degree of sterility is found in most crosses, so large numbers of crosses need to be made and recombination may be reduced because of infrequent chromosome pairing (Brar & Khush, 1986). Only in a crop such as rice, whose breeding is supported by large-scale breeding programmes in many countries, and where skilled technical personnel are available as permanent staff members, could the large-scale efforts needed to realize the potential gains from this approach be undertaken.

RFLP mapping techniques promise to improve the efficiency of wide-crossing programmes by enabling breeders to detect chromosome segments introgressed from wild species in early generations and to correlate introgressed segments with valuable traits. A model system for wide-crossing programmes is available in the material produced at IRRI in the wide-cross programme instituted by Khush. In one part of the programme, crosses were made between *O. officinalis*, a CC genome wild species, and various breeding lines of *O. sativa*. A large array of introgression lines were produced, and many characters have apparently been introgressed from the wild species, including valuable traits such as resistance to brown planthoppers and bacterial blight (Jena & Khush, 1990). Introgressed segments in many of these lines have been precisely located by RFLP analysis, giving the most complete picture of introgression available for any wide-cross programme (Fig. 10.1).

For those wild species of rice which cannot be crossed to cultivated rice with enough efficiency to justify wide-cross programmes, an alternative approach is necessary. This germplasm will be a valuable source

Fig. 10.1. RFLP map of *Oryza sativa* breeding lines showing the
location and approximate size of segments of *O. officinalis*
chromosomes which have been introgressed by wide crosses. RFLP
markers surveyed and genetic distances are shown for each
chromosome. Introgressed segments are indicated by boxes and
arrows. The map is a composite of 52 breeding lines; no lines
contained all the introgressed segments shown.

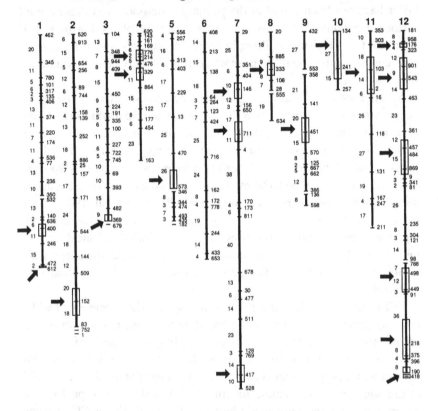

of genes for rice improvement, because many of these plants have
apparently evolved specific defences against some of the diseases and
insects that infest cultivated rice. The rice RFLP map can be transferred
to these wild species by intercrossing different accessions of the wild
species to each other to generate mapping populations. For example if a
wild species susceptible to an insect is crossed to another accession of
the wild species which is resistant, the genes for resistance can be mapped in the wild species. If closely linked RFLP markers can be mapped,
the genes of interest could then be cloned from the wild species by map-
based cloning methods and introgressed into cultivated rice by trans-

formation. This approach to cultivar improvement has never been utilized by plant breeders, who have always concentrated on trying to get crosses between wild species and cultivated plants. Thus no information is available concerning the map location of valuable genes in wild species. Before RFLP mapping and plant transformation techniques were developed, such an approach could not have been considered. Construction of genetic maps of wild species would have been too time-consuming and expensive, and if valuable genes had been mapped there would have been no way to utilize them in plant improvement. In actual fact, the gene pool available to rice breeders is now virtually unlimited. Any gene which can be cloned from any organism could potentially be transformed into rice and be expressed.

SPECIFIC AREAS FOR RESEARCH

Insect resistance
Additional sources of insect resistance are needed if rice production is to sustain the levels needed for the rapidly expanding populations in rice growing areas. Traditional approaches to this goal have utilized introgression of resistance genes from unimproved varieties and to a limited extent from compatible wild species. New approaches now possible include introduction of toxin genes cloned from *Bacillus thuringiensis* for protection against lepidopteran pests, and possible genes such as the cowpea trypsin inhibitor.

Virus resistance
Resistance to one viral disease of rice, grassy stunt virus, has been achieved by introgression from a wild species, *O. nivara*. However, many other virus diseases exist and progress has been slow in developing resistant varieties. The gene pool available for rice breeders does not appear to contain genes for complete resistance to tungro virus, for example, although genes which inhibit feeding by the insect vector are partially successful. New approaches to achieving virus-resistant plants include protection mediated by expression of viral coat protein genes in transgenic plants, the production of satellite RNA in those viruses for which this is appropriate, and the production of antisense RNA or oligonucleotide triple-helix agents (Baulcombe, 1989; Riordan & Martin, 1991). It is likely that rice will serve as a model for testing these agents in cereal crops because of the wealth of background information about these diseases.

Fungal and bacterial disease resistance

For these diseases, again much genetic and epidemiological information is available in rice, and resistance to many disease has been achieved by traditional plant breeding methods. However, for some pathogens such as *Rhizoctonia*, which causes sheath blight, no satisfactory resistance genes have been found, and a new approach is clearly warranted. Extensive study is being made of the natural defence mechanisms utilized by various plants to withstand attack by pathogens and insects. Upon challenge by a pathogen, a complex set of changes in plant metabolism occur, including synthesis of gene products such as the 'pathogenesis related proteins'. Further basic study of these responses will be necessary before genetic engineering techniques can be applied profitably.

Seed quality

Because rice has long been cultivated in small plots over a wide geographical area, there are a great number of varieties. Many of these differ greatly in grain quality, including such characters as grain shape, starch characteristics affecting cooking characteristics and texture, colour of the hull and aroma. These characteristics have become intricately intertwined with the culture of the rice growers and rice with the appropriate grain characters often commands a premium. Thus, despite the advances made in developing high-yielding varieties, local varieties are sometimes preferred. In Bali, for example, improved varieties developed at IRRI are grown for sale or export, but a local variety is grown for consumption by the producer or for sale at a premium (Eiseman, 1990). Therefore, there is much interest in developing high-yielding, semi-dwarf varieties with the various grain qualities favoured in specific geographical areas. In many areas where production of rice has (temporally, perhaps) caught up with population growth, this is the number one rice breeding priority.

Progress in improving the grain quality of improved rice varieties has been slow, however. Breeders have been attempting to transfer the grain quality of long-grained, aromatic basmati varieties into improved varieties for more than 20 years, but very little progress has been made (G. S. Khush, personal communication). This is a difficult problem for several reasons. Not much is known about the genetics of grain characters, but they appear to be quantitative in nature and to show large genotype interactions. RFLP analysis of the QTLs controlling these traits will provide information about the number and location of genes controlling these traits, and will provide a way to monitor introgression of the appropriate chromosome segments into improved varieties. Some progress has been made in identifying the compounds responsible for aroma in rice (Buttery, Turnbaugh & Ling, 1988). Basic research is needed on the chemical pathways involved in producing these compounds. Perhaps the appropriate pathways can be constructed by transformation, once the appropriate structural genes and regulatory elements are identified.

Range extension

More rice could be grown if its range could be extended to cooler, drier, or saltier environments. Again, major breakthroughs in these areas are difficult using conventional breeding, either because the traits are quantitative or because genes for the desired traits are simply not available in compatible germplasm. Selection for most of these traits in tissue culture has failed to produce stable regenerants with the desired qualities (Oono, 1984). There is not sufficient basic information available about how such characters are controlled in any plants to suggest failsafe methods of attacking these problems by gene cloning and transformation techniques. However, rice and related plants are good candidates for the study of these problems. For example, a related plant *Porteresia coarctica* is highly salt-tolerant (Finch, Slamet & Cocking, 1990). Study of the mechanisms whereby *Porteresia* achieves its salt tolerance will be extremely useful to the development of methods for the production of salt-tolerant rice.

Factory plants

Many rice growers are subsistence farmers who desperately need some sort of high-value cash crop which can be grown on their small holdings. Several studies have shown that it is possible to produce antibodies or valuable peptides in transgenic plants (Hiatt, Cafferkey & Bowdish, 1989; Krebbers & Vandekerckhove, 1990). Organizers of rice breeding programmes would be well advised to begin to develop ways to produce compounds in rice which would give a value-added component.

Summary

Rice is the most important human food crop, and an extensive amount of information is available about rice genetics, systematics, diseases, chemical, physical and ecological properties, and relationship to human culture and history. Rice has the richest and best-conserved germplasm base of any crop. Rice research is carried out in an array of university and government supported research organizations, with generous funding from international foundations. Perhaps because it is less important as a crop in the industrialized countries of Western Europe and the United States, most of the research results and experimental materials are in the public domain, and communication is good between research groups.

Rice also has many attributes needed to serve as a model organism for studies of molecular genetics and molecular biology. It has a small genome, is diploid, and extensive conventional genetic maps and molecular maps have been developed. Rice can be transformed and regenerated, so that the many research protocols which require transgenic plants can be used. The wealth of breeding lines and the background knowledge which has been developed about rice provide the raw material for rapid progress in molecular biology in rice. One of humankind's oldest plant companions can be confidently expected to achieve even greater importance in the future.

References
Barbier, P. & Ishihama, A. (1990). Variation in the nucleotide sequence of a Prolamin gene family in wild rice. *Plant Molecular Biology*, **15**, 191–195.

Baulcombe, D. (1989). Strategies for virus resistance in plants. *Trends in Genetics*, **5**, 56–60.

Beckmann, J. S. & Soller, J. M. (1990). Toward a unified approach to genetic mapping in eukaryotes based on sequence tagged microsatellite sites. *Bio/Technology*, **8**, 930–932.

Bennett, M. D. & Smith, J. B. (1976). Nulear DNA amounts in angiosperms. *Proceedings of the Royal Society of London Series B*, **274**, 227–274.

Bonierbale, M. W., Plaisted, R. L. & Tanksley, S. D. (1988). RFLP maps based on a common set of clones reveal modes of chromosomal evolution in potato and tomato. *Genetics*, **120**, 1095–1103.

Brar, D. S. & Khush, G. S. (1986). Wide hybridization and chromosome manipulation in cereals. In *Handbook of plant cell culture*, vol. 4, ed. D. A. Evans, W. R. Sharp & P. V. Ammirato, pp. 221–263. New York: Macmillan Publishing Co.

Burke, D. T. (1990). YAC cloning – options and problems. *Genetic Analysis – Techniques and Applications*, **7**, 94–99.

Buttery, R. G., Turnbaugh, J. G. & Ling, L. C. (1988). Contribution of volatiles to rice aroma. *Journal of Agricultural Food Chemistry*, **36**, 1006–1009.

Chang, T.-T. (1976a). The origin, evolution, cultivation, dissemination, and diversification of Asian and African rices. *Euphytica*, **25**, 425–441.

Chang, T.-T. (1976b). Rice. In *Evolution of crop plants*, ed. N. W. Simmonds, pp. 98–104. New York: Longman Press.

Chang, T.-T. (1984). Conservation of rice genetic resources: luxury or necessity. *Science*, **224**, 251–256.

Coulson, A., Waterston, R., Kiff, J., Sulston, J. & Kohara, Y. (1988). Genome linking with yeast artificial chromosomes. *Nature*, **335**, 184–186.

Datta, S. K., Peterhans, A., Datta, K. & Potrykus, I. (1990). Genetically engineered fertile *indica*-rice recovered from protoplasts. *Bio/Technology*, **8**, 736–740.

Eiseman, Jr., F. B. (1990). *Bali: Sekala & Niskala*, vol. 2. Berkeley: Periplus Editions.

Ellis, R. P. (1986). A review of comparative leaf blade anatomy in the systematics of the Poaceae: the past twenty-five years. In: *Grass systematics and evolution*, ed. T. R. Soderstrom, K. W. Hilu, C. S. Campbell & M. E. Barkworth, pp. 3–10. Washington: Smithsonian Institution Press.

FAO (1985). *Production yearbook*. Rome: Food and Agricultural Organization.

Feng, X.-H. & Kung, S.-D. (1991). Diversity of the protein kinase gene family in rice. *FEBS Letters*, **1**, 98–102.

Finch, R. P., Slamet, I. H. & Cocking, E. C. (1990). Production of heterokaryons by the fusion of mesophyll protoplasts of *Porteresia coarctata* and cell suspension-derived protoplasts of *Oryza sativa* – a new approach to somatic hybridization in rice. *Journal of Plant Physiology*, **136**, 592–598.

Glaszmann, J. C. (1988). Geographical pattern of variation among Asian native rice cultivars (*Oryza sativa* L.) based on fifteen isozyme loci. *Genome*, **30**, 782–792.

Glaszmann, J. C. (1989). Isozymes and classification of Asian rice varieties. *Theoretical and Applied Genetics*, **74**, 21–30.

Glaszmann, J. C. (1990). A varietal classification of Asian cultivated rice (*Oryza sativa* L.) based on isozyme polymorphism. In *Rice genetics*, pp. 83–90. Manila, Philippines: IRRI.

Gould, F. W. & Shaw, R. B. (1983). *Grass systematics*, 2nd edn. College Station, TX: Texas A & M Press.

Hiatt, A., Cafferkey, R. & Bowdish, K. (1989). Production of antibodies in transgenic plants. *Nature*, **342**, 76–78.

Hiratsuka, J., Shimada, H., Whittier, R., Ishibashi, T., Sakamoto, M., Mori, M., Kondo, C., Honji, Y., Sun, C.-R., Meng, B.-Y., Li, Y.-Q., Kanno, A., Nishizawa, Y., Hirai, A., Shinozaki, K. & Sugiura, M. (1989). The complete sequence of the rice (*Oryza sativa*) chloroplast genome: intermolecular recombination between distinct tRNA genes accounts for a major plastid DNA inversion during the evolution of the cereals. *Molecular and General Genetics*, **217**, 185–194.

Hitchcock, A. S. (1971). *Manual of the grasses of the United States*, 2nd edn. New York: Dover.

Hodges, T. K., Peng, J., Lee, L. & Koetze, D. S. (1990). *In vitro* culture of rice: transformation and regeneration of protoplasts. In *Gene manipulation in plant improvement II*, ed. J. P. Gustafson, pp. 163–183. New York: Plenum Press.

Huang, J. K., Wen, L., Swegle, M., Tran, H. C., Thin, T. H., Naylor, H. M., Muthukrishnan, S. & Reeck, G. R. (1991). Nucleotide sequence of a rice genomic clone that encodes a class-I endochitinase. *Plant Molecular Biology*, **16**, 479–480.

Huang, N., Koizumi, N., Reinl, S. & Rodriguez, R. L. (1990). Structural organization and differential expression of rice alpha-amylase genes. *Nucleic Acids Research*, **18**, 7007–7014.

Izawa, T., Miyazaki, C., Yamamoto, M., Terada, R., Iida, S. & Shimamoto, K. (1991). Introduction and transposition of the maize transposable element *Ac* in rice (*Oryza sativa* L.). *Molecular and General Genetics* (in press).

Jeffreys, A. J., Wilson, V. & Thein, S. L. (1985). Hypervariable 'minisatellite' regions in human DNA. *Nature*, **314**, 67–73.

Jena, K. K. & Khush, G. S. (1989). Monosomic alien addition lines of rice: production, morphology, cytology, and breeding behavior. *Genome*, **32**, 449–455.

Jena, K. K. & Khush, G. S. (1990). Introgression of genes from *Oryza officinalis* Well *ex* Watt to cultivated rice, *O. sativa* L. *Theoretical and Applied Genetics*, **80**, 737–745.

Jena, K. K. & Kochert, G. (1991). Restriction fragment length polymorphism analysis of CCDD genome species of the genus *Oryza* L. *Plant Molecular Biology*, **16**, 831–839.

Kao, T.-H., Moon, E. & Wu, R. (1984). Cytochrome oxidase subunit II gene of rice has an insertion sequence within the intron. *Nucleic Acids Research*, **12**, 7305–7315.

Kay, S. A., Keith, B., Shinozaki, K. & Chua, N.-H. (1989). The sequence of the rice phytochrome gene. *Nucleic Acids Research*, **17**, 2865–2866.

Kemmerer, E. C., Lei, M. & Wu, R. (1991). Isolation and molecular evolutionary analysis of a cytochrome c gene from *Oryza sativa* (rice). *Molecular Biology and Evolution*, **8**, 212–226.

Khush, G. S. (1986). Relationships between linkage groups and cytologically identifiable chromosomes of rice. In *Rice Genetics*, pp. 239–248. Manila: International Rice Research Institute.

Khush, G. S. (1987). Rice breeding: past, present and future. *Journal of Genetics*, **66**, 195–216.

Khush, G. S., Singh, R. J., Sur, S. C. & Librojo, A. L. (1984). Primary trisomics of rice: origin, morphology, cytology, and use in linkage mapping. *Genetics*, **107**, 141–163.

Kim, W. T. & Okita, T. W. (1988). Structure, expression, and heterogeneity of the rice seed prolamines. *Plant Physiology*, **88**, 649–655.

Kothari, S. L., Davey, M. R., Lynch, P. T., Finch, R. P. & Cocking, E. C. (1991). Transgenic rice. In *Transgenic plants*, ed. S. D. Kung & R. Wu. London: Butterworths (in press).

Krebbers, E. & Vandekerckhove, J. (1990). Production of peptides in plant seeds. *Trends in Biochemistry*, **8**, 1–3.

Kurata, N. (1986). Chromosome analysis of mitosis and meiosis in rice. In *Rice genetics*, pp. 143–152. Manila: International Rice Research Institute.

Lazo, G. R., Roffey, R. & Gabriel, D. W. (1987). Pathovars of *Xanthomonas campestres* are distinguishable by restriction fragment length polymorphisms. *International Journal of Systematics Bacteriology*, **37**, 214–221.

Lei, M. & Wu, R. (1991). A novel glycine-rich cell wall protein gene in rice. *Plant Molecular Biology*, **16**, 187–198.

Li, L.-C., Chen, Y.-M. & Chen, Y. (1990). Studies on protoplast culture of rice

(*Oryza sativa* L.) and plant regeneration from protoplast-derived calli. *Chinese Journal of Genetics*, **15**, 333–341.

Martin, G. B., Williams, J. G. K. & Tanksley, S. D. (1991). Rapid identification of markers linked to a *Pseudomonas* resistance gene in tomato by using random primers and near-isogenic lines. *Proceedings of the National Academy of Sciences, USA*, **88**, 2136–2140.

Masumura, T., Kidzu, K., Sugiyama, Y., Mitsukawa, N., Hibino, T., Tanaka, K. & Fujii, S. (1989). Nucleotide sequence of a cDNA encoding a major rice glutelin. *Plant Molecular Biology*, **12**, 723–725.

McCouch, S. R., Kochert, G., Yu, Z. H., Wang, Z. Y., Khush, G. S., Coffman, W. R. & Tanksley, S. D. (1988). Molecular mapping of rice chromosomes. *Theoretical and Applied Genetics*, **76**, 148–149.

McCouch, S. R., Causse, M., Fulton, T. & Tanksley, S. D. (1991). Molecular mapping of the rice genome: recent advances. In *Rice genetics II*. Manila, Philippines (in press).

Meijer, E. G. M., Schilperoort, R. A., Rueb, S., van Os-Ruygrok, P. E. & Hensgens, L. A. M. (1991). Transgenic rice cell lines and plants: expression of transferred chimeric genes. *Plant Molecular Biology* (in press).

Minami, E., Ozeki, Y., Matsuoka, M., Koizuka, N. & Tanaka, Y. (1989). Structure and some characterization of the gene for phenylalanine-ammonia-lyase from rice plants. *European Journal of Biochemistry*, **185**, 19–25.

Nakamura, Y., Leppert, M. & O'Connell, P. (1987). Variable number of tandem repeat (VNTR) genetic markers for human gene mapping. *Science*, **235**, 1616–1622.

Nayar, N. M. (1973). Origin and cytogenetics of rice. *Advances in Genetics*, **17**, 153–292.

Oono, K. (1984). Tissue culture and genetic engineering in rice. In *Biology of rice*, ed. S. Tsanoda & N. Takahashi, pp. 339–358. Japan: Japan Science Society Press.

Ou, S. H. (1985). *Rice diseases*. United Kingdom: Commonwealth Agricultural Bureaux.

Patterson, A. H., Damon, S., Hewitt, J. D., Zamir, D., Rabinowitch, H. D., Lincoln, S. E., Lander, E. S. & Tanksley, S. D. (1991). Mendelian factors underlying quantitative traits in tomato – comparison across species, generations, and environments. *Genetics*, **127**, 181–197.

Potrykus, I. (1990). Gene transfer to cereals: an assessment. *Bio/Technology*, **8**, 535–542.

Reece, K. S., McElroy, D. & Wu, R. (1990). Genomic nucleotide sequence of 4 rice (*Oryza sativa*) actin genes. *Plant Molecular Biology*, **14**, 621–624.

Riordan, M. L. & Martin, J. C. (1991). Oligonucleotide-based therapeutics. *Nature*, **350**, 442–443.

Rommens, J. M., Iannuzzi, M. C., Kerem, B.-S., Drumm, M. L., Melmer, G., Dean, M., Rozmahel, R., Cole, J. L., Kennedy, D., Hidaka, N., Zsiga, M., Buchwald, M., Riordan, J. R., Tsui, L.-C. & Collins, F. S. (1989). Identification of the cystic fibrosis gene: chromosome walking and jumping. *Science*, **245**, 1059–1065.

Sakamoto, M., Sanada, Y., Tagiri, A., Murakami, T., Ohashi, Y. & Matsuoka, M. (1991). Structure and characterization of a gene for light-harvesting Chl a/b binding protein from rice. *Plant Cell Physiology*, **32**, 385–393.

Schlessinger, D. (1990). Yeast artificial chromosomes – tools for mapping and analysis of complex genomes. *Trends in Genetics*, **6**, 248–252.

Sears, E. R. (1976). Genetic control of chromosome pairing in wheat. *Annual Review of Genetics*, **10**, 31–52.

Second, G. (1982). Origin of the genic diversity of cultivated rice (*Oryza spp.*): study of the polymorphism scored at 40 isozyme loci. *Japanese Journal of Genetics*, **57**, 25–57.

Second, G. (1985). Geographic origins, genetic diversity, and the molecular clock hypothesis in the *Oryzeae*. In *Genetic differentiation and dispersal in plants*, ed. P. Jacquard, G. Heim & J. Antonovics, pp. 41–56. Berlin: Springer-Verlag.

Shimamoto, K., Terada, R., Izawa, T. & Fujimoto, H. (1989). Fertile transgenic rice plants regenerated from transformed protoplasts. *Nature*, **338**, 274–276.

Sitch, L. A. (1990). Incompatibility barriers operating in crosses of *Oryza sativa* with related species and genera. In *Gene manipulation in plant improvement II*, ed. J. P. Gustafson, pp. 77–93. New York: Plenum Press.

Suh, S.-C., Kim, H.-I., Lee, Y.-H. & Chung, T.-Y. (1990). Plant regeneration from the transformed rice protoplasts. *Korean Society of Plant Tissue Culture*, **17**, 141–150.

Takaiwa, F., Kikuchi, S. & Oono, K. (1989). The complete nucleotide sequence of new type cDNA coding for rice storage protein glutelin. *Nucleic Acids Research*, **17**, 3289.

Takaiwa, F., Kikuchi, S. & Oono, K. (1990). The complete nucleotide sequence of the intergenic spacer between 25S and 17S rDNAs in rice. *Plant Molecular Biology*, **15**, 933–935.

Takaiwa, F., Oono, K. & Sugiura, M. (1984). The complete nucleotide sequence of a rice 17S rRNA gene. *Nucleic Acids Research*, **12**, 5441–5448.

Tanksley, S. D., Bernatzky R., Lapitan, N. L. & Prince, J. P. (1988). Conservation of gene repertoire but not gene order in pepper and tomato. *Proceedings of the National Academy of Sciences, USA*, **85**, 6419–6423.

Tanksley, S. D., Young, N. D., Patterson, A. H. & Bonierbale, M. W. (1989). RFLP mapping in plant breeding: new tools for an old science. *Bio/Technology*, **7**, 257–263.

Toriyama, K., Arimoto, Y., Uchimiya, H. & Hinata, K. (1988). Transgenic rice plants after direct gene transfer into protoplasts. *Bio/Technology*, **6**, 1072–1074.

Vergnaud, G., Mariat, D., Zoroastro, M. & Lautheir, V. (1991). Detection of single and multiple polymorphic loci by synthetic tandem repeats of short oligonucleotides. *Electrophoresis*, **12**, 134–140.

Vilgalys, R. & Gonzalez, D. (1990). Ribosomal DNA restriction fragment length polymorphisms in *Rhizoctonia solani*. *Phytopathology*, **80**, 151–158.

Wang, Z. Y. & Tanksley, S. D. (1989). Restriction fragment length polymorphism in *Oryza sativa* L. *Genome*, **32**, 1113–1118.

Wang, Z. Y., Wu, Z. L., Xing, Y. Y., Zheng, F. G., Guo, X. L., Zhang, W. G. & Hong, M. M. & Hong, M. M. (1990). Nucleotide sequence of rice waxy gene. *Nucleic Acids Research*, **18**, 5898.

Watson, L., Clifford, H. T. & Dallwitz, M. J. (1985). The classification of Poaceae: subfamilies and supertribes. *Australian Journal of Botany*, **33**, 433–484.

Wu, R., Kemmerer, E. & McElroy, D. (1990). Transformation and regeneration of important crop plants: rice as the model system for monocots. In *Gene manipulation in plant improvement II*, ed. J. P. Gustafson, pp. 251–263. New York: Plenum Press.

Wu, S.-C., Végh, Z., Wang, X.-M., Tan, C.-C. & Dudits, D. (1989). The nucleotide sequences of two rice histone H3 genes. *Nucleic Acids Research*, **17**, 3297.

Young, N. D., Zamir, D., Ganal, M. W. & Tanksley, S. D. (1988). Use of isogenic lines and simultaneous probing to identify DNA markers tightly linked to the *Tm-2a* gene in tomato. *Genetics*, **120**, 579–585.

Xie, Y. & Wu, R. (1989). Rice alcohol dehydrogenase genes: anaerobic induction, organ specific expression and characterization of cDNA clones. *Plant Molecular Biology*, **13**, 53–68.

Zhang, J.-L., Lou, X.-M., Cai, R.-Z., Huang, R.-X. & Hong, M.-M. (1991). Transposition of maize transposable element activator in rice. *Plant Science*, **73**, 191–198.

Zhu, Q. & Lamb, C. J. (1991). Isolation and characterization of a rice gene encoding a basic chitinase. *Molecular and General Genetics*, **226**, 289–296.

11

Domestication and its changing agenda

G. P. Chapman

Originally 'domestication' meant being brought within the household economy but as Darwin recognized, this redirects selection and new biotypes come to prominence. Breeding, whether of plant or animal, was to shift the emphasis still further since selection was applied to both parents. With the advent of breeding and the newer genetic techniques, domestication can be redefined as 'bringing within the household of science'. Although the changes domestication has brought about in grasses can seem impressive, genetically the links to wild grasses are very close, since, relative to the history of the Poaceae, the separation of our crops is extremely recent and, were humankind to disappear, so too would the cereals in the domesticated form we know them. Beyond our food needs, we require, increasingly, to use other grasses and thus to bring them within scientific enquiry.

An evolutionary resumé

The closest families to the Poaceae (Anarthriaceae, Centrolepidaceae, Ecdeiocoleaceae, Flagellariaceae, Joinvilleaceae and Restionaceae) are not particularly close and as Watson (1990) succinctly remarked 'it is very unlikely, therefore, that there are disguised Poaceae currently resident in any other family'. We might assume that grasses came to prominence during the 70 000 000 years since the end of the Cretaceous period but how old is the grass family? While the subfamilies through their tribes are distributed throughout the world, their genera are more localized to continents. Genera are described by Clayton & Renvoize (1986) as 'poor travellers'. The implication is that the distribution of already divergent grasses pre-dates, and generic diversification post-dates, the major episodes of continental drift. Evidence reviewed by Thomason (1987) reveals the equivalents or near-equivalents of *Festuca*, *Nassella*, *Panicum*, *Setaria* and *Stipa* in Tertiary times. Except for *Nassella* all are large genera from which many derivative genera are thought to have arisen. Among the grass subfamilies, Renvoize & Clayton (chapter 1) regard, in Arundinoideae, the Arundineae as representing the declining remnants of the earliest grasses.

From such an elusive origin, grasses have diverged so that we can identify in the five or six subfamilies broad themes of ecological adaptation: bamboos in tropical and subtropical forest, the pooids within the temperate zones and so on. Within this framework, Clayton & Renvoize (1986) have proposed a classification expressly phylogenetic, where some large genera, such as *Panicum* and *Poa*, for example, are seen as sources from which other genera are derived. For a discussion, see the review by Chapman (1990), and Renvoize & Clayton (Appendix).

Against this background has been a remarkable diversification of that most fundamental process, photosynthesis. Hattersley & Watson (chapter 2) has examined the extent to which a derivative series could be proposed. A second noteworthy feature is the *S–Z* incompatibility system, for which there are no convincing antecedents known and which therefore serves to emphasize the distinctiveness of grasses.

Apomixis (Chapman, chapter 4) has a wide distribution through the grasses and in the more venturesome grasses perhaps serves as much to sustain these as it assists stagnation among relict arundinoids.

Despite our evident uncertainty about the origin and age of the grass family, it is appropriate to emphasize how recent is the advent of domestication. While our cereals may have evolved over about 10 000 years, Manglesdorf, Macneish & Galinat (1964) implied that even so distinctive a crop as maize may have acquired its characteristic features in perhaps half that time. Of the 10 millennia that comprise our agricultural period, systematic exploitation of Mendelian genetics has taken place for less than a century, and any molecular understanding is, for crop plants, scarcely of two decades duration.

Unconscious selection?

So remarkable was the Neolithic achievement, which invented farming, that we should pause to reflect upon it. Darwin (1868) assumed that the beginnings of domestication involved 'unconscious' selection. Plants different from their wild ancestors came to provide a more convenient, regenerating food supply. Eventually, upon such a basis, settled societies began to flourish and demonstrate artistry and inventiveness. Why then should we assume that such inherent intellectual capacity had a blind spot about so crucial a resource as its food supply? Did the organization of a food supply involving cultivation arise out of a collective unconscious or did gifted individuals set an example? We have no means of knowing but surely we must hesitate to apply 'unconscious' to a selection process, the results of which have proved so durable and far-reaching. Darwin (1868) was in fact more circumspect, since he provided many examples of conscious selection by primitive peoples and expressly states that the boundary between conscious and unconscious selection is indistinct. Recently, Harlan (1989) stressed the detailed awareness of plants that hunter–gatherers have (and by implication had at the time of the origin of agriculture).

Quite frequently, one finds men and women having little or no formal training but with an inherent ability to grow or propagate 'elite' plants. There is no reason to suppose this situation has not existed indefinitely. Again, it is a commonplace among extension workers that new ideas will not be taken up generally, unless well-regarded farmers first adopt them. And who could fail to notice the minute interest farmers take in their neighbours' successes and failures?

Darlington (1969) automatically accepted Darwinian unconscious selection but, curiously, developed at some length the idea that the cereal farmer would be rewarded for thoughtful diligence and penalized by indolence and mismanagement. Zohary (1969) described in some detail the difference in appearance of brittle rachis and non-shattering wheat, and it is sufficiently striking to make one question whether the occasional perceptive individual might not have become aware of it and even pondered its significance.

It is impossible for us to conjecture the intellectual environment of these early societies. To create anything at all under such circumstances is, perhaps, astonishing. It is therefore precarious to draw on ideas much nearer our own time to help us to understand the origins of agriculture. Even so, it is worth considering the fathers of corn so sympathetically described by Wallace & Brown (1956). These, often self-contained, individuals, in pioneer America brought long-term commitment, curiosity and quiet enthusiasm to the improvement of their chosen crop. Their genetic understanding was only what they themselves could sense of the situation and of experimental training they had none. Yet, they could distinguish the better from the good and have their selections adopted by their neighbours. Who is to say that there are *no* similarities here with the first farmers? Purseglove (1968) was more cautious than Darlington and wrote of 'conscious or unconscious' selection. In those early days, then, why not credit that the best farmers grasped something of what they were doing?

However few 'nuclear' centres there were when cereal culture began, other societies were ready to accept what had been invented. Its ramifications eventually permeated society and involved not only merchants but politicians, philosophers and poets, quite apart from ordinary folk who, whatever their other privations, declined to forego bread. More cereals came into cultivation and landraces multiplied. Purseglove (1972) has described for example how the migrating Cushites are thought to have domesticated sorghum in Ethiopia when climate precluded them from growing the wheat with which they had been familiar further north.

A wider domestication

In this volume, Harlan (chapter 5) surveys domesticated grasses, de Wet (chapter 6) takes a world view of cereals and Davies & Hillman (chapter 7) concentrate on a few cereals to explore in greater detail the genetic processes underlying domestication. What emerges from these three chapters is the importance of wheat, maize and rice. They are the widespread and sought after cereals and seemingly they are somehow 'prepotent'. Whatever may be the prospects for all other cereals, for the foreseeable future they will be subordinate in importance to the major three. Other cereals are used where none of the major three will grow or, as in the case of barley, where there is a specialized requirement for brewing or for other crops, or to spread risk or where there is some deeply held food preference. Official Ethiopian policy was, for a time, to disregard t'ef, *Eragrostis tef*, since it had a negligible modern technology available, in favour of wheat, maize and barley. The local food preference has prevailed, however, resulting in a change of policy and a sustained effort to modernize the t'ef crop.

Belatedly, the importance of the fonios, *Digitaria exilis* and *D. iburua*, has been recognized in Africa and for different (gourmet) reasons so has wild rice, *Zizania aquatica*, in North America. These developments highlight a problem for the plant breeder. Clearly, with an enlarging world population, it is unwise to neglect the minor cereals but, even with our best efforts, what can be expected of them? As de Wet (1992) has shown, the prospects for pearl millet are highly encouraging but what is the judicious commitment to *Coix lachryma-jobi*, *Echinochloa crus-galli* or *Panicum miliaceum*? Is any of them a 'maize-in-waiting' or did the Neolithic farmers and their canny descendants make the right choices? Are there to be *no* surprises?

Beyond the cereals

There is a further widening of domestication for which the primary impetus is ecological decline. Numerous examples are available. Mineral spoil in industrialized areas led to the recognition of heavy metal tolerance in *Agrostis* (McNeilly, 1968), *Anthoxanthum* (McNeilly & Antonovics, 1968) and *Arrhenatherum* (Richards, 1990). Kernick (1990) stressed the need for greater interest in genera useful for desert stabilization and pasture improvement under such conditions, among them *Cenchrus*, *Chrysopogon*, *Dichanthium*, *Eleusine*, *Enneapogon*, *Lasiurus*, *Leymus*, *Oryzopsis*, *Poa*, *Sporobolus* and *Stipa*. *Miscanthus*, hitherto regarded as an ornamental, is a C_4 plant, hardy in Europe, perhaps with a future as a source of biomass from which to prepare fuel.

One curious example is *Vetiveria zizanioides*. For many years this grass has been recognized as a superb natural barrier to prevent soil erosion and create a self-terracing system in the wet tropics. The present author was familiar with it in Jamaica from 1961 onwards where it was and had been for many years a common ingredient in the management of steep hillsides. Around 1985 it was discovered by the World Bank and the United States National Research Council, both of which brought their considerable resources to popularize its use around the tropics (ASTAG, 1991; ASTAG VPN, 1991). Recently, an attempt has been made to utilize *Achnatherum (Stipa) splendens* to control spectacular rates of erosion in the Loess Plateau region of China (Grimshaw, 1991) in an area too inhospitable for *Vetiveria*.

Among crops of any consequence as cereals, apomixis is virtually unknown. As pasture grasses, several genera, *Brachiaria*, *Cenchrus*, *Chloris*, *Dichanthium*, *Panicum*, *Paspalum* and *Pennisetum*, include apomicts that are now domesticates requiring a specialized breeding technology.

A major cause of ecological decline is salt poisoning of land (salinization) and here grasses, hitherto unfamiliar, can have a beneficial effect. So important is salinization that it, together with its consequences for the agrostologist and plant breeder, will now be considered.

Salinization: alternative responses

Salinization is sometimes natural but often is attributable to agriculture. One estimate (IIEDWRI, 1987) suggests that 25–40% of the

220+ million hectares of irrigated land is subject to salinization. It is a contributory cause of the wider problem of desertification. Much of the Third World's increase in food production depends upon irrigation but, if done carelessly, salt concentration rises, productivity declines and the land may be abandoned in a situation where, through increasing population, fresh land is increasingly difficult to find. The problem is not new and the historical lessons are there for anyone prudent enough to learn them. Of all responses, surprise is the least excusable. See, for example, Jacobsen & Adams (1958), where details are quoted for salinization dating from 3500 BC to 1700 BC in southern Iraq. In this period, wheat began apparently as equal in importance to barley, the more salt tolerant crop, but in the end was abandoned. During this time, yields overall declined by two thirds. For a further account see the article by Jacobsen (1982).

THE BREEDER'S RESPONSE

Breeding for salt tolerance has been shown to be possible in *Triticum* (Gorham, 1990) for example. Lazzeri, Kollmorgen & Lörz (1990) have explored how far salt tolerance in cereals can be sought via cell culture methods, although so far the results are disappointing. Such successes as there have been from conventional breeding have sometimes induced in breeders and their sponsors the idea that because it can be done it should be done. So crucial is this issue likely to become that its rationale must be examined.

Tanton (1991) has identified four causes of salinization:

1 Poor quality irrigation water.
2 Insufficient water for leaching.
3 Impeded drainage.
4 High water table.

Theoretically, provided sufficient resources are available, each of these problems could be remedied but it is realistic to recognize that only in the first case is there a situation *unlikely to deteriorate*. Here breeding for salt tolerance could be feasible; otherwise the breeders are committed to situations where gains in salt tolerance could be progressively undermined by rising salt concentration. To allow such a situation to develop is to misuse the breeder. It would be logical (and preferable) instead to challenge the resourcefulness of the irrigation engineer.

In regard to the second of the above causes, Carruthers (1991) has pointed out that Pakistan has sufficient water for *half* its land. Politi-

cally, the situation turns on much of the land having about half the water
it needs. What is a political problem should not be handed to the
breeder disguised as a technical one.

REHABILITATION

So vast is the area subject to salinization that a shift of emphasis
is required toward techniques for rehabilitation. No single technology is
sufficient. What is needed is a carefully integrated process involving
irrigation engineers, agronomists and agrostologists. The role of the last
is especially interesting here.

Many grasses are known to possess salt glands in their leaves and
when grown in saline conditions accumulate a deposit of salt on their
leaf surfaces (for a review, see Gorham, 1992). If a mixture of (say)
potassium and sodium chlorides were supplied to such plants, sodium
chloride would be preferentially deposited. Removal of such salt-laden
foliage effectively removes salt from the supporting soil. By such means
Leptochloa fusca has been used to lower soil salt levels in Pakistan
(Islam-ul-Haq & Khan, 1971) and in India (Rana & Parkash, 1980).

In the Hoxhi Corridor region of the Gobi Desert the soil is naturally
salty. A combination of high quality irrigation water (from melted
snow) and a three-year cultivation of *Puccinellia chinampoensis* re-
habilitated the land sufficiently to grow wheat and sugar beet (Ren
Jizhou, 1992). *P. chinampoensis* lacks salt glands and excludes salt at the
root and thus functions differently from *Leptochloa*.

Technology to retrieve salinized soils at lower latitudes is at an early
stage and is considered further in the next section.

Chloridoideae

This subfamily contains only one significant cereal, namely t'ef,
but it is a group likely to increase in importance for other reasons.
Ecologically, two themes dominate. The first is adaptation to dry
habitats in low latitudes and the second is that many genera are salt
tolerant and in some cases this is allied to the ability to withstand high
pH. In soil rehabilitation, therefore, they are potentially useful in both
saline and sodic (alkaline) situations. Salt tolerance occurs here and
there throughout the Poaceae but the Chloridoideae are of topical interest, since they are adapted to regions where salinity is a problem and
because, as yet, they are poorly explored from this viewpoint. Table
11.1 sets out, for the Poaceae, by subfamily, on a world basis, currently
known salt-tolerant grasses.

Fig. 11.1. Salt glands of *Cynodon dactylon* (Bermuda grass). (a) SEM photograph showing distribution of salt glands on the upper leaf surface (Scale bar 50 μm). (b) A salt gland showing cap cell and basal cell (Scale bar 5 μm). (c) A salt gland showing secreted salt on the cap cell (Scale bar 5 μm). Photos courtesy of W. Worku.

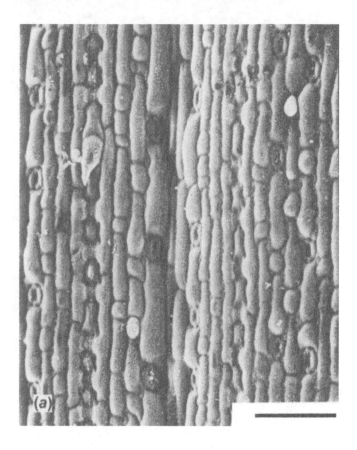

Fig 11.1 cont

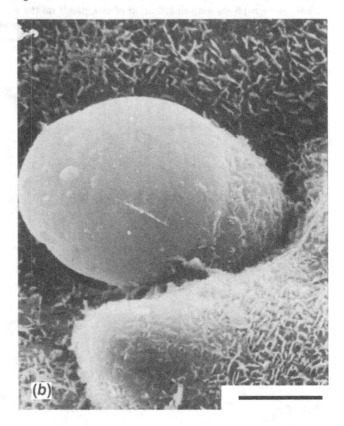

(b)

Fig 11.1 cont

Table 11.1. *Salt-tolerant grasses*

Bambusoideae	
Ehrharta calycira†	*Porteresia coarctica*

Pooideae

Agropyron cristatus	*Parapholis incurva, P. strigosa*
Ammophila arenaria	*Phalaris minor*
Arrhenatherum elatius†	*Piptochaetium napotense*†
Bromus repens, B. rubens†	*Poa barrosiana, P. bulbosa, P. lagunisa*
Cutandia memphitica†	*Puccinellia chinampoensis, P. ciliata,*
Elymus angustus, E. elongatus†,	*P. distans*†, *P. testuciformis,*
E. farctus, E. striatus, E. pungens	*P. maritima, P. peisonis*
Fetuca arundinacea†, *F. pratensis*	*Rostraria pumila*
Hainardia cylindrica	*Stipa tenuis*†
Hordeum bogdani, H. depressum,	*Trisetum flavescens*
H. patagonicum, H. secalinum,	
H. vulgare	

Arundinoideae

Aristida mutabilis†	*Phragmites australis*†
Arundo donax	*Schismus barbatus*
Danthionia richardsonii†	*Stipagrostis pernata*

Chloridoideae

Aeluropus lagopeides, A. littoralis†	*D. scoparia, D. spicata*†
Allolepis texana	*Eleusine indica*
Astrebla lappacea	*Enneapogon avenaceus*†
Bouteloua curtipendula, gracilis†	*Enteropogon acicularis*†,
Buchloe dactyloides†	*E. macrostachys*
Chondrosum brevisetum†	*Eragrostis australasica, E. curvula*†,
Chloris acicularis, C. gayana,	*E. obtusiflora, E. xerochila*
C. scariosa, C. virgata	*Jouvea pilosa*
Coelachyrum brevifolium	*Leptochloa fusca*†
Cynodon dactylon†	*Lepturus repens*
Dactyloctenium aegyptium†	*Reederochloa eludens*
D. radulans†	*Spartina coarctica, townsendii*
Desmostachya bipinnata	*Sporobolus arenarius*
Dinebra retroflexa	*Tetrachne dregei*
Distichlis humilis†, *D. palmeri,*	*Tetrapogon spathulata*
	Triodia longiceps†

Panicoideae

Andropogon gayanus†	*Echinochloa colona, E. crus-galli,*
Bothriochloa insculpta†, *B. pertusa*	*E. stagnina, E. turneriana*
Cenchrus biflorus†, *C. ciliaris*†,	*Imperata cylindrica*
C. pennisetiformis†, *C. prieurii*†,	*Lasiurus scindicus*†
C. setigerus†	*Panicum pinifolium*†, *P. turgidum*†,
Chrysopogon fulvus†, *C. gryllus*†,	*P. virgatum*
C. plumulosus†	*Paspalum scrobiculatum, P. virgatum*

Table 11.1. (*cont.*)

Dichanthium annulatum†,	*Setaria kangerensis*
D. fecundum†	*Sorghum* × *alum*†, *S. halepense*
Digitaria ciliaris†, *D. coenicola*†	*Vetiveria zizanioides*

† signifies, additionally, high pH tolerance.
Further research is likely to extend this list considerably. The genera
Cladoraphis, *Drake-Brockmania*, *Halopyrum* and *Monanthochloe* (all
chloridoids), for example, grow in habitats likely to be saline but have not
apparently been subject to physiological experiments.
Salt tolerance is not absolute, the range is considerable and the grasses listed
have not been examined by the same criteria.
Classification is that of Clayton & Renvoize in the Appendix to this volume.
Table based on Sepasal Data, Royal Botanic gardens, Kew, and augmented.

Beyond xerophytic evolution, there is centred on the chloridoids an
extreme manifestation, the 'resurrection' grasses. They have foliage
that revives after dehydration to air-dryness. In South Africa, *Eragrostis
nindenensis*, *Oropetium capense* and *Tripogon minimus* occur (Gaff &
Ellis, 1974). In Australia, these grasses include *Eragrostiella bifaria*, an
undescribed *Sporobolus* species, *Tripogon loliiformis*, all of which are
Chloridoids. Two other grass groups include six species of Australian
Micraria (Arundinoideae) and the introduced *Poa bulbosa* (Pooideae)
(Lazarides, 1992). Other resurrection plants include some ferns and
sedges, this latter group being especially palatable.

For pastoral purposes *Tripogon* in Australia is, among the grasses,
the most valuable where it is abundant on stony calcareous tablelands. It
is food for sheep, horses and kangaroos. It responds rapidly to even low
rainfall, hence its vernacular name 'five minute grass' (Lazarides, 1992).
Resurrection grasses provide a dry ground cover, prevent soil erosion
and 24 hours after rainfall of 10 mm or more yield a green fodder. These
grasses, which may have counterparts elsewhere, provide an obvious
area of both fundamental and applied interest.

Bambusoideae, Bambuseae

Any complete treatment of grasses must include bamboos but,
as Renvoize & Clayton remark (chapter 1), the Bambusoideae have
remained outside the mainstream of taxonomic research for about a
century. This near exclusion is not confined to taxonomy but holds for
the other botanical disciplines, too. Bamboos, to many plant scientists,
appear enigmatic, oriental and hardly fit subjects for enquiry. They are,

though, not so enigmatic as to be mistaken for anything but a grass and, although concentrated in the Orient, bamboos also occur widely in Africa and the New World. Of the world's several hundred grass genera only about 50 are bamboos but a view of grasses that excludes them is one that is, in every sense, diminished. Once they are included, we have to redefine grasses and add questions that, otherwise, would not have occurred to us.

The utility of bamboos is well known and has been described by McClure (1966) and Marten & Brandenberg (1980). In the West interest is stimulated by bamboo in the diet of the Giant Panda which includes *Arundinaria fangiana* and *Phyllostachys bambusoides*. This diet need not consist exclusively of bamboo but includes various roots and these animals will, for example, accept sugar cane (Schaller, 1986).

In China, bamboo sciences – cytogenetics, ecology, physiology and taxonomy – are scattered in various institutions and progress has been impeded until recently by the difficulty of travel. Bamboos are widespread through southern China but with a concentration in Yunan and Guangxi. Quite remarkable, however, are the cold-tolerant bamboos of the Tibet High Plateau. C. S. Chou (personal communication) has described *Sinarundinaria* and *Thamnocalamus* in the understorey of pine forest at about 3000 m above sea level with temperatures of −20 or −30 °C for 5–6 months of the year. For the ecology of this region see, for example, Chang (1981, 1983).

Although the common perception of bamboos is of tall, stately plants, they are not invariably so. *Arundinaria vagans* seldom grows above 30 cm. *Arundinaria disticha* forms almost a lawn and *Chusquea abietifolia* has a scrambling growth through upland Caribbean forest. *Chusquea* is interesting, too, in that monopodial and sympodial growth habits can coexist in the same genus.

Bamboo flowering and plant breeding

Bamboos are assumed to flower only at long intervals and, having done so, to senesce and die. This is an oversimplified view. Synchronous or 'mast' flowering has been reviewed by Janzen (1976), who sets out evidence for long vegetative periods and regular flowering cycles. To this must be added the experience of other field botanists with access to natural stands of bamboo. *Phyllostachys edulis* (a monopodial type) flowers infrequently but does so throughout the plant canopy. The aerial parts can senesce but over a period of about four years the rhizome sometimes rejuvenates and may re-establish a green canopy (C.

S. Chou, personal communication). *Arundinaria japonica*, a hardy ornamental in the United Kingdom, flowers occasionally and can do so without conspicuous senescence.

Irregular flowering occurs in natural stands of *Bambusa* and *Dendrocalamus*, each being sympodial, although a single clump can contain both flowering culms and others that remain vegetative. Transplanting such clumps to a breeder's garden enabled Zhang & Chen (1985) to hybridize *Dendrocalamus minor* × *D. latiflorus*, *Bambusa textilis* × *B. pervariabilis* and *B. textilis* × *B. sinospinosa*. In this garden at Guanzhou, I was shown, for example *D. latiflorus* × *B. pervariabilis*.

In 1990, Nadgauda, Parasharami & Mascarenhas demonstrated *in vitro* accelerated flowering and seeding in *Bambusa arundinacea* and *Dendrocalamus brandisii*. Informal reports from elsewhere have since confirmed flowering *in vitro*. Perhaps the means are to hand now for a more controlled approach to the study of the physiology of flowering, of intraspecies and interspecies crossing and of *S–Z* incompatibility.

Bamboos are impressively diverse, will grow in a range of habitats, serve a multitude of uses and comprise a renewable resource. At last, it seems that practical needs and scientific interests have begun to converge.

As we recognize new needs, the usefulness of more, hitherto wild, grasses will doubtless become apparent and the process of extending domestication will continue. None the less, the focus of new genetic technology is likely to remain with the major cereals. At this level, evolution under domestication might one day change not only our cereals but, in the process, eventually transform plant breeding.

Molecular biology and cereal breeding

The successes of the 'Green Revolution' are evident and while they may have drawn adverse sociological comment (see, for example, Mellor & Desai, 1985) the fact is that breeders and their colleagues contrived a technical response to human need. What is at issue is not whether the Green Revolution offered abundance in place of scarcity but, rather, have contemporary social and political arrangements optimized its results?

In the cereals, yields continue to rise partly through genetic advance and partly through better agronomy to which many disciplines make an indispensable contribution. Schmidt (1984) estimates for wheat that over 30 years the genetic contribution to yield has been about 0.74% per year. What then is the role of the molecular biologist?

Chapters 8, 9 and 10 offer a comparative approach to the three major cereals. Wheat, a polyploid inbreeder, maize, a diploid outbreeder and rice, a diploid inbreeder, are, taxonomically, quite different grasses, being, respectively, pooid, panicoid and bambusoid.

The molecular biologist has several objectives. They include (a) describing gene action in chemical terms; (b) transferring useful genetic diversity into crop plants from elsewhere; and (c) where such diversity does not exist, seeking to create it. Since systems of genetic transformation based on *Agrobacterium* are ineffective in cereals and no other substitute is readily available, we have necessarily to try to understand cereal genetic systems as a prerequisite to finding an effective transformation system. Even when it becomes possible reliably to transform a cell or protoplast, there remains the daunting problem of converting it to a stably transformed, differentiated, fertile adult.

Perhaps, at this point, it needs to be emphasized just how remarkable are our cereals. For several thousand years they have survived in monocultures (admittedly weedy) subject to a collection of pests and diseases and, it must be admitted, too, in most cases with indifferent farming. Not only this, species have been taken beyond the ecological range of their wild ancestors, resulting in unaccustomed climatic stress. Once science began to be applied, already remarkable plants revealed a new level of bounty that, in the First World, we have come to take entirely for granted. Europe's problem, undreamed of even two generations ago, is how to avoid having too much. This achievement is not that only of the plant breeder of course, but it is significant that he or she has worked within nature's confines in that the alternation of gametophyte and sporophyte has been kept entirely intact. Whether we look at wild grasses or cereals responding to our own direct needs, we are confronted by remarkably adaptive and subtle genetic systems and it is with these that the new technology seeks to engage and eventually to redirect. To be realistic, molecular biology is an investment in the future and it is evident that the heady excitement of the late 1970s has been replaced by the soberly lengthened time scales of today. The investment has to be made and progress will come but for sometime yet increased cereal yields and wider adaptability to new environments will depend on conventional plant breeding. The extension of wheat, for example, into Bangladesh described by Hess & Cassaday (1990) uses not a new but an old technology. None the less, it would be complacent to imagine, in the face of a rapidly explanding world population, that a technology, which has served us well, could not be made better.

Gametophytes

The male gametophyte, the pollen grain, is, for grasses, a relatively uniform structure and one that shows little variation in function. By contrast, the female gametophyte can arise from the archesporium and be reduced or unreduced. It may develop with or without fertilization. Alternatively, one or more nucellar cells may, aposporously, replace or coexist with the archesporially derived embryo sac. Sexual and asexual behaviour is now relatively well understood both genetically and cytologically. At an ultrastructural level there is a growing understanding, but the physiology of events around gamete fusion still require intensive investigation if, at this point in the plant life cycle, we wish to insert (or delete) genes or, in some way, radically affect heredity. We cannot, even yet, effect fertilization *in vitro*, as has been routine for some time in animal science studies. Indeed, we are still being impeded by matters of definition, as will now be explained.

Russell & Cass (1983) and Russell (1984) demonstrated for *Plumbago zeylanica* that the male gametes originating from the pollen grain were not identical and that the male gamete having a preponderance of plastids fused preferentially with the egg while that having significantly more mitochondria fused with the central cell. This finding prompted a search among other unrelated angiosperms to see if this situation was found elsewhere. Knox & Singh (1990) developed an interpretation of these events that occur around fertilization. They suggested the generative cell divided to give not two gametes but one *male gamete* and one *associate cell*, this latter being destined to fuse with the central cell to form the primary endosperm cell. This approach for grasses gains added support if, as Mogensen (1990) showed, the male gamete makes only a *nuclear* contribution to the egg while the associate cell (as a protoplast) fuses with the central cell. A fundamental question, that of 'recognition', i.e. how the two products of generative cell division fuse with their intended future partners was discussed by Knox *et al.* (1988). This acquires further relevance if, increasingly, the male gamete and the associate cell are found to be different in form or function. In the interim, we might accept two fusions in the embryo sac while suspending the term *double fertilization*. An oblique light on this is shed by pseudogamy, discussed in chapter 4. In the terminology of Knox & Singh (1990), pseudogamy required the fusion of the associate cell with the central cell but not the nuclear entry of the male gamete, and it is curious that as long ago as 1957 Snyder, studying apomixis in *Paspalum secans*, found what was interpreted as the remnant of a (the) male gamete on the outer surface of the egg. Seemingly, the only difference between pseudogamy and fertilization is that failure of the nucleus in the male cytoplasm to enter the egg cell, for fusion, to create the zygote.

This discussion provides a basis for consideration of work with isolated gametes of both sexes now brought into close proximity and even conjunction. Efforts in several laboratories have sought, enzymically, to isolate embryo sacs and male gametoplasts in a viable condition (for a comprehensive review, see Theunis, Pierson & Cresti, 1991). Provided these are available, attempts to fuse them become possible.

An absorbing study for *Zea* is that of Kranz, Bantor & Lörz (1991*a,b*). In their study, they showed that it was possible to isolate the embryo sac and, additionally, to fragment it enzymically to its constituent components. Similarly, they were able to obtain isolated male gametoplasts. Male and female elements brought into contact would not, however, fuse without the stimulus of an electrical shock. Moreover, the fusion of the gametoplasts was not confined to that with the egg cell. Several points deserve comment. First, egg/gametoplast fusion involved all constituents of both cells rather than being confined, on the male side, to nuclear entry. Secondly, the 'quasi-zygote' resulting divided but did not differentiate and, thirdly, the gametoplasts had not passed through a synergid prior to interacting with the egg.

This most interesting study highlights how intricate are the processes occurring in nature that precede zygote formation. The requirement for electrically mediated fusion presumably indicates that supposedly intact mature gametes do not fuse spontaneously and implies the need for the process equivalent to capacitation in animal sperm. Even though fusion was achieved, the fact that it did not result in differentiation raises two questions. First, what has to be done to produce not a quasi-zygote but a zygote that would go on to differentiate and, secondly, what in all of this might be the role of an intact functional endosperm?

In their second paper, Kranz *et al.* stress the advantage of including organelles to facilitate a study of cytoplasmic inheritance, utilizing cultured cell lines from which plants might eventually be regenerated.

If this kind of study were to be extended from the conventional sexual system of maize to the various sorts of apomict described, for example, in chapter 4, an area of remarkable interest would surely eventually open up.

Photosynthesis

Even photosynthesis has been appreciably diversified among the grasses with the emergence of a range of C_4 types. So radical a change must almost certainly have been a response to a changing environment and might contain the germ of further change. As Hattersley (chapter 2) points out, rising CO_2 levels may favour C_3 grasses, and rising temperatures may favour C_4 grasses, but it is not clear how these factors in some combination might operate to favour some existing or new photosynthetic variants. What seems obvious is that, if we wish to modify photosynthesis then some of the ingredients for doing so exist among present-day grasses. This is part of a larger problem, namely, how we might seek to direct the evolution of our crop grasses or indeed any other member of the Poaceae.

Directed evolution?

The founding of Rothamsted Experimental Station in 1843 marks conveniently the beginning of scientific plant agriculture. In the ensuing 150 years, such institutes have multiplied around the world, one conspicuous result being the emphasis on the importance of wheat, maize and rice. More recently, interest has begun to broaden and to re-examine neglected cereals and hardly known grasses, but the three major cereals continue to monopolize 'high science' and if they are to be model systems, it is perhaps fortunate that they are, cytogenetically and taxonomically, so different.

If, as seems likely, the world will have a population of eight billion within the life time of some of us, two concerns are that the cereal-growing environment shall be enhanced and that cereal genotypes shall continually be improved. What are the options for the latter?

Wilson reported the first rye–wheat hybrid (quoted by Borlaug, 1989). Since then sustained efforts have been made to breed both hexaploid and octaploid triticales (*Triticosecale*; Lukasewski & Gustafson, 1987), but during this time the rate of wheat improvement suggests that the addition of the rye genome is an impediment to breeding progress. Where a more piecemeal approach has been adopted involving the importation into the wheat genome of smaller chromosome components, results have been more satisfactory and formed part of mainstream breeding of wheat. Since we are not yet able to take the process to its logical conclusion and make direct molecular alterations to the genome at specified loci, there is no effective alternative so far to plant breeding, a situation that is repeated for maize and rice. (Any alternative, of course, should be not only different but better.) So pressing is the need, though, that we cannot forego the search for alternatives.

However much we might aspire to direct cereal evolution, our breeding procedures still await the outcome of meiosis and syngamy with each cycle of selection and the signs are that we must continue to do so for some time yet. And we must recognize too, the frustrating paradox that, despite immense effort, those plants, the grasses upon which humankind most depends, are among the less responsive to the new molecular genetics.

References

ASTAG. (1991). Asia Technical Department of Agriculture Division, World Bank. *Vetiver Newsletter Special Issue.*

ASTAG, VPN (1991). Asia Technical Department Agriculture Division, World Bank. *Vetiver Participants Network.*

Borlaug, N. E. (1989). Feeding the world during the next doubling of the world population. In *Plants and society*, ed. M. S. Swaminathan & S. L. Kochbar, pp. 523–554. London: Macmillan.

Carruthers, I. (1991). In *Complementary or conflicting approaches to salinity, proceedings of an ODA workshop*, ed. G. Wyn Jones & H. Gunston (in press).

Chang, D. H. S. (1981). The vegetation zonation of the Tibetan Plateau. *Mountain Research and Development*, **1**, 29–48.

Chang, D. H. S. (1983). The Tibetan Plateau in relation to the vegetation of China. *Annals of the Missouri Botanic Gardens*, **70**, 564–570.

Chapman, G. P. (1990). The widening perspective: reproductive biology of bamboos, some dryland grasses and cereals. In *Reproductive versatility in the grasses*, ed. G. P. Chapman, pp. 240–259. Cambridge: Cambridge University Press.

Clayton, W. D. & Renvoize, S. A. (1986). *Genera Graminum: grasses of the world*. Kew Bulletin Additional Series 13. London: Royal Botanic Gardens, Kew.

Darlington, C. D. (1969). The silent millennia in the origin of agriculture. In *The domestication and exploitation of plants and animals*, ed. P. J. Ucko & G. W. Dimbleby, pp. 67–78. Duckworth.

Darwin, C. (1868). *Animals and plants under domestication*, 1st edn, 2 vols. London: John Murray.

de Wet, J. M. J. (1992). Pearl millet: a cereal for the Sahal. In *Desertified grasslands: their biology and management*, ed. G. P. Chapman. London: Academic Press (in press).

Gaff, D. R. & Ellis, R. P. (1974). Southern African grasses with foliage that revives after dehydration. *Bothalia*, **11**, 305–308.

Gorham, J. (1990). Salt tolerance in the triticeae: K/Na discrimination in synthetic hexaploid wheats. *Journal of Experimental Botany*, **41**, 623–627.

Gorham, J. (1992). Grass responses to salinity. In *Desertified grasslands: their biology and management*, ed. G. P. Chapman, pp 165–180. London: Academic Press.

Grimshaw, R. (1991). Vetiver in Thailand and China. Other species with potential. *Vetiver Newsletter*, pp. 1–2.

Harlan, J. R. (1989). Self-perception and the origins of agriculture. In *Plants and society*, ed. M. S. Swaminathan & S. L. Kochhar, pp. 5–23. London: Macmillan.

Hess, E. & Cassaday, K. (1990). Third World's appetite for wheat. *Shell Agriculture*, **6**, 11–13.

IIEDWRI (International Institute for Environment and Development World Resource Institute) (1987). *World Resources 1987*, New York: IIEDWRI.

Islam-ul-Haq, M. & Khan, M. F. A. (1971). Reclamation of saline and alkaline soils by growing Kallar grass. *Nucleus*, **8**, 139–144.

Jacobsen, T. & Adams, R. M. (1958). Salt and silt in ancient Mesopotamian agriculture. *Science*, **128**, 1251–1258.

Jacobsen, T. (1982). Salinity and irrigation agriculture in antiquity Diyala Basin archaeological projects: report on essential results, 1957–58. *Biblioteca Mesopotamica*, **14**, 1–107.

Janzen, D. H. (1976). Why bamboos wait so long to flower. *Annual Review of Ecology and Systematics*, **7**, 347–391.

Kernick, M. D. (1990). An assessment of grass succession, utilisation and development in the arid zone. In *Reproductive versatility in the grasses*, ed. G. P. Chapman, pp. 154–181. Cambridge: Cambridge University Press.

Knox, R. B., Southworth, D. & Singh, M. D. (1988). Sperm cell determinants and control of fertilisation in plants. In *Eukaryote cell recognition: concepts and model systems*, ed. G. P. Chapman, C. C. Ainsworth & C. J. Chatham, pp. 175–195. Cambridge: Cambridge University Press.

Knox, R. B. & Singh, M. B. (1990). Reproduction and recognition phenomena in the Poaceae. In *Reproductive versatility in the grasses*, ed. G. P. Chapman, pp. 220–239. Cambridge: Cambridge University Press.

Kranz, E., Bantor, J. & Lörz, H. (1991a). *In vitro* fertilisation of single, isolated gametes of maize mediated by electrofusion. *Sexual Plant Reproduction*, **4**, 12–16.

Kranz, E., Bantor, J. & Lörz, H. (1991b). Electrofusion-mediated transmission of cytoplasmic organelles through the *in vitro* fertilisation process, fusion of sperm cells with synergids and central cells and cell reconstitution in maize. *Sexual Plant Reproduction*, **4**, 17–21.

Lazarides, M. (1992). Resurrection grasses (Poaceae) in Australia. In *Desertified grasslands: their biology and management*, ed. G. P. Chapman, pp . 213–234. Academic Press.

Lazzeri, P. A., Kollmorgen, J. & Lörz, H. (1990). *In vitro* technology. In *Reproductive versatility in the grasses*, ed. G. P. Chapman, pp. 182–219. Cambridge: Cambridge University Press.

Lukaszewski, A. J. & Gustafson, J. P. (1987). Cytogenetics of triticale. *Plant Breeding Reviews*, **5**, 41–94.

McClure, F. A. (1966). *The Bamboos – fresh perspective*. Cambridge: Harvard University Press.

McNeilly, T. (1968). Evolution in closely adjacent plant populations. III. *Agrostis tenuis* on a small copper mine. **23**, 99–108.

McNeilly, T. & Antonovics, J. A. (1968). Evolution in closely adjacent plant populations. IV. Barriers to gene flow. *Heredity*, **23**, 205–218.

336   G. P. Chapman

Mangelsdorf, P. C., Macneish, R. S. & Galinat, W. C. (1964). Domestication of Corn. *Science*, **143**, 538–545.

Marten, L. & Brandenberg, J. (1980). Bamboo, the giant grass. *National Geographic Magazine*, October, pp. 504–528.

Mellor, J. W. & Desai, G. M. (1985). *Agricultural change and rural poverty: variations on a theme by Dharum Narain*. Baltimore, MD: Johns Hopkins University Press.

Mogensen, H. L. (1990). Fertilisation and early embryogenesis. In *Reproductive versatility in the grasses*, ed. G. P. Chapman, pp. 76–99. Cambridge: Cambridge University Press.

Nadgauda, R. S., Parasharami, V. A. & Mascarenhas, A. F. (1990). Precocious flowering and seeding behaviour in tissue cultured bamboos. *Nature*, **334**, 335–336.

Purseglove, J. (1968). The origin and spread of tropical crops. In *Tropical crops. Dicotyledons I*, ed. J. Purseglove, pp. 9–17. London: Longmans.

Purseglove, J. (1972). Sorghum. In *Tropical Crops, Monocotyledons I*, ed. J. Purseglove, pp. 259–286. Longmans.

Rana, R. S. & Parkash, V. (1980). Karnal grass grows well on Alkali soils. *Indian Farming* (July).

Ren Jizhou (1992). The ecological role of *Puccinellia chinampoensis* on saline soil in arid inland regions of China. In *Desertified grasslands: their biology and management*, ed. G. P. Chapman, pp. 136–144. Academic Press.

Richards, A. J. (1990). The implications of reproductive versatility for the structure of grass populations. In *Reproductive versatility in the grasses*, ed. G. P. Chapman, pp. 133–153. Cambridge: Cambridge University Press.

Russell, S. D. & Cass, D. D. (1983). Unequal distribution of plastids and mitochodria during sperm cell formation in *Plumbago zeylanica*. In *Pollen: biology and implications for plant breeding*, ed. D. G. Mulcahy, G. B. Mulcahy & E. Ottaviano, pp. 135–140. Elsevier: Amsterdam.

Russell, S. D. (1984). Ultrastructure of the sperm of *Plumbago zeylanica* II. Quantitative cytology and three dimensional organisation. *Planta*, **162**, 385–391.

Schaller, G. B. (1986). Secrets of the wild panda. *National Geographic Magazine*, (March), pp. 284–308.

Schmidt, J. W. (1984). Genetic contribution to yield gains in wheat. In *Genetic contribution to yield gains in five major crop plants. Special Publication*, pp. 89–101. Madison, WI: Crop Science Society of America.

Snyder, L. A. (1957). Apomixis in *Paspalum secans*. *American Journal of Botany*, **44**, 318–324.

Tanton, T. (1991). Aspects of irrigation research. In *Complementary or conflicting approaches to salinity, Proceedings of an ODA Workshop*, ed. G. Wyn Jones & H. Gunston (in press).

Theunis, C. H., Pierson, E. S. & Cresti, M. (1991). Isolation of male and female gametes in higher plants. *Sexual Plant Reproduction*, **4**, 145–154.

Thomason, J. R. (1987). Fossil grasses: 1820–1986 and beyond. In *Grass systematics and evolution*, ed. T. R. Söderstrom, K. W. Hilu, C. S. Campbell &

M. E. Barkworth, pp. 159–167. Washington DC: Smithsonian Institution Press.

Wallace, H. A. & Brown, W. L. (1956). *Corn and its early fathers.* East Lansing: Michigan State University Press.

Watson, L. (1990). The grass family, Poaceae. In *Reproductive versatility in the grasses*, ed. G. P. Chapman, pp. 1–31. Cambridge: Cambridge University Press.

Zhang, G.-Z. & Chen, F.-Q. (1985). Studies in bamboo hybridisation. In *Recent research on bamboos. Proceedings of an International Workshop, Hangzhou, Peoples Republic of China*, ed. A. N. Rao, G. Chanarajan & C. B. Sastry.

Zohary, D. (1969). The progenitors of wheat and barley in relation to domestication and agricultural dispersal in the Old World. In *The domestication and exploitation of plants and animals*, ed. P. J. Ucko & G. W. Dimbleby, pp. 47–66. Duckworth.

Appendix

A system of classification for the grasses
W. D. Clayton & S. A. Renvoize

In the following system of classification, which is based on *Genera Graminum* (Clayton & Renvoize, 1986), the family is organized under six subfamily divisions and 39 tribes. The descriptions indicate the overall facies of the taxa, but are not intended to be comprehensive or strictly diagnostic.

Subfamily Bambusoideae
Herbs, shrubs, tall trees or climbers, commonly with lanceolate leaf blades; ligules membranaceous, sometimes with a ciliate margin. Inflorescence paniculate and often complex. Spikelets 1 to many-flowered, laterally compressed; lemmas mostly awnless, entire, several to many-nerved. Lodicules 2–3, hyaline to membranous; stamens 1–6, rarely more; stigmas 1–3. Fruit a caryopsis, rarely an achene or berry; the embryo bambusoid or oryzoid.

Chlorenchyma irregularly arranged; bundle sheaths double; microhairs usually slender, occasionally absent; fusoid cells commonly present; stomatal subsidiary cells low-dome-shaped to triangular.

Photosynthetic pathway, C_3.

TRIBE BAMBUSEAE
Trees, shrubs or climbers with hollow woody stems divided into cylindrical segments by the nodes, the lower half clad in broad leaf sheaths with a rudimentary blade, the upper half copiously branched, one or more branches grouped at each node; leaf blades usually with cross nerves and a short false petiole, deciduous from the sheath. Inflorescence sometimes a small panicle or a raceme but usually a complex bracteate structure. Spikelets of 1 to many fertile florets; glumes 2–4, shorter than the lemma; lemmas 5- or more-nerved, entire, awnless or with a short straight awn from the tip; stamens 3–6; stigmas 1–3.

Subtribe Arundinariinae
Ovary appendage absent or inconspicuous. Inflorescence usually simple with predicellate spikelets, rarely complex.
Sinarundinaria, Colanthelia, Thamnocalamus, Aulonemia, Olmeca, Myriocladus, Glaziophyton, Chusquea, Neurolepis, Arundinaria, Hitchcockella, Perrierbambus, Guaduella, Sasa, Indocalamus Pseudosasa, Acidosasa, Indosasa, Sinobambusa, Chimonobambusa.

Subtribe Bambusinae

Ovary appendage broadly conical and usually hairy. Inflorescence rarely simple, usually complex.

Racemobambos, Nastus, Pseudocoix, Hickelia, Decaryochloa, Puelia, Greslania, Semiarundinaria, Phyllostachys, Shibataea, Arthrostylidium, Criciuma, Eremocaulon, Alvimia, Apoclada, Elytrostachys, Athroostachys, Atractantha, Rhipidocladum, Actinocladum, Merostachys, Guadua, Bambusa, Oreobambos, Thyrsostachys, Gigantochloa, Dendrocalamus, Melocalamus, Sphaerobambos, Dinochloa.

Subtribe Melocanninae

Ovary appendage long, stiff, tapering. Inflorescence complex.

Schizostachyum, Oxytenanthera, Melocanna, Ochlandra.

TRIBE OLYREAE

Herbs or cane-like plants with lanceolate to oblong or ovate leaf blades. Inflorescence a raceme or panicle bearing unisexual spikelets. Female spikelets 1-flowered, disarticulating below the floret or falling entire; glumes usually longer than the floret; lemma cartilaginous to bony. Male spikelets 1-flowered, fusiform, readily deciduous; glumes lacking or minute.

Olyra, Rehia, Maclurolyra, Reitzia, Sucrea, Raddia, Raddiella, Cryptochloa, Arberella, Parodiolyra, Lithachne, Froesiochloa, Diandrolyra, Piresia, Mniochloa, Ekmanochloa, Buergersiochloa.

TRIBE PARIANEAE

Herbs, sometimes with separate vegetative and fertile culms; leaf blades linear, lanceolate or oblong. Inflorescence spiciform with a fragile rachis, bearing verticels of 4–6 male spikelets surrounding a single female spikelet, these shed as a single unit. Spikelets 1-flowered. Female spikelet sessile; glumes 2, as long as the spikelet; lemma coriaceous. Male spikelets borne upon a flattened pedicel; stamens 2, 6 or numerous.

Pariana, Eremitis.

TRIBE PHAREAE

Herbs with linear to oblong, pseudopetiolate leaf blades in which the nerves slant obliquely away from the midrib and cross-veins are present. Inflorescence a panicle, the ultimate branches bearing 1–2 female spikelets and a terminal male spikelet. Female spikelet 1-flowered, terete to inflated, disarticulating below the floret; glumes shorter than the floret, scarious; lemma involute or utriculate. Male spikelets similar to the female but much smaller and soon deciduous; stamens 6.

Pharus, Leptaspis (*Suddia* is doubtfully included here).

TRIBE ORYZEAE

Herbs with linear leaf blades. Inflorescence a panicle, the spikelets all alike or the sexes separate. Spikelets 1-flowered or 3-flowered with the two lowest florets reduced to sterile lemmas, disarticulating below the floret; glumes absent or reduced to obscure ridges at the tip of the pedicel; lemma 5–10-nerved; palea resembling the lemma; stamens usually 6.

Porteresia, Oryza, Rhynchoryza, Potamophila, Maltebrunia, Prosphytochloa, Leersia, Chikusichloa, Hygroryza, Zizania, Zizaniopsis, Luziola.

TRIBE PHYLLORACHIDEAE

Shrubby herbs with cane-like stems bearing lanceolate to ovate leaf blades. Inflorescence terminal, the primary axis with leafy margins which enfold short cuneate racemes borne along the mid-rib; racemes falling entire, comprising 1–4 unisexual spikelets on a flattened rachis. Female spikelet 2-flowered; glumes shorter than the spikelet; lower floret reduced to a many-nerved lemma, lemma of upper floret 7–11-nerved, awnless. Male spikelet similar to the female but smaller; stamens 3 or 6.

Phyllorachis, Humbertochloa.

TRIBE STREPTOGYNEAE

Herbs with linear to lanceolate leaf blades. Inflorescence a tough unilateral raceme. Spikelets several-flowered, disarticulating below each floret; glumes 2, shorter than the florets; lemma convolute, 7–13-nerved, with a straight awn; stigmas 2–3, after fertilization growing into long tendrils.

Streptogyna.

TRIBE ANOMOCHLOEAE

Herbs with ovate leaf blades, these conspicuously pseudopetiolate and with cross-nerves. Inflorescence spathate, the ultimate spatheole enclosing a single spikelet. Spikelets 1-flowered falling entire; glumes absent; lemma produced into a long herbaceous projection, lodicules absent their place taken by a dense ring of hairs.

Anomochloa.

TRIBE STREPTOCHAETEAE

Herbs with lanceolate to ovate leaf blades. Inflorescence a tough, many-sided raceme. Spikelets 1-flowered, narrowly conical, falling entire, glumes 5, short, whorled; lemma produced into a very long, coiled awn; palea split to the base with the tips reflexed; lodicules 3, long; stamens 6, the filaments united; stigmas 3.

Streptochaeta.

TRIBE PHAENOSPERMATEAE

Herbs with broadly linear leaf blades narrowed to a pseudopetiole, this twisted to bring the abaxial side uppermost. Inflorescence a panicle. Spikelets 1-flowered, falling entire; glumes 2, shorter than the floret; lemma 3–7-nerved. Fruit a globose caryopsis with thick pericarp and small apical protuberance.

Phaenosperma.

TRIBE EHRHARTEAE

Herbs with linear leaf blades. Inflorescence a panicle, a unilateral raceme or reduced to 1 or 2 spikelets. Spikelets 3-flowered with the 2 lowest florets reduced

to sterile lemmas, disarticulating above the glumes; glumes 2, shorter than or exceeding the floret; sterile lemmas or at least the upper as long as the floret, coriaceous, awned or awnless; fertile lemma 5–7-nerved, awnless; palea 0–5-nerved; stamens 1, 2, 3, 4 or 6.
Ehrharta.

TRIBE DIARRHENEAE

Herbs with narrowly lanceolate leaf blades. Inflorescence an open or contracted panicle. Spikelets of 2–5 fertile florets, disarticulating below each floret; glumes 2, shorter than the lemmas; lemmas 3(–5)-nerved. Fruit an ellipsoid caryopsis with a beak-like appendage.
Diarrhena.

TRIBE BRACHYELYTREAE

Herbs with narrowly lanceolate leaf blades. Inflorescence a panicle. Spikelets 1-flowered, disarticulating above the glumes; glumes 2, tiny; lemma awned. Fruit a linear caryopsis with a beak-like appendage.
Brachyelytrum.

Subfamily Pooideae

Herbs, usually with linear leaf blades; ligule membranaceous. Inflorescence usually a panicle, sometimes a raceme. Spikelets laterally compressed, 1–many-flowered, with a fragile rachilla; lemmas several-nerved, entire, with or without an apical or dorsal awn. Lodicules 2–3, membranous to hyaline (fleshy in Meliceae); stamens 3; stigmas 2. Fruit a caryopsis; the embryo pooid. Chromosomes large in the principle tribes, with $x=7$.

Chlorenchyma irregularly arranged with more than 4 cells between vascular bundles; bundle sheath double; microhairs absent; stomatal subsidiary cells triangular, domed or parallel sided.

Photosynthetic pathway C_3.

TRIBE NARDEAE

Tufted perennial with filiform leaf blades. Inflorescence a unilateral raceme. Spikelets 1-flowered, disarticulating below the floret; glumes reduced or suppressed; lemma 3-nerved, awned from the tip. Chromosomes small, $x=13$.
Nardus.

TRIBE LYGEAE

Perennial with tough wiry culms and involute leaf blades. Inflorescence a single spikelet enclosed by a spatheole. Spikelet 2(–3)-flowered, urn-like, falling entire; glumes absent; lemmas fused below along opposing margins to form a cylindrical tube, the upper half free; paleas fused back to back and forming a transverse septum within the lemma tube, free above. Chromosomes small, $x=10$.
Lygeum.

TRIBE STIPEAE

Mostly wiry bunch grasses with rolled or filiform leaf blades. Inflorescence an open or contracted panicle. Spikelets 1-flowered disarticulating above the

glumes; glumes usually longer than the floret; lemma terete to lenticular and often enclosing the palea, awned from an entire or bidenticulate tip. Chromosomes small, $x=11$.
Stipa, Nassella, Psammochloa, Oryzopsis, Milium, Trikeraia, Piptochaetium, Ortachne, Aciachne.

TRIBE POEAE
Herbs, usually with linear leaf blades. Inflorescence a panicle, rarely a unilateral raceme. Spikelets of 1–many fertile florets, disarticulating below each floret; glumes usually shorter than the lowest lemma; lemmas 3–13-nerved, entire, awnless or awned from the tip.
Ampelodesmos, Festuca, Scolochloa, Dryopoa, Megalachne, Podophorus, Lolium, Micropyropsis, Micropyrum, Vulpia, Wangenheimia, Loliolum, Castellia, Psilurus, Cynosurus, Lamarckia, Puccinellia, Torreyochloa, Briza, Microbriza, Rhombolytrum, Poa, Dactylis, Aphanelytrum, Austrofestuca, Parafestuca, Arctagrostis, Colpodium, Arctophila, Dupontia, Catabrosa, Phippsia, Coleanthus, Neuropoa, Eremopoa, Nephelochloa, Lindbergella, Sphenopus, Libyella, Desmazeria, Catapodium, Sclerochloa, Cutandia, Vulpiella, Erianthecium, Oreochloa, Sesleria, Ammochloa, Echinaria.

TRIBE HAINARDIEAE
Annual and perennial herbs. Inflorescence a single cylindrical, bilateral raceme, tough or fragile. Spikelets 1–2-flowered; glumes subequal and side by side, usually exceeding and covering the floret, strongly 3–7-nerved; lemma 3–5-nerved, usually awnless.
Narduroides, Agropyropsis, Pholiurus, Parapholis, Scribneria, Hainardia.

TRIBE MELICEAE
Inflorescence a panicle or raceme. Spikelets of several to many fertile florets with imperfect florets above, usually gathered into a clump of rudimentary lemmas, disarticulating below each floret or only above the glumes; glumes often papery; lemmas 5–13-nerved, awned or awnless; lodicules 2, usually connate, fleshy, truncate. Chromosomes small, $x=9$ or 10.
Glyceria, Pleuropogon, Melica, Anthochloa, Schizachne, Lycochloa, Streblochaete, Triniochloa.

TRIBE BRYLKINIEAE
Inflorescence a raceme. Spikelets of 1 fertile floret with 2 sterile lemmas below, falling entire together with the pedicel; glumes shorter than the lemma; lemmas 5–7-nerved, the sterile acuminate, the fertile awned. Fruit an ellipsoid caryopsis with a glossy umbonate cap. Chromosomes small, $x=10$.
Brylkinia.

TRIBE AVENEAE
Inflorescence a panicle, rarely racemes. Spikelets 1 to several-flowered; glumes usually membranous, longer than the adjacent lemmas and often as long as the spikelet; lemmas 3–11-nerved, usually with a geniculate dorsal awn.

Subtribe Duthieinae
Spikelets of 1 to several fertile florets; lemmas bidentate to bifid, geniculately awned from the sinus. Caryopsis with a linear hilum.
Metcalfia, Pseudodanthonia, Stephanachne, Duthiea.

Subtribe Aveninae
Spikelets with 2 to several fertile florets and usually a rudimentary floret or rachilla extension; glumes equal or unequal, the upper usually longer than the lemma but seldom enclosing the spikelet; palea sometimes gaping free from the lemma.
Helictotrichon, Arrhenatherum, Avena, Gaudinia, Relchela, Dissanthelium, Tovarochloa, Trisetum, Trisetaria, Ventenata, Koeleria, Rostraria, Graphephorum, Peyritschia, Sphenopholis, Dielsiochloa, Deschampsia, Holcus, Corynephorus, Periballia, Aira, Airopsis, Antinoria.

Subtribe Phalaridineae
Spikelets 3-flowered, the 2 lower florets male or barren, the uppermost bisexual, disarticulating above the glumes, usually enclosing the florets.
Hierochloe, Anthoxanthum, Phalaris.

Subtribe Alopecurinae
Spikelets 1-flowered; glumes commonly enclosing the floret; palea not gaping.
Agrostis, Calamagrostis, Ammophila, Simplicia, Ancistragrostis, Echinopogon, Pentapogon, Dichelachne, Triplachne, Gastridium, Lagurus, Apera, Hypseochloa, Zingeria, Mibora, Polypogon, Chaetopogon, Cyathopus, Euthryptochloa, Cinna, Limnodea, Limnas, Alopecurus, Cornucopiae, Beckmannia, Phleum, Rhizocephalus.

TRIBE BROMEAE
Inflorescence a panicle. Spikelets of several to many fertile florets, disarticulating below each floret; glumes, shorter than the lowest lemma; lemmas 5–13-nerved, usually awned rarely awnless; ovary capped by a hairy, lobed appendage bearing subterminal stigmas. Starch grains simple.
Littledalea, Bromus, Boissiera.

TRIBE TRITICEAE
Inflorescence a single bilateral raceme, the spikelets single or in groups of 2–3 at each node, rachis tough or fragile. Spikelets 1 to many-flowered, disarticulating below each floret when the rachis is tough; glumes shorter or narrower than the lemma, sometimes awn-like; lemmas 5–11-nerved, awned or awnless; ovary tipped by a small, fleshy, hairy appendage. Starch grains simple.
Brachypodium, Elymus, Hystrix, Sitanion, Leymus, Psathyrostachys, Hordelymus, Taeniatherum, Crithopsis, Heteranthelium, Hordeum, Agropyron, Erymopyrum, Secale, Dasypyrum, Triticum, Aegilops, Henrardia.

Subfamily Centothecoideae
Herbs, usually with broad leaf blades and conspicuous cross-veins; ligule membranous, sometimes with a ciliate margin. Inflorescence a panicle or racemes.

Spikelets laterally compressed, 1 to many-flowered; lemmas 5- or more-nerved. Lodicules 2, fleshy; stamens usually 3; stigmas 2. Fruit a caryopsis, the embryo centothecoid.

Chlorenchyma often forming a palisade layer below the upper epidermis, otherwise irregular; bundle sheaths double; microhairs slender; stomatal subsidiary cells triangular.

Photosynthetic pathway C_3.

TRIBE CENTOTHECEAE
Megastachya, Centotheca, Orthoclada, Chevalierella, Lophatherum, Bromuniola, Chasmanthium, Calderonella, Pohlidium, Zeugites.

Subfamily Arundinoideae
Herbs or occasionally large reeds. Leaf blades usually linear; ligule a line of hairs. Inflorescence a panicle. Spikelets 1 to several-flowered, laterally compressed, rachilla fragile; lemmas 1 to several nerved, entire or bilobed and awned from the sinus. Lodicules 2, fleshy; stamens 3; stigmas 2. Fruit a caryopsis; the embryo arundinoid.

Chlorenchyma irregularly arranged; bundle sheaths double; microhairs usually slender, rarely absent; stomatal subsidiary cells domed or triangular.

Photosynthetic pathway mostly C_3, rarely C_4.

TRIBE ARUNDINEAE
Mostly perennials. Inflorescence a panicle, sometimes spiciform, rarely a true raceme or reduced to a single spikelet. Spikelets 1 to several-flowered, usually disarticulating between the florets; glumes larger or shorter than the florets; lemmas 3–11-nerved, entire or bilobed, awnless or awned from the tip or sinus, often geniculately; palea 2-nerved, well developed. Fruit an ellipsoid caryopsis. *Tribolium, Cyperochloa, Urochlaena, Elytrophorus, Prionanthium, Spartochloa, Notochloe, Zenkeria, Piptophyllum, Styppeiochloa, Chionochloa, Danthonia, Chaetobromus, Plinthanthesis, Pentaschistis, Pentameris, Poagrostis, Pseudopentameris, Rytidosperma, Phaenanthoecium, Alloeochaete, Monachather, Schismus, Pyrrhanthera, Dregeochloa, Centropodia, Danthonidium, Anisopogon, Diplopogon, Amphiopogon, Crinipes, Nematopoa, Leptagrostis, Dichaetaria, Cortaderia, Lamprothyrsus, Arundo, Hakonechloa, Molinia, Phragmites, Gynerium, Thysanolaena.*

TRIBE MICRAIREAE
Small moss-like plants with spirally arranged leaves. Inflorescence a panicle or raceme. Spikelets 2-flowered, laterally compressed, disarticulating below each floret; glumes longer or shorter than the florets; lemmas hyaline to membranous, 1–3-nerved; palea 2-nerved and bifid to the base or undivided and 4–7-nerved.
Micraira.

TRIBE ARISTIDEAE

Inflorescence an open to densely contracted panicle. Spikelets 1-flowered, laterally compressed or terete, disarticulating above the glumes; glumes longer than the body of the lemma; lemma terete, 1–3-nerved, 3-awned, these more or less connate at the base and often elevated upon a twisted column; palea less than half the lemma length.

Sartidia, Stipagrostis, Aristida.

Subfamily Chloridoideae

Herbs, usually with linear leaf blades, often xeromorphic; ligule variable. Inflorescence a panicle or racemes. Spikelets laterally compressed, 1–many flowered, with a fragile rachilla; lemmas usually 1–3-nerved. Lodicules 2, fleshy. Fruit a caryopsis; the embryo usually chloridoid.

Chlorenchyma obscurely or clearly radiate, generally limited to a discrete layer around the bundle; bundle sheaths double; microhairs short and stout, rarely slender; stomatal subsidiary cells domed or triangular.

Photosynthetic pathway C_4.

TRIBE PAPPOPHOREAE

Inflorescence a dense panicle. Spikelets several-flowered, slightly compressed, disarticulating above the glumes but usually not between the florets; glumes 1–11-nerved, longer or shorter than the florets; lemma 9–11-nerved, the nerves produced into 7–19 awns or hyaline lobes.

Pappophorum, Enneapogon, Schmidtia, Cottea, Kaokochloa.

TRIBE ORCUTTIEAE

Inflorescence a speciform panicle or single raceme. Spikelets several-flowered, laterally compressed, disarticulating below each floret; glumes shorter than the florets or absent; lemma prominently 13–15-nerved.

Orcuttia, Tuctoria, Neostapfia.

TRIBE ERAGROSTIDEAE

Inflorescence a panicle, or of tough unilateral racemes, these digitate or scattered along an axis, rarely single. Spikelets several to many-flowered, laterally compressed, disarticulating below each floret; glumes 0–1-nerved, shorter than the lowest lemma; lemmas 1–3-nerved, entire or lobed, awned or awnless.

Subtribe Triodiinae

Leaf blades rigid, conduplicate or convolute and terete when dry, pungent; sheaths often resinous. Inflorescence a panicle or of racemes on a central axis. Spikelets with 1 to several florets, disarticulating below each floret; lemmas scarious to horny, 3–9-nerved.

Triodia, Plectrachne, Symplectrodia, Monodia.

Subtribe Uniolinae

Inflorescence a speciform panicle or of racemes on a central axis. Spikelets with several to many florets, ovate, falling entire; glumes shorter than the florets;

lemmas strongly keeled, chartaceous to coriaceous, 3–11-nerved.
Uniola, Tetrachne, Entoplocamia, Fingerhuthia.

Subtribe Monanthochloinae

Leaf blades short, stiff, flat or rolled, distichous, often pungent. Inflorescence a panicle or of racemes, often dioecious or reduced to a few spikelets. Spikelets with several to many fertile florets, disarticulating below each floret; glumes shorter than the lemmes; lemmas chartaceous to coriaceous, 5–13-nerved.
Aeluropus, Swallenia, Distichlis, Reederochloa, Allolepis, Monanthochloe, Jouvea.

Subtribe Eleusininae

Leaf blades flat or rolled. Inflorescence sometimes a panicle, more often of racemes. Spikelets with (1 or) 2 to many fertile florets, orbicular to linear, disarticulating below each floret; glumes shorter than to as long as the spikelet; lemmas keeled or rounded, membranous to cartilaginous, rarely coriaceous, usually 3-nerved, rarely 1– or 5–9-nerved.
Neyraudia, Triraphis, Habrochloa, Apochiton, Tridens, Redfieldia, Triplasis, Sohnsia, Scleropogon, Neesiochloa, Erioneuron, Blepharoneuron, Munroa, Blepharidachne, Leptochloa, Kengia, Orinus, Tripogon, Indopoa, Oropetium, Odyssea, Bewsia, Halopyrum, Gouinia, Silentvalleya, Lophachme, Leptocarydion, Trichoneura, Pogononeura, Dinebra, Brachychloa, Ochthochloa, Drake-Brockmania, Eragrostis, Cladoraphis, Richardsiella, Eragrostiella, Psammagrostis, Ectrosiopsis, Steirachne, Ectrosia, Heterachne, Pogonarthria, Harpachne, Viguierella, Coelachyrum, Vaseyochloa, Eleusine, Acrachne, Sclerodactylon, Dactyloctenium, Psilolemma, Desmostachya, Myriostachya.

Subtribe Sporobolinae

Leaf blades usually flat rarely convolute and pungent. Inflorescence a panicle. Spikelets 1-flowered, fusiform to lanceolate, disarticulating below the floret; lemma rounded on the back, membranous to chartaceous, 1–3-nerved; palea often resembling the lemma in size and texture.
Sporobolus, Crypsis, Urochondra, Calamovilfa, Muhlenbergia, Hubbardochloa, Lycurus, Pereilema.

TRIBE LEPTUREAE

Inflorescence a single cylindrical, bilateral raceme, the spikelets embedded in the fragile rachis. Spikelets 1-flowered, dorsally compressed, falling entire; lower glume minute or suppressed; upper glume appressed to the rachis, exceeding and covering the sunken floret, coriaceous, 5–12-nerved, acute to awned; lemma 3-nerved, membranous.
Lepturus.

TRIBE CYNODONTEAE

Inflorescence of tough unilateral racemes, these single, digitate or scattered along an axis, often deciduous. Spikelets with one fertile floret, with or without additional male or barren florets, cuneate to subterete, laterally or dorsally compressed, disarticulating above the glumes but not between the florets, or falling entire; glumes shorter than the floret or enclosing it; lemma 3-nerved, entire or 2–5-lobed, with or without 1–5 terminal or subapical awns.

Subtribe Pommereullinae
Inflorescence of single or subdigitate, unilateral racemes. Spikelets with 2–4 fertile florets, disarticulating above the glumes; glumes shorter than or equalling the florets; lemma 3-nerved at the base but usually branching above, emarginate to 4-lobed, awned. Grain elliptic, dorsally compressed.
Astrebla, Lintonia, Pommereulla.

Subtribe Chloridinae
Inflorescence of single, digitate or sometimes scattered unilateral racemes. Spikelets of 1 fertile floret with or without additional sterile florets, disarticulating above the glumes; glumes shorter than the floret or exceeding and enclosing it; lemma entire or bilobed, with or without a central awn or with 3 lobes or awns.
Tetrapogon, Neostapfiella, Chloris, Eustachys, Oxychloris, Austrochloris, Lepturopetium, Enteropogon, Trichloris, Afrotrichloris, Willkommia, Polevansia, Craspedorhachis, Schedonnardus, Gymnopogon, Harpochloa, Kampochloa, Ctenium, Microchloa, Daknopholis, Cynodon, Chrysochloa, Brachyachne, Lepturidium, Schoenefeldia, Pogonochloa, Spartina.

Subtribe Boutelouinae
Inflorescence of single or scattered unilateral racemes, these usually deciduous. Spikelets alike or dimorphic, of 1 fertile floret usually with additional sterile florets, disarticulating above the glumes or falling entire; glumes usually narrow, the upper as long as the floret the lower shorter; lemma 3-lobed or 3-awned.
Chondrosum, Neobouteloua, Bouteloua, Melanocenchris, Pentarrhaphis, Buchlomimus, Cyclostachya, Pringleochloa, Opizia, Schaffnerella, Buchloe, Cathestecum, Griffithsochloa, Aegopogon, Soderstromia, Hilaria.

Subtribe Zoysiinae
Inflorescence spiciform, more or less cylindrical, of numerous deciduous racemelets disposed along a central axis, the racemelets very short and sometimes reduced to a single spikelet. Spikelets alike or the upper reduced, 1-flowered, falling entire; glumes often much modified and oddly shaped; lemma membranous to hyaline; entire.
Catalepis, Monelytrum, Tragus, Zoysia, Dignathia, Leptothrium, Lopholepis, Pseudozoysia, Decaryella, Perotis, Mosdenia, Tetrachaete, Farrago.

Subfamily Panicoideae
Herbs, usually with linear leaf blades; ligule variable. Inflorescence a panicle, racemes or compound. Spikelets single or paired, usually 2-flowed, mostly dorsally compressed, falling entire, the florets usually dimorphic, the lower male or barren, the upper fertile. Lodicules 2, fleshy; embryo panicoid.

Chlorenchyma irregular or radiate; bundle sheaths single or double; microhairs slender or stout; stomatal subsidiary cells triangular or domed.
Photosynthetic pathway C_3 or C_4.

TRIBE PANICEAE
Inflorescence an open to spiciform panicle or racemes, the spikelets solitary or paired, falling entire, rarely awned; glumes membranaceous or herbaceous, rarely coriaceous, the upper often as long as the spikelet, the lower usually shorter and sometimes rudimentary; lower floret male or barren, its lemma usually membranous or herbaceous and as long as the spikelet, with or without a palea; upper floret bisexual, the lemma and palea more or less indurated.

Subtribe Neurachninae
Inflorescence a spiciform or capitate panicle. Glumes as long as the spikelet, more or less indurated; upper lemma hyaline to membranous or coriaceous with a membranous tip, the margins narrow and covering only the edge of the palea.
Photosynthetic pathway C_3 and C_4.
Neurachne, Paraneurachne, Thyridolepis.

Subtribe Setariinae
Inflorescence a panicle or of racemes. Lower glume variable, the upper glume and lower lemma as long as the spikelet; lower floret male or barren and often without a palea; upper lemma coriaceous to bony, usually with narrow inrolled margins covering only the edge of the palea but these sometimes flat.
Photosynthetic pathway C^3 and C^4.
Poecilostachys, Pseudechinolaena, Cyphochlaena, Oplismenus, Ichnanthus, Echinolaena, Lecomtella, Panicum, Ancistrachne, Lasiacis, Amphicarpum, Steinchisma, Plagiantha, Otachyrium, Triscenia, Hymenachne, Homolepis, Sacciolepis, Thyridachne, Hydrothauma, Ottochloa, Entolasia, Cleistochloa, Calyptochloa, Cyrtococcum, Streptostachys, Acroceras, Microcalamus, Echinochloa, Oplismenopsis, Chaetium, Oryzidium, Louisiella, Alloteropsis, Scutachne, Yakirra, Arthragrostis, Brachiaria, Eccoptocarpha, Urochloa, Eriochloa, Yvesia, Thuarea, Tatianyx, Whiteochloa, Acostia, Anthaenantiopsis, Paspalum, Reimarochloa, Thrasya, Thrasyopsis, Mesosetum, Axonopus, Centrochloa, Spheneria, Setaria, Paspalidium, Holcolemma, Ixophorus, Setariopsis, Dissochondrus, Plagiosetum, Paractaenum, Hygrochloa, Stenotaphrum, Uranthoecium.

Subtribe Melinidinae
Inflorescence a panicle. Spikelets laterally compressed; lower glume small or absent; upper glume as long as the spikelet; lower floret male or barren, its lemma resembling the upper glume; upper lemma cartilaginous, readily deciduous, the margins firm and flat.
Photosynthetic pathway C_4.
Tricholaena, Melinis.

Subtribe Digitariinae
Inflorescence a panicle or more often of racemes. Lower glume small or absent; upper glume less than to as long as the spikelet; lower floret usually reduced to an empty lemma; upper lemma chartaceous to cartilaginous, its flat hyaline margins enfolding and concealing half to most of the palea.

Photosynthetic pathway C₃ and C₄.

Hylebates, Acritochaete, Homopholis, Stereochlaena, Baptorhachis, Megaloprotachne, Digitaria, Tarigidia, Leptocoryphium, Anthenantia.

Subtribe Arthropogoninae
Inflorescence a panicle. Spikelets laterally compressed; glumes firmly membranous to coriaceous, awned; lower floret male with a palea, or reduced to an empty lemma; upper lemma hyaline, the palea absent or very small.

Photosynthetic pathway C₄.

Arthropogon, Reynaudia.

Subtribe Cenchrinae
Inflorescence a highly modified panicle. Spikelets borne singly or in clusters, subtended by 1 or more bristles or bracts, these often forming an involucre, all deciduous together; both glumes shorter than the spikelet, the lower often absent; lower floret usually reduced to a membranous lemma more or less as long as the spikelet, rarely male with a palea; upper lemma cartilaginous to thinly coriaceous with flat margins covering half to two thirds of the palea.

Photosynthetic pathway C₄.

Pseudoraphis, Chamaeraphis, Pseudochaetochloa, Pennisetum, Cenchrus, Odontelytrum, Paratheria, Snowdenia, Chaetopoa, Anthephora, Streptolophus, Chlorocalymma, Trachys.

Subtribe Spinificinae
Plants dioecious or bisexual. Inflorescence compound, composed of several to many single racemes or small panicles attended by prophylls and spathate sheaths, and condensed into a compact head or fascicle. Upper lemma firmly membranous to crustaceous, with flat margins covering much of the palea.

Photosynthetic pathway C₄.

Xerochloa, Zygochloa, Spinifex.

TRIBE ISACHNEAE
Herbs, sometimes climbers with semi-woody stems; leaf blades linear or lanceolate. Inflorescence a panicle or of racemes. Spikelets 2-flowered, awnless, disarticulating below the florets and sometimes tardily between them; glumes deciduous or persistent, membranous, shorter than or equalling the spikelet; florets similar or dissimilar, the lower male or bisexual, the upper female or bisexual; lemmas membranous to coriaceous, obscurely 0–7-nerved, the margins inrolled and clasping the edges of the palea.

Photosynthetic pathway C₃.

Isachne, Heteranthoecia, Limnopoa, Coelachne, Sphaerocaryum.

TRIBE HUBBARDIEAE

Inflorescence a scanty axillary panicle. Spikelets 2-flowered, awnless, disarticulating above the glumes; glumes persistent, membranous, as long as the lemma, 5–7-nerved; florets similar but the lower empty, the upper bisexual; lemmas membranous, 7–9-nerved; paleas absent.

Photosynthetic pathway C_3.

Hubbardia.

TRIBE ERIACHNEAE

Inflorescence a panicle. Spikelets 2-flowered, slightly laterally compressed, disarticulating below each floret; glumes persistent, 1–14-nerved; florets similar, both fertile, narrowly elliptic, dorsally compressed, the lemma cartilaginous, coriaceous or crustaceous, 3–9-nerved, the margins inrolled and clasping the palea keels, awned or awnless.

Photosynthetic pathway C_4.

Eriachne, Pheidochloa.

TRIBE STEYERMARKOCHLOEAE

Culms mono- or dimorphic, vegetative culms condensed, bearing a single leaf, or leaves numerous; flowering culms elongate, clad with complete or bladeless leaf sheaths. Inflorescence either a spiciform panicle of unisexual spikelets, male below, female above or the sexes in separate, hemispherical panicles. Female spikelet 2–3-flowered, terete, falling entire; glumes both shorter than the spikelet; lower floret barren, middle or upper floret female, its palea thickened, exserted from the lemma, falcate, tightly convolute around the flower, 5–13-nerved. Male spikelets 2–9-flowered, the uppermost florets rudimentary.

Photosynthetic pathway C_3.

Steyermarkochloa, Arundoclaytonia.

TRIBE ARUNDINELLEAE

Inflorescence a panicle, the spikelets often in triads. Spikelets lanceolate, slightly laterally compressed, shedding one or both florets; glumes persistent, the upper as long as the spikelet, the lower usually shorter, membranous to coriaceous, often brown and bearing tubercle-based hairs, rarely awned; lower floret male or barren, its lemma resembling the upper glume, 3–9-nerved; upper floret bisexual, subterete; lemma thinly coriaceous, bidentate or bilobed, awned from the sinus; awn geniculate and often deciduous.

Photosynthetic pathway C_3 and C_4.

Chandrasekharania, Jansenella, Arundinella, Garnotia, Danthoniopsis, Gilgiochloa, Tristachya, Zonotriche, Trichopteryx, Dilophotriche, Loudetia, Loudetiopsis.

TRIBE ANDROPOGONEAE

Inflorescence composed of fragile, rarely tough, racemes, these sometimes aggregated into large panicles but usually single, paired or digitate, terminating the culm or axillary and numerous, in the latter case each true inflorescence subtended by a modified leaf sheath and often aggregated into a leafy compound panicle. Racemes bearing the spikelets in pairs, usually one pedicelled and the other sessile, these sometimes alike but usually dissimilar, the sessile being bisexual, the pedicelled being male or barren. Sessile spikelet falling entire at maturity with the adjacent pedicle and internode, the pedicelled spikelet usually falling separately; glumes enclosing the florets and usually indurated, the lower very variable in shape and ornamentation, the upper usually boat-shaped, lower floret male or barren, the lemma membranous or hyaline and awnless; upper floret bisexual, its lemma membranous or hyaline, with or without a geniculate awn, its palea short or absent. Pedicelled spikelet sometimes similar to the sessile, but usually male or barren, awnless and smaller or even vestigial.

Photosynthetic pathway C_4.

Subtribe Saccharinae
Inflorescence terminal, of solitary, digitate or paniculate racemes with a tough or fragile rachis, and usually slender internodes. Spikelets paired, often plumose, the callus rounded or truncate; lower glume mostly thin, lower floret usually reduced to a sterile lemma; upper lemma entire or bilobed, awnless or bearing or glabrous awn.
Spodiopogon,   Saccharum,   Eriochrysis,   Miscanthus,   Imperata,   Eulalia, Homozeugos,   Polytrias,   Lophopogon,   Pogonatherum,   Eulaliopsis, Microstegium.

Subtribe Germainiinae
Inflorescence terminal, of solitary or digitate racemes with tough or fragile rachis and slender internodes. Spikelets paired, dissimilar. Sessile spikelet male or barren, sometimes enlarged or involucral at the base of the raceme. Pedicelled spikelet fertile, subterete, the callus obtuse to pungent; lower floret reduced to a lemma or suppressed; upper lemma entire or bidentate, bearing a hairy awn.
Apocopis, Germainia, Trachypogon.

Subtribe Sorghinae
Inflorescence terminal or axillary, of single or digitate or paniculate racemes; racemes with a fragile rachis and slender internodes. Spikelets paired, dissimilar. Sessile spikelet bisexual, the callus obtuse; lower floret reduced to a barren lemma; upper lemma entire or bidentate, awned. Pedicelled spikelet male or barren, well developed or much reduced.
Hemisorghum,   Sorghum,   Pseudosorghum,   Sorghastrum,   Asthenochloa, Cleistachne,   Vetiveria,   Chrysopogon,   Dichanthium,   Pseudodichanthium, Spathia, Capillipedium, Bothriochloa, Euclasta.

Subtribe Ischaeminae
Inflorescence of single, paired or digitate, terminal or axillary, rarely spathate racemes with fragile rachis and linear to obovoid internodes. Spikelets paired, dissimilar. Sessile spikelet bisexual, the callus usually obtuse; lower glume 2-keeled or rounded on the flanks; lower floret male; upper lemma bidentate, with a glabrous awn. Pedicelled spikelet variable.

Ischaemum, Thelepogon, Apluda, Kerriochloa, Triplopogon, Pogonachne, Sehima, Andropterum.

Subtribe Dimeriinae
Inflorescence terminal, of single or digitate racemes with a tough rachis. Spikelets single, shortly pedicelled, strongly laterally compressed; glumes subcoriaceous, usually keeled, often winged; lower floret reduced to a barren lemma; upper lemma bilobed, bearing a glabrous awn.
Dimeria.

Subtribe Andropogoninae
Inflorescence of single, paired or digitate racemes, these terminal, or axillary and aggregated into a compound panicle; racemes with fragile rachis and filiform to ovoid internodes, bearing paired, dissimilar spikelets; homogamous pairs inconspicuous or absent. Sessile spikelet bisexual, callus blunt and inserted into the concave tip of the internode; lower glume 2-keeled, often depressed or grooved between the keels, sometimes winged; lower floret reduced to a barren lemma; upper lemma usually bilobed and bearing a glabrous awn. Pedicelled spikelet very variable.
Andropogon, Bhidea, Cymbopogon, Schizachyrium, Arthraxon, Diheteropogon.

Subtribe Anthistiriinae
Inflorescence of single or paired racemes which are usually axillary and aggregated into a compound panicle; racemes with a fragile rachis and slender internodes, bearing paired, dissimilar spikelets, often with the lower pairs homogamous. Sessile spikelet bisexual, the callus usually pointed and applied obliquely to the top of the internode with its tip free; lower floret reduced to a barren lemma; upper lemma entire or bilobed, with a glabrous or hairy awn.
Hyparrhenia, Exotheca, Hyperthelia, Parahyparrhenia, Elymandra, Anadelphia, Monocymbium, Pseudanthistiria, Agenium, Heteropogon, Themeda, Iseilema.

Subtribe Rottboelliinae
Inflorescence of single or digitate racemes, these usually fragile, terminal, axillary or spathate, internodes variously thickened or swollen. Spikelets paired, usually dissimilar. Sessile spikelet bisexual, the callus obtuse or truncate and bearing a central peg; lower glume herbaceous to crustaceous, often sculptured; lower floret male or barren; upper lemma entire, awnless. Pedicelled spikelet variable.
Urelytrum, Loxodera, Elionurus, Phacelurus, Vossia, Hemarthria, Lasiurus, Rhytachne, Coelorachis, Eremochloa, Chasmopodium, Rottboellia, Heteropholis, Hackelochloa, Glyphochloa, Manisuris, Ophiuros, Oxyrhuachis, Mnesithea, Ratzburgia, Thaumastochloa.

Subtribe Tripsacinae
Inflorescence of single or digitate racemes with thickened internodes. Spikelets unisexual, with the sexes borne in separate parts of the same inflorescence or in different inflorescences.
Tripsacum, Zea.

Subtribe Chionachninae
Inflorescence of single axillary racemes, these often gathered into a compound panicle; internodes slender. Spikelets unisexual, the spikelets borne in separate parts of the raceme. Female spikelet sessile, enclosed by its lower glume, and accompanied by a little pedicelled spikelet borne laterally on the internode.
Chionachne, Sclerachne, Polytoca, Trilobachne.

Subtribe Coicinae
Inflorescence comprising a sessile female raceme and a pedunculate male raceme. Female raceme enclosed within a bony utricle derived from a modified leaf sheath; comprising 1 sessile spikelet and 2 pedicels, which may or may not bear vestigial spikelets. Male raceme projecting from the mouth of the utricle, the spikelets borne in pairs or triplets.
Coix.

Note: This summary differs from Clayton & Renvoize (1986) as follows:

1 *Alvimia, Criciuma, Eremocaulon, Parodiolyra, Guadua* and *Sphaerobambos* added to Bambusoideae.

2 *Cyperochloa* added to Arundinoideae.

3 *Arundoclaytonia* added to Panicoideae.

4 Tribe Thysanolaeneae subsumed in Arundineae (and *Thysanolaena*).

5 *Mildbraedochloa* removed and with *Rhynchelytrum* subsumed in *Melinis* (subtribe Melinidinae in Panicoideae).

6 *Polliniopsis* reduced to *Microstegin.*

Reference
Clayton, W. D. & Renvoize, S. A. (1986) *Genera Graminum: grasses of the world.* Kew Bulletin Additional Series 13. London, Royal Botanic Gardens, Kew.

Author index

Organism index

Note: Two systems of grass classification are widely used, those of Clayton & Renvoize (1986) and Watson & Dallwitz (1988). Watson (1990) in the Appendix to *Reproductive versatility in the grasses* provided a summary of the Australian system. To facilitate cross reference and comparison, Clayton & Renvoize have provided a matching summary of the Kew system for the present volume. Notwithstanding the apparent similarities, the reader should note that (a) in some instances there are similar but not identical spellings of some subgroupings; (b) names common to both systems need not encompass the same groups of genera; (c) an additional subfamily, the Centothecoideae, is recognized by Clayton & Renvoize.

In this index the following superscripts may assist the reader:

[W] genera or higher taxa mentioned in this volume accepted by Watson (1990) but not by Clayton & Renvoize (1986) although reference to the latter will in most cases offer an alternative assessment.

[NW] new additions to the Watson (1990) summary.

[CR] taxa currently accepted by Clayton & Renvoize but not by Watson (1990).

[PCR] taxa latterly pooled by Clayton & Renvoize with other taxa.

[+CR] taxa latterly adopted by Clayton & Renvoize.

[OA] those taxa not recognized in either of the above systems and attributable to other authorities.

References

Clayton, W. D. & Renvoize, S. A. (1986). *Genera Graminum: grasses of the world.* Kew Bulletin Additional Series 13. London: Royal Botanic Gardens, Kew.

Watson, L. (1990). *World grass genera.* Appendix. In *Reproductive versatility in the grasses,* ed. G. P. Chapman, Appendix. Cambridge: Cambridge University Press.

Watson, L. & Dallwitz, M. J. (1988). *Grass genera of the world: illustrations of characters, descriptions, classification, interactive identification, information retrieval.* Canberra; Research School of Biological Sciences, Australian National University.

POACEAE

Snowdenia 67, 349
Soderstromia 67, 347
Sohnsia 67, 346
Sorghastrum 67, 351
Sorghinae 33, 140, 145, 351
Sorghum 33, 67, 97, 141, 150, 351
  x alum 327
  bicolor 54, 158, 177, 182
  halepense 327
  (propinquum) 167
Spartina 67, 347
  coarctica 326
  townsendii 326
Spartochloa 25, 344
Spathia 35, 67
(W)(+CR)*Sphaerobambos* 339, 353
Sphaerocaryum 67, 349
Spheneria 67, 348
Sphenopholis 343
Sphenopus 342
Spinifex
  hirsutus 31
(CR)Spinificinae 31, 349
Spodiopogon 67, 351
(CR)Sporobolinae 24, 139, 346
Sporobolus 24, 67, 71, 73, 320, 327, 346
  arenarius 326
  fimbriatus 46
Steinchisma 67, 76, 79, 90, 92, 348
Steirachne 67, 346
Stenotaphrum 67, 348
Stephanachne 343
Stereochlaena 67, 349
Steyermarkochloa 27, 350
Steyermarkochloeae 8, 18, 26, 27, 350
(W)*Stiburus* 67
Stipa 21, 316, 320, 342
  splendens 320
  tenuis 326
Stipagrostis 9, 18, 19, 30, 58, 59, 67, 70, 87, 92, 345
  pernata 326
Stipeae 8, 20, 21, 92, 341
Streblochaete 342
Streptochaete 340
  crinita 15
  spicata 15
Streptochaeteae 8, 13, 340
Streptogyna 340
Streptogyneae 8, 13, 340
Streptolophus 67, 349
Streptostachys 67, 348
Styppeiochloa 25, 344
Sucrea 339
Suddia 339
Swallenia 67, 346

Symplectrodia 67, 69, 345

Taeniatherum 343
Tarigidia 67, 349
Tatianyx 59, 67, 75, 348
Tetrachaete 67, 347
Tetrachne 67, 346
  dregei 326
Tetrapogon 25, 67, 347
  spathulata 326
Thamnocalamus 328, 339
Thaumastochloa 67, 352
Thelepogon 67, 352
(W)*Thelungia* 67
Themeda 67, 141, 352
(W)*Thinopyrum*
  elongatum 233, 235
  intermedium 233
  ponticum 233
Thraysa 348
Thrasyopsis 67, 348
Thuarea 67, 348
Thyridachne 28, 67, 348
Thyridolepis 67, 69, 75, 348
(W)*Thyrsia* 67
(W)*Thyrsostachys* 339
Thysanolaena 32, 344, 353
(PCR)Thysanolaeneae 353
Torreyochloa 342
Tovarochloa 343
Trachypogon 67, 351
Trachys 67, 349
Tragus 67, 347
Tribolium 19, 25, 344
(CR)*Trichloris* 347
Tricholaena 30, 67, 140, 348
Trichoneura 67, 346
Trichopteryx 67, 70, 350
Tridens 24, 68, 346
Trikeraia 342
Trilobachne 68, 353
Triniochloa 342
Triodia 24, 48, 68, 69, 345
  longiceps 326
  pungens 46
(W)Triodieae 69, 92
(CR)Triodiinae 24, 345
Triplachne 343
Triplasis 68, 346
Triplopogon 68, 351
Tripogon 28, 48, 346
  loliiformis 327
  minimus 327
Tripsacinae 33, 35, 141, 353
Tripsacum 35, 68, 97, 141, 142, 158, 169, 353
  andersonii 171

OTHER ORGANISMS

Subject index